KV-374-862

Black's Agricultural Dictionary

Black's
Agricultural
Dictionary

D. B. Dalal-Clayton
B.Sc.,(Hons), M.Sc. (Dunelm), M. I. Biol., F.R.G.S

Adam & Charles Black London

First published 1981
by A & C Black (Publishers) Ltd
35 Bedford Row, London WC1R 4JH

ISBN 0-7136-2130-3

© 1981 D. B. Dalal-Clayton

All rights reserved.
No part of this publication may be
reproduced, stored in a retrieval
system, or transmitted in any form or
by any means, electronic, mechanical,
photocopying, recording or otherwise,
without the prior permission in writing
of A & C Black (Publishers) Ltd.

Dalal-Clayton, D. B.
  Black's agricultural dictionary.
  1. Agriculture – Dictionaries
  I. Title
  630   S411
  ISBN 0–7136–2130–3

Printed in Great Britain
by The Pitman Press, Bath

Contents

Illustrations

Photographs

Line drawings

Source of illustrations

The illustrations printed on the pages listed below are reproduced by courtesy of the following individuals, companies and publications. To them all, both author and publisher extend their thanks.

Foreword

Farmers are generally very busy people, working in our oldest industry but caught up in a tide of progress and technological change which is without parallel.

Communication is vital, but you need words to communicate—and the meaning of the words is changing all the time. I hope this book, which is long overdue, will help us all.

RICHARD BUTLER

President
National Farmers' Union

Acknowledgements

For his initial advice and encouragement the author is grateful to David Steers, Chief Press Officer of the National Farmers' Union.

He is particularly indebted to Ken Garner, Principal of Hadlow College of Agriculture and Horticulture, for his painstaking reading of the final manuscript; for making many valuable suggestions and for eliminating minor errors and drawing attention to inevitable omissions in the original text.

Many individuals and organisations have helped and have given freely of their time and advice, and have made suggestions or checked entries. Space does not permit them all to be mentioned individually, but to them a very great debt of gratitude is owed.

Authors, publishers and copyright holders have been generous in allowing the reproduction of the illustrations. The sources are listed on page viii. In the course of compiling this dictionary, the author has consulted many texts. In all instances, the sources of quotations are given.

Preface

This book is intended as a reference source for farmers, students of agriculture and all those involved in or associated with the agricultural industry in whatever capacity. It is concerned essentially with agriculture in the U.K. and embraces such horticultural, botanical, geographical, biological, physical, geomorphological and chemical terms (including formulae) as may be needed to provide adequate information and to avoid the necessity of consulting other works. Entries also include the most common and agriculturally significant diseases and pests of farm animals or crops, with their symptoms.

The extent of each entry has been determined by the need for definition of meaning, function and agricultural relevance. Statistics are quoted where they are considered important, sometimes in the form of tables. Historical terms add colour and flavour to farming language, but traditional names vary endlessly from region to region. It has therefore been difficult to choose which terms of a historical nature should be allowed in. Photographs and line drawings illustrate entries where words alone are inadequate to convey meaning. Common abbreviations, acronyms and initials used in the industry are presented as a separate glossary in the form of an appendix.

Agriculture in the U.K. today is affected by the Common Agricultural Policy of the European Economic Community and its regulations; the E.E.C. vocabulary appears increasingly in documentation and in conversation. It is therefore appropriate that a range of terms should figure in this dictionary.

Much specialist advice has been sought and offered during the compilation of this dictionary and the writing of individual entries. Nevertheless no-one is likely to agree completely with the author's selection of entries and there may be disagreement over certain definitions. The author would welcome comment and constructive suggestions.

D. B. DALAL-CLAYTON

April 1980

xi

A note on cross-referencing

Within the definitions, words set in SMALL CAPITALS indicate other entries in the dictionary.

When set within parentheses and preceded by a symbol, they signify an entry of particular relevance to the subject, thus: (◆ LUCERNE).

Relevant illustrations and tables are cross-referenced in italics, thus: (*see p. 292*).

A

A – Horizon ◗ SOIL HORIZONS.

Abaction. The stealing of cattle.

Abattoir ◗ SLAUGHTERHOUSE.

Aberdeen-Angus. A breed of black, hornless beef cattle, developed using INBREEDING from a mixture of breeds in East Scotland. They have a long, deep body and recently have increased their size very considerably. An early maturing breed, fattening on low energy feed, producing carcases much favoured by butchers. Cows yield enough milk to rear their calves. Valued in Britain as a cross for beef, and also widely used on dairy HEIFERS to produce black progeny with no calving problems (due to their small heads), the cross from this being highly favoured in the hill districts as a hill cow (*see p. 89*).

Aberystwyth Strains. Herbage plant varieties developed by the Welsh Plant Breeding Station at Aberystwyth. The most well known are the 'S' strains which were the first to be bred.

Ablactation. The cessation of milk production by mammary glands when an offspring is weaned.

Abomasum. The fourth stomach of RUMINANTS, lying close to the OMASUM or third stomach, in which digestion is completed by the supply of digestive juice. Also known as REED STOMACH, RENNET STOMACH, and sometimes called MAW.

Abort. To miscarry in birth. For plants and animals in general – to become infertile, to fail to develop or to wither away.

Abortion. Miscarriage, slipping, or premature delivery of partially developed offspring. (◗ CONTAGIOUS ABORTION)

Abreast Parlour ◗ MILKING PARLOUR.

Abscess. A swelling in some part of the body caused by an accumulation of purulent matter.

1

Absolute Drought. A period (in the U.K.) of at least 15 consecutive days during which less than 2.54 mm (0.1 in.) of rain has fallen on any day. Not an internationally accepted definition. (◗ DRY SPELL, PARTIAL DROUGHT)

Absolute Weed. One of no use to farmers under any circumstances. (◗ RELATIVE WEED)

Abstraction Licence. A licence issued by a Water Authority for the abstraction of water from a water source for domestic or commercial use, often for irrigation purposes.

Acaricide. A chemical used to kill MITES (Acarina) as distinct from insects.

Access Agreements and Orders. Access Agreements may be made by a local Planning Authority, under the National Parks and Access to the Countryside Act 1949, with landowners and occupiers, providing for public access to open country for recreation, involving the possible payment of compensation for resultant land capital value depreciation and for extra costs (e.g. extra shepherding, damage caused, etc.). There are no restrictions on ploughing. Where an Agreement cannot be negotiated an Access Order may be made, which requires Ministerial confirmation, prescribing a public right of access to open country. Compensation is payable and there may be restrictions on ploughing and other improvements. (◗ MANAGEMENT AGREEMENT)

Acclimatisation Value. The extra worth of a flock of sheep acclimatised to a farm, over an imported flock which may stray, with loss of sheep. (◗ HEFTED SHEEP)

Acclimatised Sheep. Sheep which are long accustomed to their local environment. They will normally not stray beyond the farm boundary, unlike an imported flock.

Accredited Herds Scheme ◗ BRUCELLOSIS (ACCREDITED HERDS) SCHEME.

Accredited Milk. Milk sold by a producer-retailer from a herd accredited as free from BRUCELLOSIS.

Achene. A small, dry, one-seeded FRUIT formed from a single CARPEL, which does not split open on ripening but rots gradually with seed germination. The pips on strawberries are achenes.

Acid Soil. A sour soil, having a pH of less than 7.0 due to an excess of HYDROGEN ions (H⁺). Most British soils, other than those over limestone, are more or less acid, as a result of the fairly heavy rainfall which leaches (♦ LEACHING) out the salts, especially CALCIUM. Typical acid soils are the heather moors of the north and west. (♦ ALKALINE SOIL, NEUTRAL SOIL, LIMING)

Acre. A measure of land. 4 840 sq. yds. Equal to 4 ROODS, or 10 square CHAINS. (Scottish acre, 6 150.4 sq. yds; Irish acre, 7840 sq. yds – both obsolete.) Also equivalent to 0.405 HECTARES. Originally, the amount of land that could be ploughed in a day by a team of oxen.

Acre-inch. A unit of measurement used in IRRIGATION works. The amount of water required to cover an acre to a depth of 1 in. (25.4 mm per hectare) equivalent to 101 tons or 22 460 gallons.

Actinobacillosis. An animal disease mainly of cattle and pigs, caused by bacteria (*Actinobacillus lignieres*) causing swelling and hardening of tongue and face. Also known as Wooden Tongue. The disease also occurs in sheep and is sometimes called 'Cruels'.

Actinomycosis. An animal disease caused by the microscopic 'ray-fungus' (*Actinomyces bovis*) which penetrates through small mouth wounds (caused by barley awns, foreign bodies, during change of teeth, etc.) resulting in the swelling of the jawbone with ultimate suppuration. Also known as Lumpy Jaw.

Additives. Substances added to FEEDINGSTUFFS or CONCENTRATES during manufacture, which are not principal nutrient sources. Examples include VITAMINS, minerals, TRACE ELEMENTS, certain ANTIBIOTICS, etc.

Addled Egg. A fertile egg in which the embryo died between the seventh and fourteenth day of incubation. Also a rotten egg in general terms. (♦ DEAD-IN-SHELL)

Adipose Tissue ♦ FATS.

Adsorption Complex. The group of substances in the soil capable of adsorbing other materials (such as applied fertilizers). Organic and inorganic colloidal substances form the greater part of the adsorption complex. Non-colloidal materials, such as sand and silt, exhibit adsorption to a much lesser extent than the colloidal materials. (♦ BASE SATURATION PERCENTAGE, CATION-EXCHANGE CAPACITY)

Adulteration. The adding of water to milk.

Adventitious. Arising in an abnormal position. Adventitious roots develop from parts of plants other than the roots, e.g. from stem or leaf cuttings. Adventitious buds develop from parts of plants other than in the axils of leaves, e.g. from roots.

Advisory Council for Agriculture and Horticulture. Established in 1973, succeeding the Agricultural and Horticultural Advisory Councils, to 'consider and report on agricultural and horticultural matters within the field of responsibility of the Ministry of Agriculture, Fisheries and Food which may be referred to the Council by the Minister or Ministry'. In 1979 the Government announced its disbanding. The Council, comprising twelve members eminent in agriculture and horticulture, published a number of reports, its final one being on the future water needs of the agricultural and horticultural industries.

Adze. A tool consisting of a thin arched blade with its edge at right-angles to the handle like a hoe, used for slicing off the surface of a piece of wood, e.g. in fencing work.

Aerial Spraying. There has been an appreciable increase in aerial spraying (by aeroplane and helicopter) in recent years involving pesticides, fertilizers and seed. According to the Seventh Report of the Royal Commission on Environmental Pollution, 'Agriculture and Pollution' (H.M.S.O. 1979), 628 000 hectares (c. 1 552 000 acres), excluding woodland, were sprayed in 1977. The report also roughly estimated that about 200 000 hectares (c. 494 000 acres) are sprayed annually with fertilizer. Aerial spraying of pesticides on forests is rarely practiced in the U.K.
　　Aerial applications are advantageous in various circumstances, e.g. when land is too wet, or when crop height, scale of infestation or the area of land involved is such that the use of ground machines is impractical. Techniques such as TRAM-

4

LINING may reduce the need for aerial application.

Aerial operators must be certified by the Civil Aviation Authority under the Civil Aviation Act 1971, and are required to observe various safety precautions, and have regard to such factors as the likelihood of SPRAY DRIFT, and danger to farm animals, bees, wildlife, etc. Only pesticides cleared under the PESTICIDES SAFETY PRECAUTIONS SCHEME may be used. A Code of Practice produced by the NATIONAL FARMERS' UNION and the National Association of Agricultural Contractors also covers the activities of Aerial Operators.

Wet farmland, incapable of carrying a tractor-sprayer, being aerial sprayed against liver fluke-carrying snails.

Aerobes. Organisms, especially bacteria, requiring oxygen for respiration. (◗ ANAEROBES)

Afforestation. The process of transforming an area into forest, usually when trees have not previously been grown there. (◗ DEFORESTATION, REFORESTATION)

Aflatoxin. A poisonous toxin produced by the fungus *Aspergillus flavus* and found in GROUND NUT MEAL. Causes a reduction in milk yield and growth rate in cattle, and sometimes jaundice in pigs eating the meal.

5

Afterbirth. The placenta and membranes (also called 'cleansing') which are expelled from the uterus of a mother after the birth of its offspring.

After Crop. A second crop from the same land in the same year.

After-cultivation. Harrowing, rolling and other cultivations carried out in a field after the crop has emerged.

Aftermath. Grass which springs up again after the cutting of a crop of HAY, and can subsequently be taken as a second cut. Also known as eadish, eddish, rowen and lattermath.

Agent ♦ LAND AGENT.

Agglutination Test. A blood test used mainly to detect carriers of BRUCELLOSIS in cattle and BACILLARY WHITE DISEASE in chickens. Based on the principle that if a minute dose of blood serum from a carrier (i.e. an animal with organisms of the disease in its blood, but showing no symptoms) is introduced to a suspension of the organisms, then agglutination (i.e. grouping together of the organisms) will occur. The phenomenon does not occur if the blood serum from a disease-free animal is introduced. A positive test indicates the animal has previously had the disease and still has ANTIBODIES in its blood.

Aggregate. A group of soil particles cohering so as to behave mechanically as a unit. (♦ CRUMB)

Agistment. A contract arising when one person, the agister, takes livestock belonging to another person to graze on his land for reward, payment usually being at a certain rate per week on a headage basis. Unlike grass KEEP agreements, agistment involves no interest in the land itself and is usually agreed orally. The agister is required to take care of the livestock and supply them with food, and is liable for injuries to, or losses of, any animals arising from negligence. Agisted animals are also said to be 'on tack'.

Agrarian. Relating to land or landed property.

Agrarian Revolution ♦ AGRICULTURAL REVOLUTION

Agricultural Apprenticeship Council. The forerunner of the AGRICULTURAL TRAINING BOARD. (♦ APPRENTICES)

Agricultural Chemicals Approval Scheme. A voluntary scheme under which proprietary brands of crop protection chemicals can be officially approved by the Agricultural Chemicals Approval Organisation for specific uses under U.K. conditions if label and advertising claims are fulfilled. The Scheme enables users to select, and advisers to recommend, efficient and appropriate chemicals and to discourage the use of unsatisfactory ones. A list of approved chemicals is published annually. Approved product labels are marked with the letter 'A' surmounted by a crown.

AGRICULTURAL CHEMICALS APPROVAL SCHEME

Agricultural Co-operation and Marketing Services. A producer-controlled organisation established in 1972 to expand the effective share of the market held by producer-controlled organisations by representing the co-operative viewpoint on national bodies to Government and within the E.E.C., by providing information and advisory services to co-operative managers and by undertaking development projects as an agent for the CENTRAL COUNCIL FOR AGRICULTURAL AND HORTICULTURAL DEVELOPMENT. In 1979 the marketing function was transferred to the NATIONAL FARMERS' UNION which established a marketing division.

Agricultural Credit Corporation Ltd. Established in its present form in 1964, with Government support, to extend the availability of medium term credit to farmers, growers and co-operatives, by guaranteeing the required bank borrowing.

Agricultural Development and Advisory Service. A division of the Ministry of Agriculture, Fisheries and Food providing free and impartial scientific, professional and technical advice to farmers, growers and landowners. A.D.A.S. comprises five services: AGRICULTURE SERVICE, AGRICULTURAL SCIENCE SERVICE, VETERINARY SERVICE, LAND DRAINAGE SERVICE, LAND SERVICE. In May 1980 the Minister announced proposals to merge the latter two services.

7

Agricultural Dwelling House Advisory Committees. Committees established under the Rent (Agriculture) Act 1976. They assist local Housing Authorities in considering rehousing a former agricultural worker occupying a service house (tied cottage), by assessing agricultural need and the degree of urgency for the repossession of the service house. Each Committee of three comprises an independent member and a representative of both agricultural workers and employers, and is appointed by the Chairman of the local AGRICULTURAL WAGES COMMITTEE.

Agricultural Engineers' Association. An Association representing all the principal manufacturers of agricultural and horticultural tractors, implements, machinery and ancillary equipment.

Agricultural Executive Committee ▶ COUNTY AGRICULTURAL EXECUTIVE COMMITTEES.

Agricultural Holding. Defined by the AGRICULTURAL HOLDINGS ACT, 1948, as 'the aggregate of the agricultural land comprised in a contract of tenancy, not being a contract under which the said land is let to the tenant during his continuance in any office or employment under his landlord.'

Agricultural Holdings Act (1948). An Act laying down a code of rules regulating the relationship of landlord and tenant in farming. Essentially it grants security of tenure, encouraging the tenant to make the best use of the land without fear of eviction. (▶ AGRICULTURAL HOLDING, NOTICE TO QUIT)

Agricultural and Horticultural Cooperation Scheme. A scheme applying throughout the U.K. administered by the CENTRAL COUNCIL FOR AGRICULTURAL AND HORTICULTURAL COOPERATION, providing grants for implementing approved proposals designed to organise, promote, encourage, develop or co-ordinate worthwhile co-operative activities in agriculture and horticulture.

Agricultural Land Commission. A body established under the 1947 Agriculture Act to manage and farm land vested in the Minister of Agriculture and to advise the Minister on management problems. These duties are now carried out by the LAND SERVICE.

Agricultural Land Tribunal. Set up under the Agriculture Act 1947 principally to hear appeals against decisions which affect the owner or occupier of agricultural land. The decision of a Tribunal in a case is final. The Chairman is a lawyer appointed by the Lord Chancellor, the other two members are selected from panels representative of owners and farmers. Eight such Tribunals exist in England and Wales on an area basis, each responsible for a number of counties. (♦ CERTIFICATE OF BAD HUSBANDRY)

Agricultural Lime Scheme. A scheme which operated between 1966 and 1976 providing government subsidies for the application of LIME to agricultural land.

Agricultural Marketing Acts ♦ MARKETING BOARDS.

Agricultural Mortgage Corporation. A Corporation empowered, under the Agricultural Credit Acts of 1928 and 1932, to grant medium- and long-term loans to owners or those intending on the security of agricultural land and buildings in England and Wales. Its shareholders are the major clearing banks.

Agricultural Reference Rate. The fixed exchange rate for E.E.C. national currencies against the UNIT OF ACCOUNT used for agricultural prices. Also called green rate. (♦ GREEN POUND)

Agricultural Research Council. A corporate body established by Royal Charter in 1931, concerned with the organisation and development of agricultural and food research, and the making of grants for such research. It supervises State-aided research institutes, administers its own institutes and units, and assists fundamental and applied research in universities and elsewhere.

Agricultural Revolution. A term applied to great changes in agriculture in the eighteenth and early nineteenth centuries brought about by rapid population increase requiring the production of more food. Under the open-field system of medieval times, individual cultivators had rights of cultivation on strips of land. Once the year's crops had been harvested, common grazing rights existed over all the land (♦ COMMON LAND). During the enclosure movement, which began in the twelfth century but proceeded rapidly during the eighteenth and nineteenth centuries, open fields largely disappeared and the strips of land were consolidated and hedged or fenced around with, as a result, individuals (particularly the Lords of the Manor) gaining control over particular areas of land throughout the year. This enclosure allowed sheep to be kept and later encouraged large-scale corn production. The Agricultural Revolution also saw the introduction of new methods of crop husbandry (e.g. the NORFOLK ROTATION), improved farm machinery (◗ TULL, JETHRO) and the development of selective breeding (◗ BAKEWELL, ROBERT). The Industrial Revolution in the nineteenth century created large city-based populations needing to be fed, which stimulated further improvements, e.g. in LAND DRAINAGE, the introduction of ARTIFICIAL FERTILIZERS, and the further improvement of wheat and other cereals by PLANT BREEDING.

Agricultural Science Service. A division of the AGRICULTURAL DEVELOPMENT AND ADVISORY SERVICE providing specialist advice to farmers on a range of scientific disciplines both directly and via field officers. The service has extensive laboratory facilities.

Agricultural Training Board. A body set up under the Industrial Training Act 1966 to ensure the provision of adequate training to meet agriculture's needs, now funded by Government. It consists of members appointed by the Government and others representing employers, employees and education. The Board provides grants to employers and other bodies providing approved training. There are permanent field staff aided by Area Training Committees with a similar representative structure as the Board, closely involved in assessing and monitoring local training needs and facilities. (◗ APPRENTICES)

Agricultural Wages Board. The Agricultural Wages (Regulation) Acts, 1924 to 1947, were consolidated by the Agricultural Wages Act, 1948, which established an autonomous Agricultural Wages Board and County Agricultural Wages Committees, each Committee covering one or more administrative counties. The Board comprises equal numbers of employers' representatives (nominated by the NATIONAL FARMERS' UNION) and workers' representatives (nominated by the NATIONAL UNION OF AGRICULTURAL AND ALLIED WORKERS, and the Transport and General Workers Union), and independent members and a

Chairman nominated by the Minister. The representative structure of the County Committees is the same. The Board is empowered to make Orders fixing minimum wages and holiday entitlement for agricultural workers, and certain aspects of their terms and conditions of employment. Under the 1924 Act, County Committees were able to fix minimum wages and considerable variations existed in different areas. The Board assumed these powers in 1940 to provide national wage uniformity. There are separate Boards for Scotland and N. Ireland.

Agricultural Wages Committee ♦ AGRICULTURAL WAGES BOARD.

Agricultural Wages Order ♦ AGRICULTURAL WAGES BOARD.

Agriculture. The cultivation of the land. Defined in the AGRICULTURE ACT, 1947, as including 'horticulture, fruit growing, seed growing, dairy farming and livestock breeding and keeping, the use of land as grazing land, meadow land, osier land, market gardens and nursery grounds, and the use of land for woodlands where that use is ancillary to the farming of land for other agricultural purposes.'

Agriculture Acts (1947, 1957). Acts designed to provide stability for farmers by giving guarantees of price and market, and to provide for efficient production. They also established an ANNUAL REVIEW between Government and the N.F.U. of economic conditions and prospects of the farming industry, and fixing the major farm product prices. Most farm prices are now fixed by the E.E.C. Council of Agriculture Ministers. (♦ GUARANTEE PRICES)

Agriculture Service. A division of the AGRICULTURAL DEVELOPMENT AND ADVISORY SERVICE providing specialist advice on all subjects. Responsible for a range of experimental farms and horticultural stations in England and Wales operating under differing soil and climatic conditions. These centres serve A.D.A.S. generally and are used to develop new knowledge for general application. Investigatory and promotional work is also carried out. Annual reports are available to farmers and growers.

Agronomy. The branch of agriculture concerned with the theory and practice of field-crop production and the scientific management of soil.

Agrostis. Generic name for BENT grass.

Agrostology. The study of grasses.

A.I. Centre. One of a number of centres operated mainly by the MILK MARKETING BOARD at which pedigree, proven-performance bulls are kept and their semen stored for use in ARTIFICIAL INSEMINATION. There are also a number of pig A.I. centres.

Aid ▶ GRANT AID.

Air ▶ SOIL AIR.

Aitchbone. The rump bone. The cut of beef over it.

A.I.V. A SILAGE-making process (after A. I. Virtanen, a Finnish chemist) which substitutes Hydrochloric Acid for MOLASSES as a feed for silage making bacteria.

Albino. An animal with no skin pigment, white-coated, pink-eyed. Commonly due to a single recessive gene. (▶ DOMINANT GENE)

Albumens. A group of water-soluble PROTEINS which occur in many of the tissues and body fluids of animals; e.g. egg albumen (egg white); serum albumen (in blood), lactalbumen (in milk), etc. Also called albumins.

Albuminoid Ratio ▶ NUTRITIVE RATIO.

Alderney. A term, now in disuse, for all cattle introduced into Britain from the Channel Islands, and even from Normandy and Brittany, including the now separately distinguished GUERNSEY and JERSEY breeds.

Aldrin. A persistent organochlorine insecticide, a derivative of chlorinated napthalene, harmful to fish and under restricted use. Applied as a dust, spray or seed dressing.

Alfalfa Meal. A feedingstuff comprising artificially dried and ground LUCERNE, known in America as alfalfa.

Algae. Simple, flowerless, unicellular or multicellular photosynthetic plants with unicellular organs of reproduction. They are aquatic plants, freshwater or marine (e.g. seaweeds), or plants of damp situations (e.g. damp walls, tree trunks, in soil, etc.).

Alkaline Soil. Sweet soil. Soil having a pH above 7.0 due to an excess of hydroxyl (OH⁻) ions, or having a high exchangeable sodium content, or both. Typical alkaline soils are those of the chalk and limestone hills, e.g. South Downs, Cotswolds and Craven Pennines. (♦ ACID SOIL, NEUTRAL SOIL)

Allantois. A membranous sac-like appendage for effecting oxygenation in the embryos of mammals, birds and reptiles.

Alleyways. Passage-ways between rows of hop plants in hop gardens. (♦ BROAD ALLEY)

Allopolyploid. A POLYPLOID organism to which two different species have each contributed one or more sets of CHROMOSOMES (e.g. probably cultivated wheats). (♦ ALLOTETRAPLOID, AUTOPOLYPLOID)

Allotetraploid. An ALLOPOLYPLOID arising when an ordinary HYBRID between two species which contains a CHROMOSOME set from each parent, doubles its chromosome number. Ordinary hybrids are usually sterile since their chromosomes cannot pair during MEIOSIS. In an allotetraploid each chromosome can pair with its homologue, thus overcoming sterility, and immediately creating a new species. Tobacco probably originated in this way. Allotetraploidy is only known in plants at present. Artificial production of allotetraploids is important in creating new agricultural and horticultural varieties. Also called amphidiploid.

Allotment. A small plot of land mainly used for vegetable growing, often let for spare-time cultivation under a public scheme.

Allotment Garden. Defined by the Agriculture Act, 1947 as 'an ALLOTMENT not exceeding forty POLES in extent which is wholly or mainly cultivated by the occupier for the production of vegetables or fruit for himself and his family.'

Alluvium. Material such as clay, silt and sand deposited by rivers and streams. (♦ COLLUVIUM)

Alpha Amylase. An ENZYME naturally occurring in wheat seed at germination, causing starch breakdown to sugar. If unduly active in MILLING WHEAT it reduces the quality of bread and cakes. Tested for by either the HAGBERG TEST or the FARRAND TEST.

13

Alpine. A breed of goat native to the Swiss Alps, from which the BRITISH ALPINE breed developed.

Alsike Clover. A species of clover, *Trifolium hybridum*, named after Alsike near Uppsala in Sweden, characterised by pale pink or white flowers, a smooth fairly upright stem, toothed leaflets with long pointed stipules, and a pod containing 1–3 small seeds. It tolerates adverse soil conditions better than RED CLOVER or WHITE CLOVER, and is less susceptible to stem rot.

Alternate Host ◗ HOST.

Alternate Husbandry ◗ LEY FARMING.

Alternate Leaves. Leaves attached individually to the stem of a plant to opposite sides at each successive NODE. In some plants the leaves may be arranged in a progressive spiral. (◗ OPPOSITE LEAVES)

Amel Corn. An inferior variety of wheat (*Triticum vulgare dicoccum*), the larger SPELT. Also called French Rice.

Amelioration. The process of enhancing the agricultural value of land by drainage, tillage, liming, manuring, etc.

American Bronze. A breed of turkey, green-bronze in colour, at one time the commonest to be found in the U.K.

Amino Acids. The chemical units which link together into polypeptide chains to form PROTEINS, and of fundamental importance to life. 'Essential' amino acids are those which an organism cannot synthesise and must be obtained from its environment.

Ammonia. (NH_3). A pungent colourless gas used in the manufacture of FERTILIZERS (e.g. AMMONIUM NITRATE, SULPHATE OF AMMONIA). Anhydrous ammonia is a very concentrated nitrogenous fertilizer (82% nitrogen) consisting of pure ammonia in liquid form under pressure and applied by injection into the soil, usually when moist, where it vaporises and combines with soil COLLOIDS. Aqueous ammonia, a by-product from gas works called gas liquor, comprising ammonia in solution, used to be commonly used as a fertilizer, applied to the soil surface.

Ammonium Fixation. The adsorption or absorption of ammonium IONS by the mineral or organic fractions of the soil so that the ions become relatively insoluble in water and unexchangeable by CATION EXCHANGE.

Ammonium Nitrate. (NH_4NO_3). The commonest nitrogen FERTILIZER currently in use in Britain either as a STRAIGHT FERTILIZER or in COMPOUND FERTILIZERS. Now available as a pure salt, safe for agricultural use, but prior to 1965 it was only available as an unstable synthetic salt, liable to caking, and was usually blended with CHALK or GROUND LIMESTONE. (♦ NITRO-CHALK)

Ammonium Phosphates. Mono- and di-ammonium phosphates are occasionally used in the pure state as fertilizers but, more commonly, they are constituents of concentrated COMPOUND FERTILIZERS. Both the phosphate and the nitrogen are soluble in water. (♦ FERTILIZERS, ARTIFICIAL FERTILIZERS)

Amnion. A protective membranous sac containing amniotic fluid cushioning the developing embryo of an animal.

Amphidiploid ♦ ALLOTETRAPLOID.

Amylase ♦ ALPHA AMYLASE.

Amylopsin. An ENZYME found in PANCREATIC JUICE which breaks down STARCH into sugars.

Anaerobes. Organisms capable of living in absence of free oxygen (gaseous or dissolved).

Anaerobic Conditions. Environmental conditions in which free oxygen (gaseous or dissolved) is absent.

Anaesthetic. A chemical substance producing loss of sensation, removing all sensitivity to pain. A local anaesthetic produces lack of feeling in a limited area of the body. A general anaesthetic acts on the whole body producing total unconsciousness. Used in surgical operations by veterinary surgeons.

Analysis. A term usually applied to chemical or physical methods of determining the composition of substances. (♦ SOIL ANALYSIS, CHEMICAL ANALYSIS and PHYSICAL ANALYSIS)

15

Anaplasmosis. An infectious disease of cattle, characterised by anaemia.

Anchor Wire. Strong wire for supporting wire-work, one end attaching to the top of a straining pole, the other to a fixture in the ground (a deadman). Common in hop gardens. Also known as guy wire.

Ancona. A light breed of fowl laying white eggs; has black feathers with white tips, a single comb and yellow legs.

Andalusian. A light breed of fowl laying white eggs; has blue feathers with darker edges, a single comb, and slate-coloured legs. At one time commonly called the 'Blue Spanish'.

Anglo-Nubian. A breed of GOAT first established by crossing imported oriental lop-eared goats with English or other breeds. It produces milk relatively rich in BUTTERFAT.

Anglo-Nubian-Swiss. The former name for a breed of goat. (♦ BRITISH GOAT)

Angora. 1. A breed of rabbit which produces wool of various colours (white, grey, fawn, smoke or blue).
2. A breed of goat with long silky wool (the true mohair).

Anhydrous Ammonia ♦ AMMONIA.

Animal Fats. FATS derived from animals and characterised by a high proportion of saturated FATTY ACIDS, as distinct from plant fats or oils which are characterised by a high proportion of unsaturated acids.

Animal Harmonisation Standards. Standards concerning animal health for harmonising the differences in legislation between existing E.E.C. member states and those nations which joined in 1973.

Animal Manure. The excreta of farm stock. (♦ FARMYARD MANURE)

Animal Movement Order ♦ MOVEMENT RECORD.

Animal Starch ♦ GLYCOGEN.

Animal Unit ♦ LIVESTOCK UNIT.

Animals Act (1971). An Act, applying in England and Wales, concerned with strict liability for damage done by animals, detention and sale of trespassing livestock, animals straying on the highway, and the protection of livestock against dogs.

Anion ▶ ION.

Annelida. An animal phylum comprising ringed or segmented worms such as EARTHWORMS, leeches and bristleworms.

Annual. A plant which completes its life cycle in one year, i.e. grows from seed, flowers, fruits, and dies. (▶ BIENNIAL, PERENNIAL)

Annual Labour Units. The estimated number of full-time worker equivalents on a farm. One annual unit represents the activity of a person working normal full-time hours throughout the year. Part-time workers are converted to full-time equivalents in proportion to their actual working time.

Annual Review. An annual review, also called the Price Review, of the conditions and prospects of the agricultural industry, introduced by the AGRICULTURE ACT, 1947, consideration being given to the action needed to promote and maintain stability and efficiency in agriculture for food production, and setting farm prices (excluding those for poultry and horticultural produce) for each year. Initially the review involved direct negotiations between the Government and the NATIONAL FARMERS' UNION. This process was later broadened to encompass wider consultation with other interest groups such as landowners, farmworkers and consumer organisations, and was accompanied by the publication of the 'Annual Review of Agriculture' White Paper. Since Britain joined the E.E.C. in 1973, most farm prices (except those for sheepmeat, wool, retailed milk and potatoes) have been determined by the Council of Ministers, and the various interest groups have made their representations, less directly, through their respective E.E.C. umbrella organisations (the N.F.U., for instance, through C.O.P.A.). The Government continues, however, to publish an annual White Paper reviewing the state of the industry.

Annual Ring. The annual growth of secondary wood (XYLEM) in stems and roots of temperate woody plants, seen as a series of concentric rings in a stem cross-section due to the distinction in size between xylem elements formed in spring and autumn. Annual rings are used to estimate the approximate age of a tree. Growth is constant throughout the year in the tropics and no rings are seen. Annual rings may also be seen in cow horns.

Anther. The part of a STAMEN containing POLLEN.

Anthony. The smallest pig in a litter. Derived from St Anthony, the patron saint of swineherds.

Anthrax. A NOTIFIABLE DISEASE of animals caused by a bacterium, *Bacillus anthracis*, which lingers in the soil and in infected carcases. It occurs world-wide but is endemic in Europe. In Britain it occurs in cattle and pigs, rarely in sheep and horses, and the chief source of infection is imported infected feed – usually from India and African countries. The disease is characterised by high fever, sometimes diarrhoea, and death in 24–48 hours with blood usually emanating from the body orifices. There is a legal requirement for carcases to be burnt.

Antibiotics. Antibacterial drugs obtained from living organisms, such as moulds, bacteria or other micro-organisms, which act by inhibiting the growth and multiplication of germs and other bacteria in the body. Examples are penicillin and the tetracyclines. Antibiotics have no effect on illnesses caused by viruses.

Antibody. A natural substance produced by the body in response to the presence of foreign substances (antigens), such as bacteria or viruses, to protect itself against disease or infection. The antibody neutralises the antigen by combining with it. Most VACCINES work by stimulating the formation of antibodies against a particular disease.

Antigen ♦ ANTIBODY.

Antiseptic. A substance which inhibits the growth of, or kills, germs. Commonly applied as disinfectants to skin wounds or to timber to counteract infection.

Antitoxin. A type of ANTIBODY that acts against a poison or toxin that has entered the body.

Aorta. The main artery leaving the heart, through which passes the arterial blood supply to an animal's body.

Aphid. A plant louse, also called greenfly. Aphids are small, soft-skinned, often green plant-bugs. They suck sap mainly from young leaves and shoots, causing reduced growth and vigour, and leave a sweet, sticky excretion (honeydew) which is eagerly eaten by ants and which occludes the leaf and shoot pores. Often found in large numbers on wild and cultivated plants. Many are pests, e.g. BEAN APHIS.

Apiary. A place where BEEHIVES are kept.

Apiculture. Bee-keeping.

Apple. Many edible varieties of apple have been cultivated from the wild crab apple (*Pyrus malus*). Both culinary and dessert types are grown, the latter being the most important fruit crop in the U.K. with the main production areas being the South-East (particularly Kent), Herefordshire and Worcestershire. Cider apple production is mainly restricted to the West Midlands and South-West England. The area of apple ORCHARDS (excluding cider apples) has been declining in recent years and in 1979 stood at an estimated 30 000 hectares (*c.* 75 000 acres).

Apple and Pear Development Council. A producer-financed organisation, established in 1966 by Order in Council, responsible for promoting consumption of English apples and pears. It is involved in all areas of publicity activity, in market and consumer research, new variety development, and in marketing policy. In 1980 it was assigned additional functions related to the certification of products and to the promotion of scientific research, of inquiries into methods of production and materials, and of cooperation or co-ordination in supply, production and marketing. The Council is financed by an annual charge on growers with 2 hectares or more of land planted with 50 or more apple or pear trees (other than cider apple or perry pear trees).

Apple Blossom Weevil. An insect (*Anthonomus pomorum*), approximately $\frac{1}{2}$ cm long, laying its eggs in apple and pear flower buds, on which the larvae feed so that no fruit develops.

Apple Sawfly ♦ SAWFLY.

Apple Sucker. A plant louse, *Psylla mali*, able to jump with its large hind legs. Its pale yellow, stump-winged offspring suck sap from opening apple flowers, causing fading and browning.

Appointment Grade. An agricultural worker, not less than 20 years of age, appointed to a managerial position with two other whole-time workers under his control (Appointment Grade I), or to a position with responsibility for day-to-day supervision of farming operations normally with one other whole-time worker under his control (Appointment Grade II). Terms defined in the Agricultural Wages Order. (◗ AGRICULTURAL WAGES BOARD)

Apprentices. Formal apprenticeship in agriculture began in Great Britain after the Second World War. The Scottish Apprenticeship Council for Agriculture and Horticulture began its scheme in 1948, and in 1953 the Agricultural Apprenticeship Council introduced a scheme in England and Wales. In 1969 the AGRICULTURAL TRAINING BOARD took over administration of both schemes which were amalgamated in the early 1970s with the dissolution of the two Councils.

Apprenticeship in agriculture is the traditional route to CRAFTSMAN status. Apprentices undergo training and supervised practice at work on farms, attend classes of associated further education and take specified PROFICIENCY TESTS.

Approved Bull. A bull approved by the Ministry of Agriculture providing semen for ARTIFICIAL INSEMINATION. Health, genetic qualities and physical characteristics of the bull and its progeny are conditions of the 'approval', which may be varied or withdrawn if appropriate.

Approved Chemical ◗ AGRICULTURAL CHEMICALS APPROVAL SCHEME.

Aqueous Ammonia ◗ AMMONIA.

Aquifer. A bed of rock which holds water and allows water to percolate through it. (◗ ARTESIAN WELL)

Arable Crops. Annual crops which require the land to be ploughed and cultivated (unless direct drilled) and seeds to be sown, as distinct from those which yield a product without such annual activities (i.e. fruit, grass, etc.). The main arable crops grown in the U.K. include the CEREALS, POTATOES, ROOTCROPS, and various VEGETABLES.

Arable Land. Land capable of being ploughed, that is, fit for TILLAGE, as opposed to that only suited for PASTURE. The term generally means land ploughed at regular intervals.

Arachnida. A class of ARTHROPODS including spiders, ticks, mites, harvestmen and scorpions. They are mainly carnivorous, usually with six pairs of limbs, four for walking and the first pair as jaws. They have no antennae, simple eyes and the head and thorax are fused together.

Arbitration. The hearing or determination of a dispute by means of an arbiter, an unprejudiced person, rather than through a court of law. Both parties agree in advance to accept the arbiter's decision. In agriculture, arbitration can be general, e.g. over the sale of cereals, or statutory under the AGRICULTURAL HOLD-INGS ACT, e.g. to determine the rent of agricultural land.

Arboriculture. The systematic culture and care of trees and shrubs.

Arbour. A grass-plot, garden, herb-garden, or orchard. Term now obsolete.

Are. A unit of metric land measure, 100 sq. metres. (♦ HECTARE)

Area of Outstanding Natural Beauty. An area (in England and Wales) designated under provisions of the National Parks and Access to the Countryside Act, 1949, by the COUNTRYSIDE COMMISSION, as one with high landscape qualities. The local planning authorities in A.O.N.B.s are obliged to consult the Countryside Commission when preparing development plans. In contrast to National Parks, the powers of the local authority in an A.O.N.B. are concerned with conservation of natural beauty, grant aiding tree planting, clearing eyesores, providing public access to the countryside and setting up warden services.

Arenaceous. Sandy. Arenaceous soils have a high proportion of sand particles. Sandstone is an arenaceous rock. (♦ ARGILLACEOUS)

Argente. A breed of rabbit of French origin, with a silvery coat. There are several shade varieties, e.g. Argente Blue, Argente Silver, etc.

Argillaceous. Clayey. Argillaceous soils have a high proportion of clay particles. Mudstones and marls are examples of argilla-ceous rocks. (♦ ARENACEOUS)

Arid. A term applied to land that is dry, parched and often devoid of vegetation. Also means deficient in rainfall – usually applied to a climate or region where rainfall is barely sufficient to support vegetation.

Ark. 1. A corn bin.
2. A poultry house, usually mobile, and used for the out of doors rearing of birds. Triangular in cross section, sometimes with a slatted floor, usually with one end covered for roosting purposes. Less common than they used to be. Also a movable shelter for pigs kept under FREE-RANGE. (◗ PIGGERY)

Arpent. An old French measure for land. Still used in Quebec and Louisiana, varying from $1\frac{1}{4}$ acres to $\frac{5}{6}$ of an acre.

Arris ◗ COMB.

Arrish. Stubble or a stubble field.

Arsenic. A chemical element, steel-grey, brittle and crystalline. Found naturally in a number of minerals. Compounds are very poisonous and are used as weed-killers and pesticides. (◗ LEAD ARSENATE)

Arterial Drain. A watercourse which carries water draining from smaller ditches.

Arterial Drainage. Drainage work on major rivers and watercourses which are under the control of Regional Water Authorities, INTERNAL DRAINAGE BOARDS or local Authorities. (◗ FIELD DRAINAGE)

Artesian Well. A well sunk into an AQUIFER sandwiched between impermeable strata in which the hydrostatic pressure normally forces a continuous flow of water up the well. The pressure is due to the well outlet being below the level of the water source. Normally occurs in basins (e.g. the London Basin).

Arthropod. The Arthropoda form the most numerous phylum of the animal kingdom. Characterised by an external skeleton of hardened cuticle, divided into segments some of which bear jointed paired limbs (the word arthropod means jointed feet). Includes insects, spiders, millipedes, centipedes, crabs and shrimps, etc.

Artificial Fertilizers. FERTILIZERS made by chemical processes or mined. Also known as inorganic (mineral) fertilizers, chemical fertilizers, artificial manures. (♦ COMPOUND FERTILIZER, ORGANIC FERTILIZERS, MANURES, STRAIGHT FERTILIZER)

Artificial Insemination. The introduction, by an instrument, of semen collected from a male into the uterus of a female for animal breeding purposes. Called A.I. for short. First begun on a commercial basis in the U.K. in 1942. A bull produces 50–100 times more semen in a normal mating than is needed for a cow to conceive. Thus, by the collection, dilution, and refrigerated storage of a bull's semen, it is possible for that bull to sire, via A.I., far more progeny during and even after its lifetime than would be possible by normal mating. A.I. also reduces the spread of venereal disease and enables farmers to economise by dispensing with their own or a commercial bull. Diluted semen is stored at − 196° C., after the addition of glycerol, and may be kept for years. A.I. of cattle is strictly controlled in England and Wales by the Ministry of Agriculture, with bulls providing semen requiring 'approval' (♦ APPROVED BULL) and the handling of semen is licenced.

A.I. plays an important part in cattle breeding, most of the dairy cattle in England and Wales, and a significant proportion in Scotland, being bred in this way, through centres operated by the MILK MARKETING BOARD and other organisations. There are also a number of pig A.I. centres. In N. Ireland the A.I. service is run by the Department of Agriculture. Substantial sales of semen are made to overseas buyers seeking to improve their herds. (Source: 'Agriculture in Britain', Central Office of Information, R5961/1977).

Artificial Manures ♦ ARTIFICIAL FERTILIZERS.

Ascaris. A genus of Nematode worm (♦ ROUNDWORMS) which infests the small intestine.

Ascomycetes. A class of fungi, known as Sac Fungi, which carry spores in a sac or ascus. They range in size from very small to quite large, the fruiting bodies of the latter often being rounded or tuberous, not toadstool-shaped. Many are parasites; many cause plant diseases; others 'ripen' cheeses.

Asexual Reproduction. Reproduction without GAMETES. In plants it consists of either spore-formation or VEGETATIVE REPRODUCTION.

Ash. 1. Constituent of MILK comprising various salts in solution, of which calcium and potassium phosphates and sodium chloride predominate, and also various VITAMINS (A, B, C, D and E). In normal milk the ash proportion remains fairly constant at about 0.75% (◗ MILK COMPOSITION). A significant reduction may indicate added water.
2. A term used in the analysis of FEEDINGSTUFFS for the residue left after thorough burning and the destruction of ORGANIC MATTER. Otherwise called mineral matter.

Aspergillosis. A fungal disease of chicks resulting from inhalation of spores of *Aspergillus* from damp or mouldy litter or feedingstuffs, causing the growth of white nodules in the lungs, and resulting in breathing difficulties.

Ass. A small, usually grey, long-eared animal of the horse genus. Also known as donkey. Used as a draught animal and crossed with the horse to produce a mule.

Assart. To reclaim for agriculture by grubbing. Assarted land is that cleared of trees and bushes.

Association. A major plant COMMUNITY in which more than one plant species is dominant, as in mixed deciduous woodland. Also applied in modern usage to any small unit of natural vegetation. (◗ SOIL ASSOCIATION)

Association of Agriculture. A voluntary body established in 1947 and recognised as an educational charity. It provides an educational advisory service to schools, issues publications, promotes school-farm links, arranges farm visits, organises conferences and acts as an agent of the Countryside Commission in the promotion of farm open days for the public. Its aim is to promote a better understanding of farming and the countryside.

Atavism. The appearance in an animal of ancestral, but not parental, characteristics, usually after an interval of several or many generations. Also called reversion and loosely known as 'throw-back'.

At-foot. Suckling. Thus a ewe with lambs at foot has suckling lambs.

Atrophic Rhinitis. Nasal inflammation in young piglets with rapid uneven growth of the snout, causing deformity.

Attenuated Strains. Strains of pathogenic micro-organisms, mainly bacteria and viruses, which have lost their virulence.

Attested Area. An area in which a programme of specific animal disease eradication is operating, declared as Attested, the disease having been reduced to a very low level. Specifically applied to BRUCELLOSIS and BOVINE TUBERCULOSIS.

Attested Herd. A herd certified by the Ministry of Agriculture as being free of TUBERCULOSIS.

Attested Herds Scheme. A scheme introduced in 1935, to replace the earlier Tuberculosis Order, as a measure to control BOVINE TUBERCULOSIS. A Register of ATTESTED HERDS was established, with inclusion dependent on herds receiving a Certificate of Attestation, having been subjected to a number of tests and showing no REACTORS. Area eradication began in 1950 extending the scheme on a voluntary basis, later introducing compulsory slaughtering of reactors. Bovine TB was virtually eradicated from British herds by 1960 when the whole U.K. was declared an Attested Area.

Auction. A sale at which property (in agriculture, generally a farm, equipment or animals) is offered for sale on the understanding that the highest bidder obtains it. The seller may put a reserve price on the property below which it may not be sold. A DUTCH AUCTION is one in which the auctioneer starts with a high price and reduces it until a bid is made.

Auger. 1. A tool for boring holes. A soil auger is used to bore into the soil and withdraw a small sample for observation.
2. A mechanism for conveying granular materials (mainly grain) or slurry, comprising a power-driven helical screw in a close-fitting tube.

Aujeszky's Disease. A virus disease of cattle and pigs made notifiable (♦ NOTIFIABLE DISEASE) in 1979, characterised by intense itching of the hindquarters. Named after the Hungarian, Aujeszky, who first described it in 1902. Cattle lick the infected part bare and may bite it. Paralysis occurs within 24 hours and death within 48 hours. In pigs the disease is milder with less mortality, characterised by loss of appetite, vomiting, diarrhoea, convulsions, acute salivation, and throat paralysis. Sows may abort. The disease also affects dogs and rats which can transmit the disease by biting. Also called Pseudo-rabies, Infectious Bulbar Paralysis, and mad itch.

Auricle. 1. A small, ear-like, extension at the base of a leaf, especially in grasses.
2. One of the chambers of the heart which receives blood from the veins, passing it to the VENTRICLE.

Auroch. The now extinct wild ox of central Europe (*Bos urus*), regarded as the progenitor of modern cattle.

Autopolyploid. A POLYPLOID organism in which all the CHROMOSOMES come from the same species. (♦ ALLO-POLYPLOID)

Auto Recorder. A milking machine comprising a container on a spring balance which records the weight of milk from each cow.

Auto-sex-linked. Pure breeds of poultry in which day-old male and female chicks respectively exhibit different colours.

Autotrophic. Organisms able to manufacture their own food from inorganic materials using energy from an outside source. Most green plants are completely autrotrophic, manufacturing organic foods from carbon dioxide, water, and mineral salts, using sunlight energy fixed by CHLOROPHYLL. The process is known as PHOTOSYNTHESIS. Some bacteria are autotrophic, using energy obtained by oxidising inorganic substances. NITRIFYING BACTERIA in the soil are examples.

Autumn Cleaning ♦ STUBBLE CLEANING.

Autumn Fly. A non-biting fly (*Musca autumnalis*) which irritates cattle in large numbers and feeds on nose and eye secretions and open wounds. They cause cattle to huddle together and cease feeding.

Auxins. A group of plant growth-regulating substances or hormones, produced by growing tips of stems and roots. Artificial or synthetic auxins have also been produced which are important in agriculture and horticulture, e.g. inhibiting sprouting of potato tubers and thus lengthening the storage period, and preventing fruit drop in orchards. Some synthetic auxins exhibit differential toxicity, 2, 4-D is toxic to DICOTYLEDONS, but not to MONOCOTYLEDONS, and is used successfully to control weeds in cereal crops and lawns. (◗ GROWTH REGULATOR)

Available Nutrients. Chemical elements or compounds in the soil that can be readily absorbed and assimilated by growing plants ('available' should not be confused with 'exchangeable'). (◗ PHOTOSYNTHESIS)

Available Water. The portion of water in the soil that can be readily absorbed by plant roots. Most soil scientists consider it to be the water held in the soil against a pressure of up to about 15 bars. (◗ FIELD CAPACITY)

Avena. The generic name for OATS.

Average All Pigs Price. The estimated average price of all types of pig marketed in the U.K. each week, based on representative DEADWEIGHT and LIVEWEIGHT quotations.

Away-going Crops. A crop sown by a tenant farmer before leaving the farm at the end of the tenancy, which he is permitted to return to, harvest and remove. Also known as FOLLOWING CROP or OFF-GOING CROP. (◗ HOLDOVER)

Awn. A small bristle at the tip of a flower or fruit, especially in the grasses, or of a leaf. In barley and oats the awn is attached to the husk and encloses the grains.

Awner. The part of a THRESHING MACHING which cuts off the awns, consisting of knives rotating on a fast-moving spindle.

Axil. The angle between the upper side of a leaf and the stem on which it is borne; the normal position for LATERAL BUDS.

Axillary. Growing from an axil. Thus an axillary bud grows from the axil between stem and leaf.

Axle-tree. The beam on a cart or wagon connecting two opposite wheels.

Aylesbury. A breed of large duck, white, generally with white to pinkish white skin and bill, bred for the table.

Ayrshire. A breed of hardy, relatively disease-free, dairy cattle, with MILK YIELDS second only to the FRIESIAN breed. White-and-brown, or occasionally white-and-black in colour. The distinctive udder is symmetrical with well positioned, often small, teats. The slender, wide set horns grow out and incline upwards. They are usually removed shortly after birth. The milk contains small fat globules and is very suitable for cheese making (*see p. 91*).

Azotobacter. Nitrogen fixing bacteria present in the soil. (◆ NITROGEN CYCLE, NITROGEN FIXATION)

B

B-Horizon ♦ SOIL HORIZON.

Babcock Test. A test for determining the BUTTERFAT content of milk and cream, involving the use of sulphuric acid to break down the proteins which surround the fat globules in the milk, releasing the fat for measurement. The test is still used in the U.S.A., but in Britain the MILK MARKETING BOARD now uses the GERBER TEST.

Baby Beef. The meat from cattle sold after fattening between the ages of 12 and 18 months.

Bacillary White Diarrhoea. An often fatal bacterial disease of young chicks caused by *Salmonella pullorum*, which causes serious losses between 4 and 14 days of age. It is inherited from parent stock which themselves have survived an outbreak. Carrier parents can be detected by a simple blood test. Known as B.W.D. for short. Also called Pullorum disease.

Bacillus. A genus of spore-producing bacteria. Also a general term for rod-shaped bacteria.

Back Band. A strap, rope or chain passing over a cart saddle, and holding up the shafts of the cart.

Back Board. A board at the back of a cart.

Back Chain. A chain fastened across the back of a DRAUGHT HORSE to the ridger rods on the shafts of a cart or wagon. Also called a ridger.

Back Country. Rural areas not thickly populated.

Back End. The later part of a season; the late autumn.

Backwoods. Remote, uncleared forest land.

Bacon. The meat from the back and sides of a pig, preserved by CURING with brine. Bacon may be sold 'green' (♦ GREEN BACON) or after wood-smoking which produces strong-flavoured 'smoked bacon'. (♦ WILTSHIRE CURE)

Bacon and Meat Manufacturers' Association. A limited company formed in 1976 from the British Bacon Curers' Federation and the Meat Manufacturers' Association. It represents the interests of its members which are companies in the U.K. meat processing industry. These interests are principally to ensure adequate supplies of good quality raw materials at reasonable prices and to support farmers by maintaining a viable industry to ensure a ready market for meat products. Advice is provided to members, particularly on interpreting and dealing with U.K. and E.E.C. legislation.

Bacon Pig, Baconer ♦ PIG.

Bacon Side. A half carcase of a BACON PIG.

Bacteria. Minute, single-celled organisms which cannot be seen by the naked eye. Most are rod-shaped from *c.* 0.0005–0.005 mm long (1/50 000–1/5 000 in.). They are commonly called 'germs' or 'microbes' and inhabit soil, water, air and other living organisms in millions. Some are responsible for diseases, e.g. BRUCELLOSIS in cattle. Others are essential for maintaining soil fertility and aid in the decomposition of ORGANIC MATTER. Amongst the most important are those capable of NITROGEN FIXATION in the soil and root nodules of certain plants (♦ AZOTOBACTER, NITROGEN CYCLE)

Bactericide. A substance capable of killing BACTERIA.

Bacteriolysis. The destruction of a bacterium by an ANTIBODY.

Bacteriophage. A VIRUS that parasitises BACTERIA. Also known as a phage.

Badderlock. An edible seaweed (*Alaria*) resembling tangle. A Scottish term.

Badger. The largest British mammalian carnivore (*Meles meles*), with a striking black and white striped face, nocturnal and living in holes in the ground (sets). Widespread and fairly common. The Badgers Act 1973 introduced controls for the killing of badgers and in 1977 certain 'control areas' were established to deal with and destroy, if necessary, badgers infected with BOVINE TUBERCULOSIS.

Badging Hook ♦ BAGGING HOOK.

Badland. A type of land generally devoid of vegetation and broken as a result of serious erosion. Named after the so-called 'Bad Lands' in Dakota, U.S.A.

Bag. The udder of a cow.

Bag Muck. An old term for ARTIFICIAL FERTILIZERS, due to the fact that they are applied from a bag.

Bag Up. A descriptive term for a cow with an enlarged udder prior to the birth of a calf, due to development of milk vessels.

Bagging Hook. A sickle-like tool with a square tip and smooth cutting edge. Also known as fagging-hook, bag-hook, badging-hook. Used for the cutting of corn, pulse crops or grass. (♦ SACK)

Bail. 1. A division between the stalls of a stable.
2. An Australian term for a frame holding a cow's head during milking.
3. A small mobile shed for milking cows in the fields, mounted on skids for towing by tractor and divided into 6–8 stalls of a walk-through type. (♦ HOSIER SYSTEM)

Left: A herringbone bail in a fixed site, fitted with bulk loft.
Right: Accompanying dairy shed.

Bail Milking. The milking of cows in a portable shed or BAIL, either moved from field to field frequently (♦ HOSIER SYSTEM) or kept in a relatively permanent and sheltered position. Sometimes the cattle are kept in temporary or permanent yards during the winter and milked in an adjacent bail.

Bailiff. A person employed to look after or manage a farm on behalf of the owner, receiving either a salary or share of the profits.

Bait. A mixture of hay, oats, chaff and other dry food used as horse feed.

Bakers. Potatoes sold for human consumption, which are of WARE POTATO standard, and which will not pass through a 65 mm horizontal mesh when manipulated, but will pass through a 90 mm mesh. (♦ MIDS)

Bakewell, Robert (1735–95). An English agriculturist who farmed in Leicestershire. A pioneer of livestock breeding and growing improved foodstuffs. The originator of the improved LEICESTER sheep and the improved LONGHORN cattle, and breeder of cart horses and pigs.

Balance plough. An old type of one-way plough with the right- and left-hand bodies (♦ BODY) attached to separate beams and handles, the beams being connected together at an angle, and having a common axle and pair of wheels. While one plough operated the other was carried more or less vertically in front of it. At the end of a furrow the working beam was raised and the other lowered. Also called a tipping plough.

Balanced ration. A ration in which the nutrient ingredients are correctly proportioned for the needs of the type of stock eating it. (♦ MAINTENANCE RATION, PROTEIN EQUIVALENT, STARCH EQUIVALENT, NUTRITIVE RATIO)

Bald Faced. Having white on the face, as a horse.

Bale. 1. A compressed package of hay or straw for easy handling and storage, tied usually with wire, sisal or plastic twine. Bales may be square, rectangular, round or flat depending on the type and normal operating density of the BALER used.
2. A portable field milking shed. Also called BAIL or Hosier bale. (♦ HOSIER SYSTEM)

Bale Loader. A device attached to the front loader arms of a tractor to lift or clamp groups of BALES for loading onto a trailer or stack, or into a barn. Commonly used in conjunction with a BALE SLEDGE drawn behind a BALER. Also called BALE SLAVE.

Bale Slave ▶ BALE LOADER.

Bale Sledge. A sledge towed behind the delivery point of a ram BALER, on which the operator stacks bales (usually 6 or 8), although many are now unmanned. These stacks are removed from the sledge periodically. This process allows bales to be grouped rather than scattered indiscriminately about a field by the baler, and facilitates loading (▶ BALE LOADER) and transport to the barn.

A ram baler and bale sledge.

Baler. A machine that picks up HAY or STRAW from a SWATH or WINDROW, compressing and tying it into BALES. The density to which the bales are compressed can be adjusted according to the crop condition or other requirements. The most popular balers produce rectangular bales between 45 and 90 cm long, but normally 35 × 45 cm in cross section, and weighing 9–18 kg. Balers producing large 500 kg bales are becoming increasingly popular when large quantities of straw need handling.

Baler Twine. Twisted cord used in a BALER to tie bales. Nowadays polypropylene twine has largely replaced sisal and wire.

Balk ◗ BAULK.

Ball. A BOLUS for a horse.

Balling Gun, Balling Pistol. An instrument which 'shoots' large pills down the throat of a sick horse.

Bampton Nott. An ancient breed of sheep which existed in the Bampton district, crossed with the LEICESTER in the early nineteenth century and improved to become known as the DEVON LONGWOOL.

Banding. 1. The tying of bands of grease round fruit trees to trap insects.
2. The fastening of corn in SHEAVES. (◗ BANDSTER)

Banding-in. The tying together, at breast height, of the several BINES arising from a hop plant.

Bandster. A person who binds sheaves of corn after the reapers (obsolete).

Bane ◗ LIVER FLUKE.

Bantam. A small variety of domestic fowl with feathered legs. Said to originate from the town of Bantam in Java.

Bar Pig. A HOG; a castrated male pig. Also called barrow pig.

Barb. 1. A swift breed of horse.
2. A dark-coloured fancy pigeon.

Barban. A TRANSLOCATED HERBICIDE, used either alone or in mixtures to control WILD OATS and other grass and broad-leaved weeds.

Bare Fallow. Land, particularly heavy land, left FALLOW for a year, during which it is ploughed several times so that the sun and wind dry the soil during the summer months, desiccating and killing perennial weeds. Success depends on summer rainfall. Bare fallowing is used less frequently nowadays due to the availability of HERBICIDES.

Bare Land Holding. Farm land lacking the basic buildings necessary for agricultural production.

Barley. A CEREAL (*Hordeum sativum*), probably the first to be cultivated by man, grown for its grain which is used either as stock feed or for beer making. The bearded seeds in the ear are arranged in 2, 4, or 6 rows. Two sub-species are important in the U.K.:
  (*a*) Two Rowed Barley (*H. distichon*), usually sown in spring, used for malting.
  (*b*) Six Rowed Barley (*H. hexadistichon*), usually sown in autumn, used for stockfeeding.
  Many varieties of 'winter' and 'spring' barleys are commercially available (*see p. 95*). (⬥ MALTING BARLEY)

Barleycorn. 1. The grain of barley from which malt is made (personified as John Barleycorn). Also a term for a single grain of barley.
  2. A measure of length equal to ⅓ inch.

Barm. Yeast. Also the froth of fermenting liquor.

Barn. A building in which grain, hay and implements, etc., are stored.

Barn Dried Hay ⬥ BATCH DRYING, HAYMAKING, STORAGE DRYING and TUNNEL DRYING.

Barn Hay Drying ⬥ STORAGE DRYING and TUNNEL DRYING.

Barn Machinery. A general term for implements or machines which are not used in the fields, e.g. HAMMER MILL, GRAIN DRIER, weighing machines, chaff cutters, etc.

Barnevelder. A heavy breed of fowl, with black and white varieties, although similar in colour to a partridge. It has a single comb, yellow legs, and the hens produce dark brown eggs.

Barnyard. The area adjoining a barn. The farmyard.

Bar-point Share. A plough SHARE, which was mostly used in Scotland, for stony conditions, consisting of a 50 cm (20 in.) bar of high quality steel taken through the wing to form the point. As the point wore it could be adjusted and turned over. It lasted longer than other shares in abrasive conditions.

Barred Feathers. Poultry feathers exhibiting a horizontal pattern of black and white stripes or bars. Other colours may be combined with black such as buff or gold, thus giving buff-barred and gold-barred feathers.

Barrel Bulk. A measurement of five cubic feet.

Barren. (*a*) (Of animals) incapable of producing offspring, due to infertility or sterility.
(*b*) (Of plants) not producing fruit or seed.
(*c*) (Of land) bare of vegetation, incapable of producing useful crops; arid.

Barrener. A female farm animal adjudged to be BARREN. A term usually applied to cows.

Barrow Pig ▶ BAR PIG.

Barton. A farmyard.

Base Exchange Capacity ▶ CATION EXCHANGE CAPACITY.

Base Saturation Percentage. The extent to which the ADSORPTION COMPLEX of a soil is saturated with exchangeable CATIONS other than Hydrogen and Aluminium. It is expressed as a percentage of the total CATION EXCHANGE CAPACITY.

Basic Price. A price fixed annually to indicate levels of E.E.C. support for certain fruit and vegetables (apples, pears, cauliflowers and tomatoes in the U.K.). For fruit and vegetables INTERVENTION BUYING actually takes place at Buying-In Prices set at between 40 and 70% of the basic price. Provision exists in theory for market support of pigmeat with intervention at between 75 and 92% of the basic price. (▶ COMMON AGRICULTURAL POLICY)

Basic Seed ▶ SEED CERTIFICATION, SEED POTATOES.

Basic Slag. A by-product of steel manufacture used as a FERTILIZER. Fine, powdery, grey-black, and rich in PHOSPHATE and LIME.

Basidiomycetes. A large group of FUNGI, containing most of the familiar types such as mushrooms, toadstools, bracket and jelly fungi and puff balls, and also parasitic forms such as RUSTS and smuts. Spores are produced by club-shaped cells called basidia.

36

Basil. 1. An aromatic herb used in cooking.
2. A sheepskin, roughly tanned and undressed.

Bast. PHLOEM. The inner bark of a tree, used in long strips for basket- and mat-making, and for tying plants to stakes.

Bastard Fallow. Land left FALLOW for half the summer, between the harvesting of one crop (usually SILAGE or HAY) and the sowing of the next crop, during which it is ploughed to kill perennial weeds by desiccation. The procedure is the same as for a BARE-FALLOW except that it does not waste a year. Success is very dependent on favourable weather conditions. Also known as half fallow, pin fallow, short fallow or summer fallow.

Bastard Trenching. A method of deep digging in which the soil layers are replaced in their natural sequence, as opposed to ordinary trenching in which the sequence of soil layers taken with each spit are reversed when replaced.

Bat. A pole used in hop gardens for supporting wire-work, usually Sweet Chestnut or Larch. (◆ INSIDE BAT, OUTSIDE BAT)

Batch Drying. 1. A method of drying BALES of HAY in batches in a barn, usually using heated air blown through a ventilated floor. Batches of about four layers are dried at a time until moisture content is reduced below 20%, are then removed for storage, and the next batch introduced. This method is less popular nowadays than STORAGE DRYING and TUNNEL DRYING. (◆ HAYMAKING)
2. ◆ GRAIN DRIER.

Batfowling. Catching birds at night by showing a light and beating the bushes.

Batology. The study of brambles.

Batt. 1. The long wooden handle of a SCYTHE.
2. A faggot of stout sticks.

Battery Hens. Hens kept for their laying lives indoors in CAGES, normally in three-tiered rows. Laying batteries may be static, semi- or fully-automatic, depending on the method of feeding, watering and manure clearance. Battery housed birds achieve a higher rate of egg production than those under other systems (*see p. 72*). (◆ DEEP LITTER, FREE RANGE)

Baulk. 1. An unploughed strip of land left to mark the boundary between fields.
2. A ridge created by ploughing and then left unploughed.
3. A beam of timber.
4. A ridge (balk) in which potatoes are planted (◗ RIDGER).

Baulking Plough ◗ RIDGER.

Bay. 1. The space between two columns in a barn, or any recess in a farm building.
2. A stall in a stable.
3. The combined cry of hounds hunting an animal, or the last stand of an animal cornered by hounds.
4. A description of a horse, reddish brown to chestnut in colour.

B.C.P. Test. A test used to determine the presence of antibiotics in milk (◗ MILK QUALITY SCHEMES) which superseded the T.T.C. test in 1976. It involves the addition of a bacterial culture to a milk sample together with the colour indicator Bromo-Cresol-Purple (B.C.P.). If no antibiotics are present the culture grows, produces LACTIC ACID, and causes the B.C.P. to change colour through green to yellow. If antibiotics are present no growth occurs, no acid is produced, and there is no colour change.

Beam. 1. A large straight piece of iron or timber usually square in cross-section, forming one of the main structural members of a building, normally supporting rafters.
2. The main shaft of a plough, wooden or metal, to which are fixed the MOULDBOARD, COULTERS and wheels, and the handles of horse drawn ploughs (*see p. 338*).
3. The main trunk of a stag's horn.

Beamage. An allowance on an animal carcase for loss of weight by evaporation.

Bean(s). Various leguminous plants, and their seeds borne in pods, distinguished from VETCH, CLOVER and PEAS by having a square, hollow stem. They are tall, erect plants, varying in length according to variety, and develop a strong tap root bearing NODULES containing bacteria capable of NITROGEN FIXATION. When ploughed-in, valuable NITROGEN and ORGANIC MATTER is returned to the soil. The seeds or beans are rich in PROTEIN. The main types are:

Vicia faba, dry seeds of which (field beans) are used for stock-feeding, and the green seeds of which (broad beans) are used for human consumption.

Field beans are usually grown as a break crop in cereal ROTATIONS, particularly on heavy land. Both autumn- and spring-sown varieties are used. Autumn-sown beans, generally known as winter beans, ripen 2–4 weeks earlier than spring beans but are more susceptible to CHOCOLATE SPOT. Spring beans include the small, round or flat-oval tick bean (grown for the pigeon trade as a cash crop) and the larger, flat-oval horse bean (used for stock-feeding). Field beans were widely grown in the Second World War, after which the area grown declined. There is now renewed interest in the crop. Broad beans are mainly frozen and canned, although some are picked green.

Phaseolus vulgaris, the French bean, of which climbing and dwarf types exist. Only the latter is important in the U.K. It is mainly grown in the warmer south and east of England, under contract, for artificial drying, canning or freezing. It is harvested unripe with the pods, known as green beans, sometimes string beans. The dry ripe seeds are known as haricot or kidney beans. The dry seeds of one dwarf type, the navy bean, is imported from the U.S.A. for manufacturing baked beans. They have been grown in Britain but seem unlikely to become commercially successful.

Phaseolus coccineus, the runner bean, grown mainly as a market garden crop for its pods. Climbing and dwarf types exist.

Other types not grown in the U.K. include *Phaseolus lunatus,* the green and dry seeds of which are respectively called lima and butter beans, and *Glycine max,* the SOYA BEAN.

Bean Seed Fly. A purple insect pest (*Delia spp.*) which attacks French and Runner Beans, particularly seedlings.

Bean Straw. The HAULM of bean plants.

Bear. BARLEY. In Scotland now the little-grown four-rowed variety. Also called Bere or Bigg.

Beard. 1. The hairy tuft on the lower part of a goat's jaw, or on a turkey's neck (also called BRUSH).

2. A term for an AWN or threadlike spike, as on the ears of barley.

Beastings ♦ COLOSTRUM.

Beaters. 1. Steel bars on the drum of a THRESHING MACHINE or COMBINE HARVESTER which dislodge the grain out of the ears of cereals as they rotate. Also known as beater bars.
2. Those who rouse game in shooting or hunting.

Beaumont Period. A period of 48 hours during which temperatures do not fall below 10°C (50°F) and relative humidity remains above 75%. Outbreaks of POTATO BLIGHT may be expected within 3 weeks of such a period.

Beck. 1. A brook or small stream in the north of England. Equivalent to a BURN in Scotland.
2. A type of hand-hoe used to chop the soil around a hop plant.

Bed. 1. A garden plot. (♦ SEED BED)
2. A layer or stratum of rock.
3. A sleeping place for an animal. (♦ BEDDING)

Bedded Set. A young hop plant, having rooted in a nursery bed from a cutting.

Bedding. Litter for farm animals to sleep on, usually straw, shavings, sawdust, etc.

Bedrock. The solid rock beneath the soil.

Bee. A furry insect of the Order Hymenoptera. There are solitary and social bees. The latter live in colonies. Each colony has one Queen bee together with large numbers of female 'workers' and a fewer male 'drones'. The most highly socialised is the Honeybee (*Apis mellifera*) kept in hives by beekeepers for commercial HONEY production. Bumblebees (*Bombus spp.*) have much smaller colonies. A growing practice is the hiring and placing of BEEHIVES in ORCHARDS to promote pollination and fruit setting.

Bee Bread. 1. The pollen of flowers collected by bees and fed, together with honey, to the larvae.
2. The local name of several plants yielding nectar.

Bee Fold. An enclosure for beehives.

Beef. The meat of a bull, cow or steer. Beef breeds and beef cattle are those reared specifically for meat production rather than for milk.

Beef Bull. A bull reared and fattened to produce beef. Male cattle for fattening are usually castrated to produce STEERS. Bull carcases are leaner than those of steers, with darker muscle and a heavier forequarter. The advantage of beef bull production is that bulls grow faster, convert food more efficiently, and achieve greater carcase weight than steers.

Beef Carcase Classification Scheme. A scheme begun in 1972 and administered by the MEAT AND LIVESTOCK COMMISSION and operated at abattoirs. It is designed to identify carcases with common characteristics for the benefit of meat suppliers. Carcases are described by code in respect of five main factors; weight, sex, fatness (♦ FAT CLASSES), CONFORMATION, and age group.

Beef Cows. Cows and heifers kept mainly for rearing calves for beef production, as distinct from DAIRY COWS.

Beef Premium Scheme. An E.E.C. scheme, applied only in the U.K., which replaced the U.K. Fatstock Guarantee Scheme for cattle. A seasonally varied scale of weekly target prices is set for each marketing year and if the average U.K. market price for CLEAN CATTLE falls below the weekly TARGET PRICE in any week, a variable premium is paid to bring the producer's total return to the target price level. The premium is paid on a unit weight basis on all certified sales either LIVEWEIGHT or DEADWEIGHT and is limited to a maximum figure. Provision exists for separate rates to be paid in Northern Ireland and Great Britain and for INTERVENTION buying-in prices to be reduced by the amount of any premium paid.

Beef Shorthorn. A breed of compact, short-legged beef cattle. Variously coloured including red, white, red and white, and roan, with a short, broad head and short horns, usually curling down, although some types are polled. The breed has been exported widely throughout the world. Continuous crossing with HIGHLAND cattle has resulted in a new breed, the LUING.

Beehive. A box in which bees are kept to produce HONEY. Several types of single- and double-walled hives are in common use, the former favoured for commercial production. Hives contain two separate chambers, a lower brood chamber in which the Queen lays and rears her BROOD, above which is the honey chamber or supers containing frames in which the bees store surplus honey which is periodically removed.

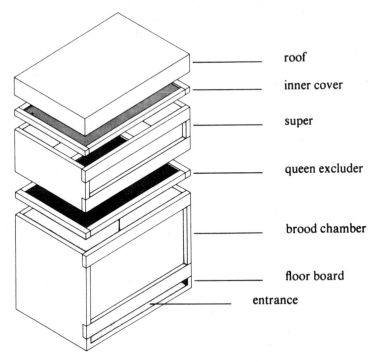

roof

inner cover

super

queen excluder

brood chamber

floor board

entrance

Exploded view of modern hive.

Beestings ♦ COLOSTRUM.

Beeswax. Wax secreted by bees and used to construct the cells of the HONEYCOMB. Used in polishes and cosmetics.

Beet. 1. A plant (*Beta spp.*) of the goosefoot family with a succulent root, used as a food and as a source of sugar. (♦ FODDER BEET, SUGAR BEET and MANGEL)
2. A sheaf of harvested flax.

Beet Cyst Nematode. A cyst-forming, plant-parasitic eelworm or roundworm (*Heterodera schactii*) which attacks BEET.

Beetle. 1. An insect of the Order Coleoptera, in which the fore-wings are reduced to hard and horny protective covers for the hind-wings and body.
2. A heavy wooden hammer with a long handle used for driving wedges and posts, or crushing paving stones, etc.

Beeves. Cattle.

Belgian Hare. A breed of rabbit with the appearance of a hare, having an abnormally long head, ears and legs. Rich red or chestnut in colour with dark, wavy ticking.

Bell. 1. To bellow or roar. The cry of a stag at rutting time.
2. The catkin containing the female flowers of the HOP. Also to be 'in flower'.

Bell Wether. The leading sheep of a flock, on the neck of which a bell is hung.

Bellows. An instrument producing a current of air to blow up a fire, commonly used by blacksmiths.

Belly Band. A saddle-girth. Also a strap fastened to the shafts of a cart and passing under the belly of a horse towing it, so that the cart is prevented from tipping backwards.

Belt. 1. A zone of country. (♦ GREEN BELT)
2. A term for SHEEP SCAB.

Belted Galloway. A variety of GALLOWAY cattle with a white belt round the body. Also called Beltie.

Beltie ♦ BELTED GALLOWAY

Benazolin. A TRANSLOCATED HERBICIDE, used only in mixtures, for controlling broad-leaved weeds.

Beneficial Cultivations. Those cultivations which enhance the condition of the land, or are good for crops.

Benefit Ratio. The ratio of the value of benefit to the cost of providing land drainage or flood alleviation works; applied to the agricultural potential of land to be protected.

Benefits. Benefits or advantages which, for the purposes of minimum wage rates fixed by the AGRICULTURAL WAGES BOARD, may be reckoned as payment of wages in lieu of payment in cash. Such benefits and their value, which may from time to time be prescribed, are defined in the AGRICULTURAL WAGES ORDER and include a house, whole milk, potatoes, board and lodging, and casual meals.

Bennet. A term used in the South of England for dry grass stalks. Also called BENT.

Benomyl. A SYSTEMIC fungicide.

Bent. A general term for short flowering stems of grasses projecting above the turf and old dried grass-stalks. Also applied to the genus *Agrostis*, some species of which are used for PASTURE and HAY. Also a term for a heath, hillside, slope, grassy area or unenclosed pasture.

Bere ♦ BEAR.

Berkshire. A breed of pig. Rusty black in colour but the feet, snout and tail tip are white. Short-faced with heavy jowls.

Berkshire Knot. A breed of sheep, also called Nott, found on the Berkshire, Hampshire and Wiltshire Downs which, together with the WILTSHIRE HORN breed, was crossed with the SOUTHDOWN to develop the HAMPSHIRE DOWN breed.

Berry. A fleshy fruit usually containing many seeds buried in pulp. Examples include tomato, gooseberry, currant (but not strawberry, raspberry or blackberry).

Betaine. A colourless, tasteless, crystalline substance present in sugar beet, producing a strong fishy taste on decomposition.

Beveren. A breed of rabbit. Most are coloured but some are white. They have well developed backs and hindquarters.

B.H.C. Benzene hexachloride, a persistent organochlorine insecticide used mainly in sprays and dusts. Particularly used to control WIREWORMS and other soil pests.

Bible. The third stomach of a RUMINANT.

Bid In. At an AUCTION, to overbid the highest offer.

Biddy. A hen.

Bident. A two-year-old sheep.

Bield. A Scottish term for a naturally protected spot such as an overhanging bank where sheep sometimes shelter.

Biennial. A plant that flowers and bears fruit only in its second year, then dies. Examples include carrot and beetroot. (◆ ANNUAL, PERENNIAL)

Biestings ◆ COLOSTRUM.

Biffin. A variety of cooking apple, deep red in colour, Also a baked apple of this kind, flattened into a cake.

Bifoliate. Having two leaves or leaflets.

Big Bud. A swelling of currant buds due to a gall-mite.

Bigg. Four-rowed BARLEY. Also called BEAR.

Bile. A bitter, yellowish green secretion of the liver, important in the digestion of FATS, and stored in the gall bladder. Also called gall.

Bill Hook. A hatchet with a curved point, used in hedge trimming, etc.

Billy Goat. A male goat. Also known as a BUCK.

Bin. 1. A large receptacle for storing FEEDINGSTUFFS, corn, fruit, etc.
2. A canvas container on a collapsible wooden frame used initially for storing hops when picked. Also called a crib.

Binder. A machine (full name: reaper-binder) which cuts and binds corn in sheaves. Powered by p.t.o., although older and heavier machines obtained power from a large BULL WHEEL. It produces long, unbroken straw, used in thatching. Binders are less frequently seen in Britain nowadays, harvesting by COMBINE HARVESTER becoming more universal. Sometimes called self-binder.

Bine. The slender flexible stem of a climbing plant. Especially the HOP where the term refers to the growth of stem, leaves and cones developed from a single basal shoot, and is sometimes applied generally to the entire aerial growth of the hop plant.

45

Bing. A feed passage, particularly for sheep.

Bink. A wasps' or bees' nest.

Biological Control. The control, sometimes destruction, of pests and parasites using other organisms, often natural predators.

Bird Damage. TOP FRUIT buds and ripe SOFT FRUIT are often taken by birds, particularly bullfinches. Farm crops suffer especially in winter and spring when they may be the only food source. BRASSICAS, CLOVERS and CEREALS are destroyed during winter and newly sown seeds of cereals, peas and beans may be eaten, particularly by Wood Pigeons, in the spring. Grain from lodged crops and grain shed during harvest is also eaten, and sugar beet can be damaged in summer when the fleshy parts of leaves are eaten. Controls include shooting, bird scarers and narcotic baits.

Bird Minding. Scaring birds away from fields and ORCHARDS by creating noise, e.g. shouting, firing guns, etc. Also called bird scaring or bird starving. (◆ BIRD DAMAGE)

Birdsfoot Trefoil. A perennial plant (*Lotus corniculatus*), with small yellow pea-flowers named from the claw-like appearance of its seed-pods. Grown for FODDER. Also known by several folk names, e.g. Tom Thumb and Eggs and Bacon.

Bit. The metal part of a bridle placed in a horse's mouth, for control.

Bite. Grazing. EARLY BITE is grazing in the early spring. LATE BITE is grazing at the end of the growing season.

Bitter Pit. A disease of apples, characterised by brown spots and depressions.

Black Aphis ◆ BEAN APHIS.

Black Dolphin ◆ BEAN APHIS.

Black Fly ◆ BEAN APHIS.

Black Grass ◆ SLENDER FOXTAIL.

Black Land. Heather moors, dark in appearance. Also a term for the dark, humus-rich, fenland soils. (◆ FEN)

Black Mustard. A cruciferous plant (*Brassica nigra*) grown for its seeds to produce table mustard. Seeds also contain up to 22% oil, extractable under pressure. Grown mainly in Lincolnshire, Norfolk, Huntingdon and Cambridgeshire. (◗ WHITE MUSTARD)

Black Rust. A fungal disease (*Puccinia graminis*) of wheat, oats, barley, rye and several grasses. Forms reddish brown (becoming black) spots or lines on stems and leaf sheaths. Develops late in Britain and seldom causes damage. Its ALTERNATIVE HOST is common barberry (*Berberis vulgaris*).

Black Welsh Mountain. A breed of sheep very like the WELSH MOUNTAIN but with a rich dark-brown wool.

Blackcurrant. A shrub (*Ribes nigrum*) grown for its small black berry. (◗ SOFT FRUIT)

Blackface ◗ SCOTTISH BLACKFACE.

Blackhead. A disease of turkeys caused by a protozoan, *Histomonas meleagridis*, infecting the liver and intestines. Symptoms include ruffing of feathers, loss of appetite and a mustard-yellow diarrhoea. A common cause of loss in turkeys, particularly when young.

Blackleg. 1. A bacterial disease (*Erwina atroseptica*) of potatoes which causes blackening and rotting of the stem base. Leaves lose colour, turn brown, and stems die. Brown rot infects the tuber. Principally a seed borne disease.
2. A fungal disease (*Phoma batae* and *Pythium spp.*) of sugar beet and mangold seedlings, which become black at ground level, and threadlike, then wilt and die.
3. A bacterial disease (*Clostridium spp.*) of sheep (also called gas gangrene) and cattle (also called black quarter or felon). In sheep *C. chauwaei* infect ewes via wounds, often at lambing, dipping or shearing, and lambs at castration or docking. It causes hot painful swellings in muscles which darken and may be gassy, and death within hours. In cattle, bacteria from soil infect minute wounds, particularly in calves, producing symptoms similar to sheep.

Blackquarter ◗ BLACKLEG.

Blade. 1. The flat or expanded part of a leaf. Cereals are 'in blade' when the flat, thin, long leaf is formed but before the corn ears have developed.

2. The thin cutting edge of a knife, axe, scythe, etc.

Blade Bone. The scapula, the flat bone at the back of an animal's shoulder.

Blae. Hardened clay or carbonaceous shale, often blackish or dark bluish, and spread on arable land to improve soil texture.

Blanket Peat Bog. One of Britain's three main soil types covering large tracts of Ireland and northern Britain. Dominated by *Sphagnum* mosses, developed where rainfall exceeds evaporation, and thus leached and acidic. Human activity has in many parts caused erosion of blanket bogs and they have tended to dry out and become moors dominated by HEATHER. (♦ BROWN EARTH, PODSOL, PEAT)

Blast ♦ BLOAT.

Blaze. A white mark on an animal's face.

Bleb ♦ STURDY.

Blet. Incipient internal decay in fruit.

Blight. A plant disease caused by fungal parasites and various insects, APHIDS, MILDEW, RUST, SMUT, etc. Usually a blight will attack a whole crop and sometimes a particular crop throughout a region. Nowadays the term is usually restricted to POTATO BLIGHT.

Blind. A plant which fails to flower or produce fruit.

Blind Gut. The CAECUM.

Bloat. The swelling of a cow's rumen due to gas from fermentation of green food, particularly lush grass containing white clover, or from frosted or mouldy food, and obstructions in the gullet. It causes respiratory distress and, in acute cases, death. Also called hoven, rumen tympany or blast.

Blood Mare. A thoroughbred MARE.

Blood Meal. A residue from a slaughterhouse, rich in PROTEIN, low in mineral content, used as a FEEDINGSTUFF, particularly in rations for pigs and poultry.

Blood Sports. Those sports involving the killing of animals, e.g. fox-hunting. (♦ FIELD SPORTS)

Blood Stock. Thoroughbred horses, bred through generations for their excellent qualities.

Bloodline. Succeeding generations of animals of a species which are specially bred to maintain the genetic composition responsible for specific desirable features or qualities. Maintaining a bloodline is also known as breeding true. (♦ CROSS-BREEDING)

Bloom. 1. A flower or flower BLOSSOM.
2. The delicate powdery dust on fruit.
3. The shine of the coat of an animal or on well-tanned leather.

Blossom. A flower or bloom, especially preceding the development of edible fruit.

Blossom Honey. Honey produced wholly or mainly from the nectar of blossoms.

Blossom Wilt. A fungal disease (*Sclerotinia laxa*) of apples, pears, plums and cherries, causing fruit to rot and blossom and young leaves to wilt and die.

Blow. 1. To develop BLOOM or BLOSSOM.
2. The condition of a cow with its stomach swollen by gases, mainly METHANE. (♦ BLOAT)
3. A term for soil blown by the wind. (♦ BLOWN OUT LAND)

Blow Fly. 1. A general name for a number of species of fly which deposit their eggs in flesh and carcases. Also called flesh fly.
2. The Bluebottle (*Calliphora spp.*), a two-winged fly with metallic blue wings. The larvae feed on animal matter and dung.

Blower ♦ FORAGE BLOWER.

Blown. 1. A descriptive term for sheep suffering from attack by maggots, particularly of the BLOW FLY.
2. A descriptive term for an animal suffering from BLOAT.

Blown Out Land. An area from which all or almost all the soil has been removed by wind erosion. Usually barren, and unfit for crop production.

Blue. A descriptive term of pigs with a blue-black patchy skin and white hairs.

Blue Albion. A hardy, rare, dual-purpose breed of cattle mainly found in the Peak District of Derbyshire. Blue, blue and white, blue roan, or blue roan and white in colour. It resembles the SHORTHORN from which it was bred.

Bluefaced Leicester. A large, long bodied, prolific breed of SHEEP, descended from the DISHLEY LEICESTER. Characterised by dark pigmented skin on the top of the head, ears, and down the back, and by a Roman nose, upright ears and fleece of tightly twisted wool. Rams are popularly used to produce crossbred ewes.

Blue-Grey. A hornless crossbred type of cattle resulting from mating a white SHORTHORN bull with a GALLOWAY or ABERDEEN ANGUS COW.

Bluestone ♦ COPPER SULPHATE, BORDEAUX MIXTURE.

Boar. An uncastrated male pig used for breeding.

Board of Agriculture. A chartered society existing between 1793 and 1822 to promote improved husbandry and to encourage increased agricultural production. It received Government grant aid and carried out local surveys of agricultural conditions. It became a statutory Government department in 1889, after pressure from the CHAMBER OF AGRICULTURE, assuming all Government responsibility for agriculture, land and animals in the U.K. Power was exercised by its President. In 1903 it took over the administration of fisheries laws. In 1911 the Board of Agriculture for Scotland was established, assuming all Scottish duties (except animal health). In 1919 forestry duties were transferred to a new FORESTRY COMMISSION and the Board was reconstituted into the Ministry of Agriculture and Fisheries. (♦ MINISTRY OF AGRICULTURE, FISHERIES AND FOOD)

Boarding. The practice of tilting a plough towards the ploughed land to increase the pressure on the MOULDBOARDS. (♦ CHECKING)

Bobby Calf. A term for the unwanted male offspring of a milk-producing breed of cattle, usually a CHANNEL ISLAND BREED, used for veal.

Bodge. An old measure for oats, approximately a gallon, or half a PECK.

Bodkin ▶ WHIPPLE-TREE.

Body. The operating part of a PLOUGH comprising a COULTER, SHARE and MOULDBOARD. Most modern ploughs have several bodies.

Bog. In general terms a spongy, usually peaty, wet area of marshy land. The term is applied more strictly to wet, very acid PEAT, characterised by *Sphagnum* moss, found particularly in areas of heavy rainfall (e.g. central Ireland). Distinct from a FEN which has an alkaline to slightly acid peaty soil. (▶ BLANKET PEAT BOG)

Bogey. A low heavy cart used in the past for carrying HAYCOCKS.

Bogle. A SCARECROW.

Boiler. A PULLET or HEN which is sold for the table at the end of the LAYING PERIOD.

Bole. The trunk of a tree.

Boll. 1. An old measure of capacity for grain. In northern England varying from 2 to 6 BUSHELS; in Scotland usually 6 bushels.
2. An old measure of weight for grain equal to 140 lbs (63.5 kg).

Bolt. 1. A bundle of willow shoots or reeds about 3 ft in circumference.
2. To break away, as when a horse suddenly dashes off.
3. A sieve for separating BRAN from flour. (▶ BOLTING CLOTH)

Bolter. 1. A term used of biennial plants, which normally take two years to flower and set seed, but which get out of hand and produce seed instead in the first year, thus reducing the harvest yield. Examples of crops which sometimes bolt include cabbage and sugar beet.
2. A BOLTING CLOTH.

Bolting. A bundle of straw.

Bolting Cloth. A fine cloth used for sifting meal. Also called bolter.

Bond. A twine made of twisted cornstalks once used to tie corn sheaves by hand.

Bondager. An old Scottish term for a COTTAR bound to render certain services to a farm. Also a female worker paid by a cottar to do fieldwork on his behalf for a farmer.

51

Bone. To remove the bones from a carcase.

Bone Flour. Finely ground bones, containing phosphate, used as FERTILIZER and also as a FEEDINGSTUFF.

Bone Meal. Bones, coarsely ground, after removing most of the fat. Used as a FEEDINGSTUFF or FERTILIZER. (♦ MEAT-AND-BONE MEAL)

Bonnet. The second stomach of a RUMINANT.

Bonney-clabber. An Irish term for milk naturally clotted on souring.

Boosey Pasture. A field in which cattle are allowed to be kept whilst they finish a farmer's stock of FEEDINGSTUFF when he quits a farm. (♦ HOLDOVER)

Boost. An instrument used to mark sheep, often with the owner's initials, usually using hot tar.

Bordeaux Mixture. A mixture of copper sulphate (bluestone or blue vitriol), LIME, and water, used as a fungicidal spray on crops. (♦ BURGUNDY MIXTURE)

Border Leicester. A breed of sheep evolved from BAKEWELL'S improved LEICESTER sheep, with long, close, shiny wool in separate locks or 'purls', soft white hair on the face and legs, a long bald head held high, with aquiline nose, and ears carried high. Very hardy and prolific, frequently crossed (*see p. 394*). (♦ GREYFACE)

Borecole ♦ KALE.

Boron. (B). A chemical element present in borax and boric acid, and essential to crops, especially roots, in trace quantities only (♦ TRACE ELEMENT). Excess LIME can induce deficiency, particularly in LIGHT SOILS, and result in DEFICIENCY DISEASES (e.g. BROWN HEART, HEART ROT).

Bosk. A thicket or little wood.

Bot Fly. A term applied to several species of two-winged flies parasitic on ungulates. Examples are *Oestrus ovis* which infests sheep, and *Gastrophillus intestinus,* the Common Horse Bot, the larva of which develops in the horse's intestine.

Bothy. A barely furnished dwelling for lodging farm workers on Scottish farms.

Bottle. A bundle of HAY or STRAW.

Bottle Feeding. The feeding of young animals with liquids, mainly milk, from a bottle fitted with a teat.

Bottom Growth. Those grasses and clovers in a pasture growing close to the ground, as opposed to the taller plants, or top growth.

Bottom Land. Alluvial deposits such as a floodplain.

Bound Stock. Those animals, which usually have been bred on a farm, such as a flock of sheep or a dairy herd, which remain with the farm when it is sold or changes hands. Also used to mean sheep acclimatised to a HIRSEL.

Bout. A spell of work, particularly applied to ploughing and cultivating where it means a return trip across a field.

Bovine. Concerning cattle.

Bovine Leucosis. A cancerous disease, Leukaemia, of cattle. It may remain benign throughout life, or become malignant usually in cows 4 to 8 years old, resulting in fatality.

Bovine Tuberculosis. A chronic contagious bacterial disease (*Mycobacterium tuberculosis*) of cattle, virtually eradicated from Britain in 1960, following the operation of the ATTESTED HERDS SCHEME introduced in 1935, and a programme of area eradication (with compulsory slaughter of REACTORS) begun in 1950. (♦ TUBERCULOSIS)

Bovine Typhus ♦ CATTLE PLAGUE.

Bowery. An American term for a farm.

'Boy's Land'. A colloquial term for LIGHT SOIL, easily cultivated, as distinct from man's land. (♦ HEAVY LAND)

Brace. A pair of dead (shot) GAME birds.

Bracken. A tall fern (*Pteridium aquilinum*), also called Brake Fern, common on hillsides, HEATHS, MOORS and in woods, where burning or overgrazing allows it to establish itself extensively excluding heather or grasses.

Bract. A small leaf or scale with a relatively underdeveloped blade, in the axil of which a flower stalk develops.

Bracteole. A small BRACT.

Bradford Worsted Count. The traditional method for measuring the fineness of worsted spun yard, represented by the number of hanks of 560 yds (512 m) that are required to give a weight of 1 lb of wool. Other count systems in common use include metric count – the number of kms of yarn required to give a weight of 1 kg, and tex – the weight in grams of 1 km yarn. (◆ WOOL GRADES)

Brae. A Scottish term for a slope or hillside usually bounding a river or stream.

Braird. The first sprouting shoots of corn or other crop. Also called breer or brere.

Brahman. Indian or ZEBU cattle characterised by loose skin on the throat and dewlap, with well developed sweat pores, a muscular hump over the neck and shoulders, and large drooping ears. Commonly steel-grey in colour, but variations from black to white are found, and some red strains. They are long-lived cattle, tolerant of high temperatures, and resistant to insects and various tropical diseases. Imported in large numbers into Brazil and Texas in the late nineteenth and early twentieth centuries.

Brake. 1. A harrow. (◆ BRAKE HARROW)
2. A framework in which a restless horse is confined during shoeing.
3. A light carriage for breaking a horse to harness.
4. Brushwood or thicket. (◆ BRAKY)
5. An implement for crushing or braking flax or hemp. Also called a flax brake.
6. A scissor-like tool for stripping willow bark for basket-making.

Brake Harrow. A heavy HARROW for breaking up clods.

Braky. Full of bracken or brake, or rough and thorny.

Bramble. The blackberry (*Rubus fruticosus*), a thorny shrub, or clamberer, with edible fruits (blackberries). Often found as impenetrable thickets. In general terms, any allied thorny shrub.

Bramley. A well known variety of apple.

Bran. The husks of ground corn separated from the flour by bolting (♦ BOLT). A palatable FEEDINGSTUFF with a useful PROTEIN and FIBRE content, especially for cattle and poultry. It is often fed to sows prior to farrowing and, as bran mash, is given to sick animals with no appetite. Also used in making brown bread. (♦ WEATINGS)

Brand. 1. A mark burned onto an animal's hide for identification purposes by means of a hot branding iron.
2. General term for BLIGHTS or fungal diseases of grain crops.

Brands. A type of wool oddment purchased by the BRITISH WOOL MARKETING BOARD, comprising those parts of fleeces bearing brand marks which have been clipped off separately. Mainly derived from sheep in Wales.

Brank ♦ BUCKWHEAT.

Brash. 1. Loose disintegrated rock and rubble, or a soil containing many stone chippings or rock fragments.
2. Pruned branches of conifers. (♦ BRASHING)

Brashing. The pruning of the lower branches of young conifers between 15 and 20 years old up to a height of about 5 ft to allow better access into a plantation.

Brassica. The generic name for cabbage (*B. oleracea*), and its related plants including cauliflower, broccoli, kale, savoy, Brussels sprouts, and for turnip (*B. rapa*) and swede (*B. rutabaga*).

Bratting. A now obsolete method for protecting sheep during severe weather by use of a 'brat' or cloth tied round the body.

Brawn. 1. A BOAR.
2. A meat preparation from cut, boiled and pickled pig's head and ox-feet.

Brawner. A male pig castrated after serving the sows. Also called a stag.

Braxy. A bacterial disease (*Clostridium septique*) of sheep causing severe inflammation of the stomach, dizziness, exhaustion and loss of appetite. Sometimes induced by indigestion caused by eating frosted herbage.

Braxy Mutton. Meat from a sheep that has died of BRAXY. Also a general term for meat from a sheep having died of disease or accident.

Braxy Pasture. Grassland infected with *Clostridium septique*. (♦ BRAXY)

Breadcorn. Corn, from the flour of which bread is made.

Break. 1. To tame or train a horse to wear a saddle or for draught work.
2. A change of crop in a ROTATION programme. Thus an arable break could be corn grown for several years in a field following usually a root crop and to be succeeded by perhaps a LEY.
3. To cut up an animal's body.

Breakaway. A stampeding animal.

Break-furrowing. The ploughing out of alternate FURROW slices which are turned over onto the unturned slices. At one time practised on light stubble land (called ribbing).

Breaking Ground. Ploughing uncultivated or fallow land.

Breast. The MOULDBOARD of a plough.

Breast Plough. An ancient but simple plough, spade-like, with a long shaft and cross bar which was pressed against the chest. Used for paring turf.

Breast Wire ♦ BUTCHER SYSTEM.

Brecham. A Scottish term for a horse collar.

Breeching. A strong leather strap on a harness passed round a horse's haunches, attached to the saddle and the shafts of a cart, allowing the horse to reverse the cart.

Breed. 1. To reproduce (both animals and plants).
2. To promote the reproduction of animals and plants, often under control, in order to select certain characteristics for transmission to offspring. (♦ BLOODLINE, CROSS-BREEDING, PLANT BREEDING)
3. A strain, race, variety, stock or kind of plant or animal. Mostly the result of a continuous cycle of hybridisation followed by INBREEDING of an isolated pocket of plants or animals.

Breed in, Breed out. To introduce or remove a characteristic from an animal breed or plant variety by continuous breeding of those individuals lacking or having the characteristic, respectively, until it becomes fixed in or is lost from the breed or variety. (◗ IN-BREEDING, CHARACTER)

Breed Societies. Clubs established for individual livestock breeds by breeders, to compile and maintain a HERD BOOK or FLOCK BOOK of registered animals conforming to the breed type, and to look after the breed's general welfare. Societies usually organise annual sales, competitions and quality tests, mainly at agricultural shows, to judge the performance and merit of breed animals.

Breeder. A person who controls and promotes the reproduction of plants or animals.

Breeding Crate. An apparatus designed to take the weight of bull which is too heavy for the heifer or cow with which it is to be mated. Also called a service pen or service crate. Alternatively, ARTIFICIAL INSEMINATION may be used.

Breeding Stock. Farm animals selected for producing offspring, as opposed to being fattened for slaughter, etc., in order to maintain or increase the size and quality of a herd or flock.

Breer, Brere ◗ BRAIRD.

Breeze Fly, Breese. A blood-sucking, two-winged, tabanid fly (*Chrysops spp.*: Common Breeze Fly, *C. caecutiens*). It is large, dark, with conspicuous triangular yellow marks on its upper abdomen surface, the wings having a dark band across the middle and another dark mark near the tip. Particularly troublesome to horses. Related to GADFLIES and CLEGS.

Brewers' Grains. The residue of barley after being used in brewing beer, consisting mainly of PROTEIN and FIBRE, used either wet or dried as an animal FEEDINGSTUFF. Also called Draff. (◗ DISTILLERS' GRAINS, MALTING)

Bridle. An apparatus fitted to a horse's head to control it, including the BIT and reins.

Bridlepath, Bridleway. A track or path along which there is a right to ride horses if shown on a County DEFINITIVE MAP.

Bright Parts. Those parts of a tool or implement which keep themselves bright by use and require no painting, e.g. mould-boards, tines.

Brim. 1. A sow on HEAT.
2. To put a boar to a sow for mating.

Brindled. A descriptive term of animals and plants meaning marked with spots or streaks.

Bring Off. To hatch out eggs.

Brisket. The breast of an animal. The term is used for the cut of meat from next to the ribs.

Bristles. Short, stiff, coarse hair, particularly on the back and sides of a pig.

Britch. The thigh and twist part of a sheep, and also the wool from it which is coarse and of low quality.

British Agricultural Export Council. A non-commercial export promotion organisation established in 1966, financed by the Government and agricultural exporting industries.

British Agrochemical Supply Industry Scheme Ltd. An independent compulsory registration scheme introduced in 1978 and administered by a Board of Management, for U.K. distributors of crop protection products cleared by the PESTICIDES SAFETY PRECAUTIONS SCHEME (excluding rodenticides and wood preservatives). The scheme is aimed at raising the standards of product storage, distribution and application, and the training and competence of advisers and selling agents in pesticide safe usage and efficacy. The scheme introduced a Certificate in Crop Protection (by examination) for field, sales and technical staff, and the periodic inspection of registered premises.

British Alpine. A breed of GOAT developed from Swiss and Old English stocks. Black, with white markings on the face and sometimes on the belly. Short-haired, although males have a mane of long hair along the backbone and a well-developed beard. Females produce plentiful milk containing about 4% BUTTERFAT.

British Dane. A breed of cattle established by the RED POLL Cattle Society following the importation of Danish Red cattle.

British Farm Produce Council. An independent organisation launched in 1960 to promote and extend the consumption and use, in the U.K. and elsewhere, of British farm produce.

British Friesian ♦ FRIESIAN.

British Goat. A variously coloured breed of goat (formerly called Anglo-Nubian-Swiss), bred from Swiss, Nubian and Old English stock by British breeders.

British Herbage Seeds Committee. A committee formed by the BRITISH SEEDS COUNCIL to advise and assist the NATIONAL SEED DEVELOPMENT ORGANISATION in such matters as seed growers' prices, contracts, cleaning charges, etc. Such matters were formerly dealt with by the Aberystwyth Committee. (♦ ABERYSTWYTH STRAINS).

British Poultry Federation. A trade association founded to bring together specialist interest and expertise within the egg and poultry meat industries for negotiations with the Government, E.E.C., MPs, Local Authorities, etc., on all matters affecting their livelihoods. Full members of the Federation include:
British Egg Association.
British Poultry Breeders and Hatcheries Association.
British Poultry Meat Association.
British Turkey Federation.
Duck Producers' Association.

British Romagnola. A breed of beef cattle first imported into the U.K. in 1974 from North-East Italy, and well adapted to hot summers and cold wet winter conditions. Interest is increasing in Romagnola bulls for crossing (♦ CROSS-BREEDING) to produce suckler calves (♦ SUCKLER COW) with enhanced growth potential. Cows calve easily and produce vigorous calves.

British Saddleback. A dual-purpose breed of pig recently derived by the amalgamation of the Essex and Wessex breeds, both bearing similar markings, being mainly black, with white front legs and a white belt or saddle extending over the shoulders, and with white hind feet and tail tip. Offspring may be all black or the white belt may vary in width, covering nearly the whole body in some cases. The breed is traditionally used as a basis for cross bred sows for outdoor pig keeping systems (*see p. 329*).

British Seeds Council. A forum, with various specialist committees, established in 1955, through which seed growers and users and the seed trade co-operate and discuss common problems. At its inaugural meeting its aims were described as 'the improvement of quality and the expansion of the production of seed (other than cereal seed) in the U.K. and sales at home and abroad, in so far as this is economically sound and in the best interests of British Agriculture and Horticulture'.

British Sugar Corporation. A public company established in 1936 by Act of Parliament, in which the Government currently has a 24% shareholding. Each year the Corporation contracts to buy the entire sugar beet crop grown by British farmers under E.E.C. quotas, which it then processes, markets and sells, currently supplying half of the nation's needs. Agricultural research is also undertaken and the results passed on to growers.

British White. A breed of dual-purpose, polled cattle, directly descended from large wild cattle, and also linked to similar cattle introduced into Britain by the Vikings. Vivid white in colour, with black or red ears, muzzle, teats and eyelashes, and with spots on the leg fronts above the hooves. Previously called Park cattle. Very few herds remain.

British Wool Marketing Board. A MARKETING BOARD established in 1950, assuming the previous powers of the Ministry of Agriculture, to grade ($ WOOL GRADES), value and purchase home-grown raw wool offered by registered producers, and subsequently to market it. All persons with more than 4 adult sheep may sell their wool only to the Board, with whom they are required to register.

The U.K. is divided into ten 'wool' regions. Producers in each region elect one 'Regional Member' to the Board and producers throughout the nation elect two 'Special Members'. Three members are appointed by the Government. The operation of the Wool Marketing Scheme is monitored by regional committees comprising elected wool producers. All Board and Regional Committee members serve a three year term of office. Various Board committees exist for liaison and negotiation with the wool trade and industry.

Each year, in the ANNUAL REVIEW, the Government fixes an

overall GUARANTEED PRICE for fleece wool (expressed in pence per kg). Based on this price, after deduction of marketing and other costs, the Board prepares its annual schedule of prices for all wool grades. Reduced prices may be payed for inferior wool. Producers sell their wool through merchants, who grade it and make up bulk lots (averaging 5 000 kg) for each grade, which are later sold at auctions. The latter are held approximately each fortnight in Bradford, Edinburgh and Exeter, and are conducted for the Board by the Committee of London Wool Brokers (C.L.W.B.). About 1.5 million kg of British wool is offered at each sale.

The Board also organises training in sheep shearing and wool grading, offers a ram fleece assessment service (to help producers select the best rams for breeding), and operates wool product development, sales and promotion departments.

Brize ♦ BREEZE FLY.

Broad Alley. A double-width ALLEY in a hop garden for allowing vehicular access.

Broad Bean. The common BEAN (*Vicia faba*).

Broad Red Clover ♦ RED CLOVER.

Broadcast Fertilizer Distributor. A machine which broadcasts fertilizer. Various distributive mechanisms exist including spinning discs, revolving fingers, rollers, brushes and chains, etc. (♦ BROADCASTER, FERTILIZER DISTRIBUTORS)

Broadcaster. A machine for sowing seeds, mainly grass and clover, consisting usually of a hopper supplying seed to one or more revolving brushes by which it is scattered. (♦ FIDDLE)

Broadleaf. 1. A term applied to non-coniferous trees, almost coterminous with deciduous. Leaves are usually broad, flat and thin, as opposed to linear needles, and veins are networked as opposed to parallel. Broadleaf trees or hardwoods which are slower growing than CONIFERS and produce compact hard wood, take longer to provide a profit when grown commercially. They are suited best to lower hill slopes and deep fertile lowland soils.
2. A term applied to dicotyledonous (♦ COTYLEDON) plants, usually weeds, e.g. dock, FAT HEN, etc.

Broadshare(r). A heavy type of CULTIVATOR fitted with broad flat SHARES, used to work very hard ground and for stubble cleaning. Also called stubble-breaker. (♦ THISTLE-BAR)

Broccoli. Hardy forms of cauliflower harvested in winter and early spring as opposed to autumn and summer cauliflowers. SPROUTING BROCCOLI is quite distinct, lacking the tight heads of cauliflowers, with numerous small curds borne on side branches developed in leaf-axils. (♦ BRASSICA)

Brock. Food scraps, pigswill. In some areas a term for potatoes unfit for sale. Also a badger.

Brockbar. A pure breed of fowl, AUTO-SEX-LINKED, with BARRED FEATHERS, a single comb, yellow legs, and laying tinted eggs.

Brocket. 1. A stag in its second year with its first horns, un-branched and dagger-like.
2. Mottled.

Broiler. A chicken reared for the table under intensive conditions for 9–11 weeks and killed weighing 1.6–1.8 kg (3.5–4 lb) liveweight.

Broken Furrow. A type of furrow slice turned by a DIGGER PLOUGH, rough and broken, as opposed to the smooth con-tinuous furrow left by a LEA PLOUGH. (♦ UNBROKEN FURROW)

Broken Horse. A horse trained to the saddle or BRIDLE.

Broken Work. Ploughing which leaves the furrows broken. Also called broken rib work. (♦ BROKEN FURROW, WHOLE WORK)

Broken-mouthed. A descriptive term of an old sheep unable to deal with its food requirements due to loss of some teeth.

Broken-winded. Having short breath or defective respiratory organs, and thus incapable of hard work. Particularly applied to horses.

Brokes. A type of wool oddment purchased by the BRITISH WOOL MARKETING BOARD comprising large pieces of wool torn from or rubbed off a fleece by sheep. (♦ LOCKS)

Brome Grass. A large genus (*Bromus*) of long-awned grasses strongly resembling oats, but more flowery, mostly unimportant as fodder and generally regarded as weeds in Britain and becoming increasingly troublesome. Also called lop-grass.

Brood. 1. A family of birds hatched together.
2. To sit (as a hen) on eggs to hatch them.
3. The eggs, larvae and nymphs of bees in a brood chamber.
(♦ BEEHIVE)

Brood Chamber ♦ BEEHIVE.

Brood Comb. A wooden frame enclosing a wax sheet from which bees build the cells of the HONEYCOMB, placed in the brood chamber of a hive. (♦ BEEHIVE)

Brood Mare. A female horse kept for breeding.

Brooder. A unit containing a heat source in which newly hatched chicks are kept under controlled, gradually reducing, temperature conditions throughout the BROODING STAGE.

Brooder Pneumonia ♦ ASPERGILLOSIS.

Brooding Stage. The period, 4–8 weeks, from hatching until poultry chicks cease to need some form of heat for survival, provided by a BROODY HEN under the protection of its wings, or artificially in a BROODER.

Broody Hen. A hen inclined to sit on its eggs to hatch them. Also called a sitter.

Brook. A small stream.

Brose ♦ OATMEAL.

Brow Antler, Brow Tine. The first tine of a deer's horn.

Brown Calcareous Soils. Well-drained, brownish soils developed over LIMESTONE or calcareous sandstone, of medium to heavy texture, with slightly acid surface layers but calcareous in the subsoil.

Brown Earth. One of the three main British soil types including much well-drained, agriculturally desirable land in England, particularly in the south and midlands, typical of the mild climate with moderate rainfall. Normally supporting deciduous woodland. It consists of a dry brown topsoil and usually lighter coloured brownish subsoil. Also called brown soil. (◗ PODSOL, BLANKET PEAT BOG)

Brown Heart. A DEFICIENCY DISEASE of swedes due to lack of BORON, causing browning or mottling of the root making it unpalatable.

Brown Oil of Vitriol (B.O.V.). Sulphuric acid, formerly used as a weed-killer, particularly against broadleaves such as charlock.

Brown Rot. A fungal disease (*Sclerotinia fructigena*) of apples, pears and plums, which turns the ripe fruit brown and rotten. They become mummified, hanging on the trees.

Brown Soil ◗ BROWN EARTH.

Browse. 1. To feed and nibble on young shoots, buds and twigs of plants.
2. The tender shoots of shrubs and trees which cattle may feed on.

Brucellosis. A bacterial disease of the reproductive system, mainly of cattle (*Brucella abortus*), sheep (*B. melitenis*) and pigs (*B. suis*), although all mammals are susceptible. In cattle it causes infection of the womb and placenta and abortion of calves. It is spread mainly via aborted calves, afterbirth, discharges and milk which contaminate pasture, food and water. It causes undulant fever in humans. Also called contagious abortion.

Brucellosis (Accredited Herds) Scheme. A voluntary scheme introduced in 1967 as a first step towards eradicating BRUCELLOSIS. Its main aim was to identify and register Brucellosis-free herds, with periodic testing and appropriate management to maintain such herds as a nucleus of stock to replace REACTORS in other herds which were slaughtered (with compensation). Replaced by the BRUCELLOSIS INCENTIVES SCHEME in 1970.

Brucellosis Incentives Scheme. Introduced in 1970 to replace the BRUCELLOSIS (ACCREDITED HERDS) SCHEME and to facilitate an area eradication programme. Incentive payments are provided to owners of clean or relatively clean herds to seek voluntary accreditation in advance of compulsory eradication in an area, with compensation for slaughtered REACTORS.

Brush. 1. Loppings and trimmings from trees, shrubs and hedges. Also to lop or trim such plants, or to cut down weeds using a BRUSHING HOOK.
2. ♦ BEARD.

Brushing Hook. A sharp sickle-shaped tool used to trim hedges and clear undergrowth.

Brushwood. Low, rough, scrubby, undergrowth or thicket.

Brussels Sprouts. A variety of cabbage in which the axillary buds on the upright main stem develop to produce sprouts having the appearance of tiny cabbages. (♦ BRASSICA)

Brutting. A method of summer pruning used mainly by nut-growers, but also applied to old fruit trees, in which side shoots of the bushes are snapped off near to their base by hand, and the ends left hanging, preventing SECONDARY GROWTH.

Bryophyta. A group of plants comprising mosses and liverworts.

Bubbly-Jock. A Scottish term for a turkey cock.

Buchts. A term used in the Borders for sheep gathering pens.

Buck. 1. The body of a cart.
2. A male deer, goat, hare or rabbit.

Bucket-feeding. Feeding a young animal with milk and gruel from a bucket, rather than allowing it to suckle its mother.

Buckhound. A small staghound used for hunting bucks.

Buckling. A male goat, 1–2 years old.

Buckrake. A simple implement with close-set long tines, usually rear-mounted on a tractor hydraulic lift linkage, used for collecting and transporting cut grass and GREEN CROPS for SILAGE making. Also used for other purposes such as transporting straw and hay bales.

Buckshot. A large type of shot used in deer-shooting.

Buckwheat. Two frost-sensitive cultivated plant species, common buckwheat (*Fagopyrum esculentum*) and Tartarian buckwheat (*F. tartaricum*), often classed as CEREALS, but actually of the dock family. Grown mainly in patches, providing grain for pheasants, and sometimes as a FORAGE crop. Common buckwheat has white or pink flowers and dark brown seeds, sharply triangular in cross section. Also called brank.

Bud. 1. A compact, undeveloped shoot or unexpanded leaf or flower, encased in protective scales, arising from a leaf axil or terminal on a branch.

2. In bud; about to flower or put forth leaves.

3. A rudimentary horn on the head of a young animal.

Budding. A form of propagation involving the joining of a bud or a shoot (the SCION) of a desired variety on to the healthy stem of another cheaper or more readily available related variety (the STOCK). Roses are commonly budded onto wild briar stocks and cherry varieties onto wild cherry stocks.

Bug. One of a variety of plant sap-sucking or blood-sucking insects. Also a general term for any beetle or grub.

Buhr ♦ BURR.

Buist. An identification mark on sheep or cattle made by the owner (e.g. with tar), usually after SHEARING.

Bulb. The swollen underground stem of some plants, comprising thick, fleshy scales or leaves with thickened bases, surrounding a short stem and enclosing a bud which develops into main leaves and a flowering stem. A food storage and vegetative reproductive structure. Examples include onion and lily.

The production of bulbs for planting in flower gardens, as well as bulb flowers, takes place particularly in the Holland area of Lincolnshire, Norfolk, the east of Scotland, Cornwall, and in the Isles of Scilly where more than 50% of the cultivable land is devoted to bulbs which flower there from November to May.

Bulb Dips. Chemical compounds in which bulbs are immersed to provide protection against pests and diseases.

Bulk. The dry matter in animal rations.

Bulk Milk. Milk from a number of cows collected into a CHURN, or more usually from a herd into a large container or vat. The collection of bulk milk by tankers (bulk collection) for transport to dairies has become almost universal in the U.K. (♦ CHURN COLLECTION, MILK COOLER)

Bulk Storage. The storing of harvested grain, animal feedingstuffs, fertilizers, etc., loosely, in large containers, covered bays, or silos, rather than in sacks.

Bulky Feedingstuff. A FEEDINGSTUFF containing little nutritive value relative to its volume, e.g. hay, straw or chaff, in contrast to CONCENTRATES.

Bull. An uncastrated adult male bovine animal.

Bull Beef. Beef from a bull as opposed to a STEER. A bull carcase provides more lean meat than a steer's.

Bull Calf. A male calf.

Bulldog. 1. A deformed calf with a bulldog-like appearance due to its short legs and jaws and swollen abdomen.
2. A split ring which grips a bull's nostrils through which a rope can be passed for holding or leading.

Bullet. A dose of a mineral, e.g. cobalt or magnesium, introduced to an animal by a special 'gun', providing it with a long-lasting supply, preventing DEFICIENCY DISEASES.

Bullimong. A mixture of oats, peas and mixed grains sown as a FORAGE CROP. The custom was confined to certain limited areas and has probably now ceased.

Bulling Heifer. A maiden HEIFER which has reached the right size and age for mating.

Bullock. A castrated bull. A bullock calf is a castrated male calf.

Bull-wheel. A large wheel providing drive to the machinery of a BINDER or REAPER.

Bumble Foot. A condition of poultry characterised by abscesses between the toes, caused by thorns, stones, glass fragments, etc., penetrating the soft tissue, or by bruising, and resulting in lameness.

Bunch. 1. A handful of HAY.
2. A herd.

Buncher ▶ TRUSSER.

Bunt. A fungus disease (*Tilletia caries*), mainly of wheat but also of rye, filling the ears with a mass of black, fish-smelling, greasy spores. Bunted grains burst when threshed, the spores contaminating and discolouring healthy grain, so that it is useless as seed or for milling. Also called stinking smut.

Bunt Order. The social hierarchy or PECK ORDER of a herd determined by aggression, and sometimes fighting, until positions in the order are established. Thus, there will be a dominant animal (often the largest) to which all others are submissive; a second animal submissive to the first but dominant over the rest, and so on down the order to the least aggressive animal which is submissive to all in the herd. The bunt order determines, for instance, which animals take precedence in feeding or drinking at a small trough. Mixing of, or additions to, herds upsets the order temporarily, causing stress and affecting productive performance.

Bur, Burr. 1. The prickly seed-case or fruiting head of certain plants, particularly Burdock, which adhere to animal fur and clothing.
2. A knob at the base of a deer's horn.
3. A knotting growth on a tree, leaving marks in timber.
4. The catkin or CONE of the hop.
5. Arenaceous rock from which millstones (burr-stones) are made. Also one of the corrugations of a millstone.

Burgundy Mixture. A mixture of copper sulphate and sodium carbonate (washing soda) used as a fungicide. (♦ BORDEAUX MIXTURE)

Burn. A Scottish term for a small stream or brook.

Burnt Lime. A form of LIME with the highest NEUTRALISING VALUE. Mainly CALCIUM OXIDE and various residues produced by burning lumps of CHALK or LIMESTONE, and usually kibbled or ground before use. It readily absorbs water to form CALCIUM HYDROXIDE and cannot be stored for very long. Caustic and irritant, it is also used to destroy carcases of animals which have died from infectious diseases. Also called quicklime, lump lime and shell lime.

Burrow. An underground tunnel excavated by rabbits, foxes or other animals.

Burry Wool. Fleeces which have caught seed BURS in them.

Bury ◗ CLAMP.

Bush. A small shrub, usually woody with several stems.

Bush Drain. A type of field DRAIN constructed by packing bushes into a trench which is then refilled.

Bush Fruit. Fruit grown on bushes as opposed to trees or CANE, e.g. blackcurrants, gooseberries.

Bush Fruit Tree. A fruit tree, the lowest branches of which are below 75 cm (30 in.) above the ground.

Bush Harrow. An old type of light harrow with a barred wooden frame with bushes or branches woven through it, used for covering grass seeds. Developed from the primitive harrow with a log tied on top of a bush.

Bushel. A dry measure by which grain was once generally computed, and also used for fruit, containing 1.28 cubic feet or 8 gallons. Many farmers still DRILL seed in bushels per acre or refer to grain in terms of lbs per bushel when discussing its quality, but, following metrication, grain traders now use kg per hectolitre. Fruit is now usually picked into bulk bins rather than bushel boxes. Also called sieve.

Butcher System. A design, once popular, now rare, for hop garden wire-work, with three horizontal wires attaching to the poles, the lowest at about 15 cm (6 in.) and the middle (breast wire) at about 120 cm (4 ft).

Butt. 1. A tree trunk.
2. The base of a leaf stalk.
3. One of the short furrows ploughed where a field runs to an angle or at an awkward corner.

Butter ◗ MILK PRODUCTS.

Butterfat. The fatty substances contained in milk, mainly glycerides of palmitic and oleic acids, present as minute fat globules, the size of which vary according to breed, and diminish from time of calving. The largest are present in JERSEY milk, churn easily and are best for butter-making. The smallest are present in AYRSHIRE and FREISIAN milk and are best for cheese. The butterfat content of milk is determined by the GERBER TEST and nowadays also by a number of semi-automatic instruments (e.g. Foss Milkotester) which are themselves calibrated using the Gerber test.

Butterfly. A breed of rabbit. (◗ ENGLISH)

Buttermilk. The residual liquid following the churning of cream into butter.

Buttor, Buttor Board. The part of a BALER against which the HAY or STRAW is compressed when being formed into bales. Also the part of a BINDER that neatly levels the cut ends of corn as the sheaf is built up.

Buying-in Price. Support buying of fruit and vegetables takes place in the E.E.C. by reference to a buying-in price which is set between 40 and 70% of the BASIC PRICE. (◗ COMMON AGRICULTURAL POLICY)

Byre. A cow-house.

C

C-Horizon ♦ SOIL HORIZONS.

Cabbage. A biennial cruciferous plant (*Brassica oleracea*) grown both as FODDER and for the table. Characterised by a short stem with a very large head comprising tightly overlapping leaves. Closely allied to other BRASSICAS such as broccoli, Brussels sprouts, cauliflower, kale, and savoy cabbages, all of which have been developed by long selection from the wild cabbage. Cabbages are commonly fed to cattle, usually after milking, and pigs.

Cabbage Lettuce. A type of lettuce resembling a cabbage both in shape and its possession of a distinct heart. (♦ COS LETTUCE)

Cabbage Root Fly. A fly (*Erioischia brassicae*), resembling the housefly, which lays white eggs in April–early May in the soil surface near the stems of seedling BRASSICAS. The maggots, which hatch within a week, feed on the young roots and stems, causing plant water loss and frequently death.

Cabbage White Butterfly. A common butterfly (*Pieris brassicae*) with white wings, and yellowish-green, black-spotted caterpillars which feed on cabbages and other BRASSICAS. Also called Large White Butterfly.

Cable Cultivation. Drawing a plough or other implement across a field and back by cables, engines sited on opposite sides of the field providing the power. An obsolete practice.

Cadastre. The official register of the ownership of land.

Cade. 1. A lamb or colt reared on the bottle. Also called COSSET LAMB.
2. ♦ KEG.

Caecum. The blind, sac-like end of the large intestine, connected to the small intestine. Also called blind gut.

71

Cage Rearing (of piglets). A recently introduced system of pig management involving early weaning at 7–10 days and transfer to cages kept in darkness (except during twice-daily feeding) under very warm conditions. Piglets are transferred to normal pens at about 5 weeks.

Cages. Cages constructed of wood, metal, or a mixture of both are used for holding BATTERY HENS in numbers from 1 to 30. Nowadays cages holding 3–7 birds are the most popular. Cage design is fairly standard; 45 cm (18 in.) high at the front sloping to 35 cm (14 in.) at the rear, and 45 cm (18 in.) deep. Width varies from 37.5 cm (15 in.) for 3 birds to 80 cm (32 in.) for 7 birds. Battery cages are normally in three tiers approximately 2 m (6 ft) high.

Battery cages.

Cain. In old Scottish law, rent paid in kind, especially poultry.

Cake. The seeds of various plants, mainly tropical (e.g. groundnut, cotton, soya bean, coconut, linseed and palm nut kernel) compressed (to extract oil for cooking, soap manufacture, etc.) into flat slabs, or straights, rich in PROTEIN, containing oil and FIBRE in varying amounts, used as cattle food until the 1940s. In recent years more efficient solvent extraction has replaced compression, and seed residue is now available as flakes or meal, containing less oil, mixed by feedingstuff compounders with cereals and their by-products, together with molasses and steam, to produce extruded 'nuts' or 'pencils', commonly called cattle cake or meal. Oil seed rape has now become a useful constituent of cattle cake in Europe. Cake is now a generic term for processed livestock feedingstuffs. ($ CONCENTRATES, CUBER, DECORTICATION, HOME-MIXING, FEEDINGSTUFFS, MEAL)

Cake-breaker. A machine once used by farmers to mince CAKE into smaller pieces for feeding to cattle.

Calcareous. Containing CALCIUM, usually in the form of CHALK or LIMESTONE.

Calcium. (Ca). A soft white chemical element, essential to life, and an important constituent of bones and teeth, and of the several compound forms of LIME used in agriculture (CALCIUM CARBONATE, CALCIUM HYDROXIDE and CALCIUM OXIDE).

Calcium Carbonate. ($CaCO_3$). A white insoluble solid occurring naturally as CHALK, LIMESTONE, marble and calcite. ($ GROUND LIMESTONE, LIME)

Calcium Cyanamide. (CaN_2). Nitrolime. An ARTIFICIAL FERTILIZER, powdery or granular, blue-black in colour, supplying LIME and NITROGEN to crops. Converted by soil water to AMMONIA. Now rarely used.

Calcium Hydroxide. ($Ca(OH)_2$). Slaked lime or hydrated lime, formed by the action of water on CALCIUM OXIDE. Used in a finely divided condition by horticulturists to correct soil acidity.

Calcium Oxide. (CaO). Quicklime. A lumpy white powder formed by burning CHALK or LIMESTONE. Usually ground up for use as LIME (ground lime). Caustic and irritant, it is used to destroy carcases of animals which have died from infectious diseases. Also called burnt lime, and lump lime. ($ LIME REQUIREMENT, NEUTRALISING VALUE)

Calcium Sulphate $ GYPSUM.

Calf. The offspring (in its first year) of a cow. A male is called a bull calf, a female is called a heifer calf, quey calf or cow calf.

Calf Scours. An important bacterial disease (often *E.coli*) of both dairy and beef calves, often fatal. Also called neonatal diarrhoea.

Calk, Calkin. A pointed, turned-down piece on a horse-shoe preventing slipping.

Calluna. The generic name for HEATHER.

Callus. 1. The protective tissue which grows, as a swelling, over wounds in woody plants.
2. The cartilaginous growth, containing COLLAGEN, which first grows to heal a bone fracture, later becoming bone tissue.
3. Any thickened tissue, often hardened or leathery.

Calorie. The quantity of heat needed to raise the temperature of 1 g of water by 1°C. A kilo-calorie (K. cal) or Large Calorie (Calorie) is 1 000 calories.

Calve, Calve Down. To give birth to a calf. (♦ DOWN CALVER)

Calyx. The outer ring of the parts of a flower, composed of sepals, usually green, sometimes coloured. (♦ COROLLA, PERIANTH)

Cambar. An AUTO-SEX-LINKED breed of poultry, with barred feathers, a single comb, yellow legs, and producing white eggs.

Cambium. A layer of actively dividing cells in plants, producing XYLEM, or wood, to its inside and PHLOEM, or bast, to its outside in the process of SECONDARY THICKENING. Usually located just beneath the bark of trees and shrubs.

Cambridge Roll. A type of ROLLER, the cylinder consisting of a number of narrow ribbed rings, loosely mounted side by side on the axle. Generally heavier than the FLAT ROLL and effective as a clod crusher. Also called a RING ROLL.

Canadian Holstein ♦ HOLSTEIN.

Candling. A process by which eggs are passed over a candling machine in which a light source detects blood spots and shell cracks. The apparatus is usually combined with an egg grading machine. (♦ EGG WEIGHT GRADES)

Cane. The stem of raspberry, blackberry and other similar plants, and of the larger grasses, such as sugar cane or bamboo.

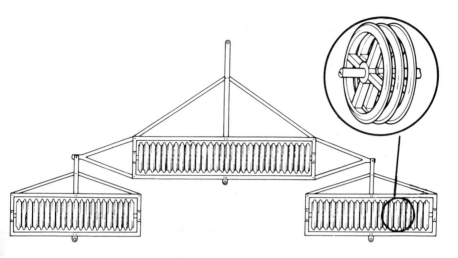

Cambridge Roll.

Cane Sugar. A type of storage sugar (Sucrose) present in sugar cane, sugar beet, ripe fruits and tree sap (e.g. maple sugar). It is broken down by intestinal ENZYMES into GLUCOSE and FRUCTOSE which can then be absorbed by the bloodstream to provide energy. (◊ MOLASSES)

Canker. 1. A general term for a number of fungal diseases of trees (often due to *Nectria sp.*), especially those affecting fruit trees, in which growing tissues are destroyed and cortical tissues malformed, forming cancerous growths and open wounds.
2. A fungus disease causing softening of the horn of a horse's hoof, inflammation and the production of a soft cheesy material.
3. A colloquial term for ear mange in dogs and for the pussy conditions which can develop if not treated.

Cannibalism. The practice of an animal eating its own kind. It sometimes occurs with indoor-housed poultry due to overcrowding, boredom (scratching for insects is precluded in cages) or protein deficiency, and is controlled by removing the point of the upper beak (debeaking) or the use of soft red lighting. Sows will also occasionally eat their own or other sows' piglets under intensive breeding conditions.

Cannon Bone. A bone supporting the limbs of horses, between the knee and FETLOCK of the forelegs, and between the HOCK and fetlock of the hind legs.

Cannula. A tube containing a sharp-pointed plunger (a TROCAR) which is inserted to puncture an animal's rumen. The cannula is left in position when the trocar is withdrawn to allow gas to escape.

Canson Kale ♦ THOUSAND-HEADED KALE.

Cant. 1. A sloping bank.
2. An area of woodland clear-felled at one time. Often the unit of sale (by auction) for sweet chestnut (*Castanea sativa*), commonly about an acre in extent.
3. An area designated to a ploughman to plough in a ploughing competition.

Canterbury Hoe. A type of hoe with three long, flat, sharp prongs. Used to break up clods.

Canterbury Lamb. A meat trade term for lamb imported from Canterbury province in New Zealand.

Cantle. The raised, curved hind part of a saddle.

C.A.P. An abbreviation for COMMON AGRICULTURAL POLICY.

Cap. 1. ♦ CAPPED BLOSSOM.
2. ♦ CAPPING.

Capability Class, Soil or Land. A rating indicating the general suitability of a soil or area of land for agricultural use. (♦ LAND CLASSIFICATION)

Capillary. A minute hair-like blood vessel, one of many permeating the body's tissues.

Capillary Action. The phenomenon of a liquid rising up a narrow tube, or the formation of drops, bubbles and films, due to high inter-molecular attraction within the liquid. Important in soil as the natural forces of capillary attraction cause water to rise through the minute 'capillary space' between soil particles and through soil pores. (♦ SOIL WATER, CAPILLARY WATER)

Capillary Water. The water held in the soil in pores, and capillary spaces, and as a thin film surrounding soil particles, held under tension against the pull of gravity. Capillary water does not, therefore, drain away or evaporate but may move from wetter to drier parts of the soil under hydrostatic gradient. It is effectively the 'permanent' soil solution containing dissolved substances such as applied plant nutrients. (♦ IMBIBITIONAL WATER, GRAVITATIONAL WATER)

Capital Grant Schemes ♦ FARM CAPITAL GRANT SCHEME, HORTICULTURAL CAPITAL GRANT SCHEME.

Capon. A castrated cockerel. Nowadays caponisation is practised chemically by injecting female hormones (oestrogens) which cause the testes to regress. It also produces a more tender carcase and stops cockerels crowing and fighting.

Capped Blossom. Apple blossom in which the petals are unable to open and form a brown 'cap' over the buds, as a result of the Apple Blossom Weevil eating the heart of the blossom.

Cappie. A disease of sheep (mainly old lambs and young sheep), generally those grazing hill areas in autumn and winter, considered to be due to phosphorus deficiency. Characterised by lack of thrift (♦ THRIFTY) and thinning of the skull bones. In bad cases sheep cannot eat or close the mouth. Also called double scalp.

Capping. The formation of a crust on a soil during a hot, dry spell, usually after heavy rain which causes the surface soil particles to slake down, blocking up the soil pores.

Capriform, Caprine. Goat-like.

Capsid Bug. One of a large family of greenish or brownish plant BUGS, slender and elongate, about 6 mm ($\frac{1}{4}$ in.) long. The common Green Capsid (*Lygocoris pabulinus*) is a pest of fruit and vegetables. The Apple Capsid (*Plesiocoris rugicollis*), which preys on the stem tips, fruit and foliage of apple trees, is now largely controlled by D.D.T. and other insecticides.

Capsule. 1. A type of hard, dry, dehiscent fruit, opening in various ways to release the seeds, e.g. poppy.
2. A gelatinous envelope around certain kinds of bacteria.
3. A soluble gelatinous case containing a dose of medicine, administered orally to an animal.
4. The fibrous membrane enveloping the various organs of the body, e.g. liver, spleen.
5. A thin metal envelope in an incubator, containing a liquid (e.g. ether) which expands and contracts, maintaining an even temperature regime in the incubator.

Captan. A fungicide used in seed dressings for peas and vegetables. Also used effectively against apple scab.

Carbamates. A group of alkaloid INSECTICIDES which are selective and to some extent SYSTEMIC in their action, and are non-persistent in the environment, e.g. carbaryl, methiocarb.

Carbohydrates. A large group of compounds containing carbon, hydrogen and oxygen only, of the general formula $Cx(H_2O)y$. They are produced by plant PHOTOSYNTHESIS and are essential to METABOLISM in all living organisms. Classified into (*a*) *monosaccharides*, simple single-molecule sugars such as GLUCOSE and FRUCTOSE, (*b*) *disaccharides*, sugars composed of two condensed simple sugar molecules such as SUCROSE, and (*c*) *polysaccharides*, compounds composed of several, frequently very many, condensed simple sugar molecules. Examples include STARCH, the principal form in which energy is stored by plants; CELLULOSE, the main plant structural material, and GLYCOGEN, the main energy storage form of animals.
Carbohydrates form the largest part of the food of animals which have only small amounts of glycogen and sugars in their bodies. The complex polysaccharides in plants are broken down by animals into simpler sugars by digestion before they can be absorbed and used. The energy stored in carbohydrates is released, to power living processes and provide heat, by 'burning' with oxygen during cellular RESPIRATION.

Carbon. (C). A chemical element. The principal constituent of all ORGANIC compounds and essential to life. (◗ CARBON CYCLE)

Carbon Assimilation ◗ PHOTOSYNTHESIS.

Carbon Cycle. The circulation of carbon atoms between living organisms and the atmosphere. CARBON DIOXIDE is assimilated during PHOTOSYNTHESIS by plants to produce CARBOHYDRATES and other complex carbon compounds. Animals feed directly on plants or on herbivores to obtain these compounds which are broken down during both animal and plant RESPIRATION to release carbon dioxide back to the atmosphere. Carbon dioxide is also released from the decaying remains of plants and animals by the action of bacteria and fungi.

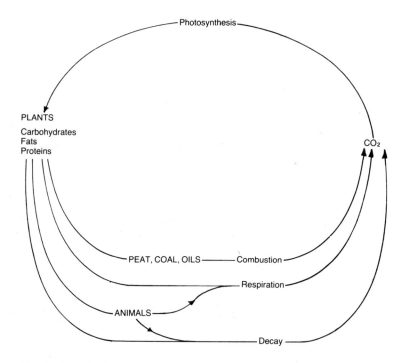

The Carbon Cycle.

Carbon Dioxide. (CO$_2$). A colourless, odourless, tasteless gas occurring in the atmosphere. Utilised by plants in PHOTOSYNTHESIS to produce CARBOHYDRATES. A by-product of animal and plant RESPIRATION and combustion (e.g. straw burning). (◆ CARBON CYCLE)

79

Carbon-Nitrogen Ratio. The ratio of carbon to nitrogen in organic material and soils. The ratio is about 40:1 in unharvested cereal straw added to the soil as residue, and narrower in leguminous residues due to their higher nitrogen content. British soils have a C:N ration of about 10 or 12:1. This represents the approximate soil humus content since little unaltered plant residues remain in the soil. Additions of organic residues with high C:N ratios may temporarily deplete a soil of ammoniacal and nitrate nitrogen as high micro-organism activity uses nitrogenous compounds for protoplasm formation. Additions of residues with C:N ratios below 15:1 result in a release of ammonia by micro-organism in excess of their requirements, which is converted to nitrate available to plants.

Carbon Tetrachloride. (CCl_4). A sweet smelling, colourless liquid used against liver flukes and round worms in animals (excluding cattle for which it is not safe). Usually administered orally in gelatin capsules.

Carcase, Carcass. A dead body. In butchers' terms an animal's body after removal of the head, limbs, hide and offal. Also called a dressed carcase.

Carcase Classification. A system operated at abattoirs mainly by the MEAT AND LIVESTOCK COMMISSION but also by some wholesalers, designed to categorise carcases of cattle, sheep and pigs with common characteristics such as weight, fatness (♦ FAT CLASS) and CONFORMATION, to assist potential buyers in the selection of carcases to meet their requirements.

Carex. A genus of SEDGE.

Carincle. The red and fleshy part of a turkey's neck below the WATTLE.

Carotenes. A group of reddish yellow pigments. They are unsaturated hydrocarbons ($C_{40}H_{56}$), occurring in carrots and butter, providing their distinctive colours. Carotenes are widely distributed in plants, acting as photosynthetic pigments (♦ PHOTOSYNTHESIS) in cells lacking CHLOROPHYLL, and are converted into vitamin A by animals.

Carpel. The female part of a flower, consisting of an OVARY (containing OVULES which develop into SEEDS after FERTILIZATION) and the STYLE, the tip of which receives the POLLEN. The carpel(s) after fertilization develops into the FRUIT. Plants may have single carpels (e.g. plum), several joined carpels (e.g. orange), or many separate carpels (e.g. buttercup).

Carpus. The human wrist, or corresponding part of an animal's fore-limb containing the animal's carpal bones, usually called the knee.

Carr. Boggy ground or land reclaimed from such areas by drainage (◗ LAND DRAINAGE). Also a fen wood, common in East Anglia, dominated by alder.

Carrier. An animal possessing or 'carrying' the germs of a disease, having itself recovered and showing no symptoms, but nevertheless able to pass the infection on to other animals.

Carrot. A root crop bred from the wild carrot (*Daucus carota*), with a long, tapering, reddish or yellowish root, which is sweet and edible. Grown mainly for human consumption, often on contract, for canning or freezing. Surplus or unmarketable roots are used as stock feed, mainly for cattle.

Carrot Fly. A small pestilent fly (*Psila rosae*), green-black in colour, attacking carrots, celery, parsley and parsnips. Eggs are laid in the soil near the plants and the slender white LARVAE burrow into the roots near the growing tips. Growth is retarded and foliage loses its freshness. With serious infestation roots may fork or show distorted growth, and in some cases crops may fail.

Carry-on-period. The stage in poultry rearing following BROODING and before chicks are able to survive without assistance.

Carse. A Scottish term for an ALLUVIAL river-side plain.

Carse Bean. A hardy variety of spring bean. The name is derived from the CARSES of Stirling where it originated. Similar to the Kilbride or Scotch horse bean. (◗ BEANS)

Cart Lodge. A shed in which carts are kept. Also called cart house.

Carter. 1. A person operating a cart to transport produce or goods. Also called waggoner.

2. A local term for the person in charge of a stable.

Cartilage. The 'gristle' of meat. Hard, pliable, skeletal tissue in animals, especially when young. Often later converted into bone with the deposition of calcium. The discs separating the vertebrae are cartilage, acting as 'shock absorbers'.

Cartridge. The 'mole' of a MOLE PLOUGH.

Carucate. An old measure of land, varying from place to place, equivalent to the amount of land a team of oxen could plough in a season.

Caryopsis. The one-seeded fruit of grasses in which the OVARY wall (pericarp) is united with the seed coat (testa).

Cascade Mechanism. Part of a hop-picking machine which uses an inclined belt or belts to extract unwanted pieces of stem or bine.

Caschrom. An ancient wooden foot-plough, with a long curved handle, a flat blade narrowing to a point, and a foot bar by which the blade was forced into the ground, after which it was twisted to turn the soil to one side. It was still used in the remote areas of Scotland up to a few years ago.

Case. A term applied to eggs. Equal to 3 LONG HUNDREDS, i.e. 30 dozen, or 360.

Casein. The main PROTEIN of milk, present in colloidal solution. It is coagulated by the addition of acid or RENNET. This is the principle of cheese-making.

Cash Crop. A crop grown for sale rather than for feeding to farm animals.

Cassava ♦ MANIOC.

Cast. 1. Undigestible matter ejected by a bird, earthworm, etc.

2. To throw off or shed, e.g. when a horse loses or casts a shoe.

3. Offspring born or dropped prematurely.

4. To drop an animal onto its side, often using a rope, to facilitate some manipulation.

5. A sheep unable to get up, having fallen on its back.

6. In apiculture, a second or subsequent swarm of bees, with one or more virgin Queens.

7. ◗ SPLIT.

Cast Ewes. Those old breeding ewes sold from hill to lowland farms where they are used to produce further crops of lambs.

Casting. A method of SYSTEMATIC PLOUGHING in which the tractor and plough are turned to the left each time they come out or go into work, so that already ploughed land is circled in an anti-clockwise direction. Usually carried out alternatively with GATHERING when ploughing marked LANDS in a field. Also called splitting (*see p. 339*).

Castock. A Scottish term for a cabbage stock. Also called custock.

Castor Oil. A foul-tasting oil from the seeds of the tropical castor plant (*Ricinus communis*) used as a medical agent by veterinary surgeons, mainly as a laxative, and also as a lubricant.

Castration. The removal of, or atrofication of, the testicles. Male cattle are castrated to prevent fighting and indiscriminate breeding when housed with HEIFERS, and to make bulls less bad tempered and easier to handle. Carcases from older castrated cattle carry more weight, mainly fat, especially in the forequarters, than when uncastrated. Young male pigs, sheep and horses are also usually castrated unless being kept for breeding purposes. (◗ CAPON)

Casual Workers. Those employed on a temporary basis, usually seasonally, as and when work is available and often for harvesting purposes. There are no fixed or regular hours of employment; work may be accepted or refused at the employee's discretion, and payment is usually on a piece-work basis but can be hourly. Sometimes called day-workers.

Casualty. In butchery, an animal slaughtered for meat prematurely as a result of an accident, e.g. a broken limb.

Catch Crop. An extra crop, usually quick-growing, grown between two main crops in a ROTATION.

Catch Drain, Catchwater Drain. A ditch on a hillside to catch or check surface water run-off and channel it away to a watercourse. Catch drains are characteristic of the FENS, preventing water flowing into them from clay 'upland' areas.

Catchment. The area which drains the rainwater falling on it, via streams and rivers, eventually to the sea or into a lake. Separated from an adjacent catchment area by a ridge of high land or WATERSHED. Also called a drainage area. (♦ RIVER BASIN)

Caterpillar. 1. The LARVA of a butterfly (e.g. CABBAGE WHITE) or moth, often causing damage to crops by feeding on foliage. 2. ♦ TRACKLAYER.

Cation. A positively charged ION.

Cation Exchange. The interchange of a cation (♦ ION) in solution and another cation held on the surface of clay, and organic, colloids in the soil. For example, in ACID SOILS subject to leaching, calcium ions have largely been displaced and replaced by hydrogen ions. The addition of LIME corrects the acidity by exchanging calcium ions for hydrogen ions. Temporary sea water inundation of farmland replaces soil calcium ions with sodium ions. Applications of gypsum (calcium sulphate) corrects the alkalinity by exchanging calcium ions for the sodium ions. Cations applied in the form of FERTILIZERS are adsorbed by soil colloids, usually being exchanged for hydrogen ions. Nutrient ions in the soil solution taken up by plant roots are replaced by exchangeable cations from the soil colloids.

Cation Exchange Capacity (C.E.C.). The total amount of exchangeable cations (♦ ION) that a soil can absorb, expressed in milliequivalents per 100 g of soil or of other adsorbing material such as clay. It is a measure of the potential of a soil to hold nutrient cations for plant absorption. Clayey or organic soils usually have a large C.E.C., whilst the opposite applies to sandy or weakly organic soils. Agriculturally productive soils require application of FERTILIZERS rich in cations and the C.E.C. is a guide to the quantity and frequency of application. Also called Total Exchange Capacity, Base Exchange Capacity, and Cation Adsorption Capacity.

Cat's Tail ♦ TIMOTHY.

Cattabu. A cross between a common European breed of cattle (*Bos taurus*) and a ZEBU breed (*Bos indicus*). In Africa there have been many such crosses, e.g. Africander. (♦ CATTLE)

Cattle. Bovine animals. There are 250 major breeds and nearly 1 000 breeds worldwide. They fall into two groups, those mainly European breeds developed from *Bos taurus*, and those developed from *Bos indicus* (Indian cattle or ZEBUS). Cattle breeds are classified as beef, dairy or dual-purpose. The main breeds of each type currently kept in Britain are indicated below. Their characteristics are defined alphabetically elsewhere.

Beef Breeds: Aberdeen-Angus, Beef Shorthorn, Belted Galloway, Blonde d'Aquitaine, British Romagnola, Charolais, Chianina, Devon, Galloway, Gelbvieh, Hereford, Highland, Limousin, Maine-Anjou, Marchigiana, Murray Grey, Simmental, South Devon, Welsh Black.

Dairy Breeds: Ayrshire, British Dane, Dairy Shorthorn, Friesian, Guernsey, Jersey, Kerry.

Dual-purpose: Dexter, Meuse-Rhine-Ijssel, Red Poll.

Beef Cattle. The U.K. produces about 85% of its beef requirements. Beef cow numbers in 1979 totalled about 1.5 million, 41% of which were in herds of 50 and over, although average herd size was 18.

Cattle Numbers in the U.K. (at June each year). ('000 head).

	Average of 1968–70	1975	1976	1977	1978	1979
Total cattle and calves	12 393	14 717	14 069	13 854	13 625	13 534
of which: Dairy cows	3 248	3 242	3 228	3 265	3 270	3 278
Beef cows	1 227	1 899	1 764	1 680	1 580	1 537
Heifers in calf	837	903	939	824	859	865

(Source: Annual Review of Agriculture, H.M.S.O. 1980)

About two-thirds of home-fed beef production is derived from the national dairy herd. Many dairy cows are crossed with beef bulls for beef production, and in recent years many beef breeds have been imported from continental Europe.

In hill and upland areas cows are kept for the production of single suckled calves (▶ SINGLE SUCKLING). The usual breed of cow is a beef/dairy cross, often Hereford cross Friesian, with hardier cattle such as the Galloway and Welsh Black on the higher farms. The predominant sires used are the Hereford or Charolais with smaller numbers of Aberdeen-Angus, Limousin and Simmental. The hill and upland areas are important in the production of STORE cattle which are moved to lowland farms for fattening.

The development of intensive and semi-intensive systems of production has led to slaughtering at ages from 15–24 months. The majority of calves from the dairy herd, either pure-breeds or beef crosses, are reared and fattened on semi-intensive systems.

Some of the more important breeds of beef cattle now kept in the U.K.

Simmental.

Welsh Black.

Galloway.

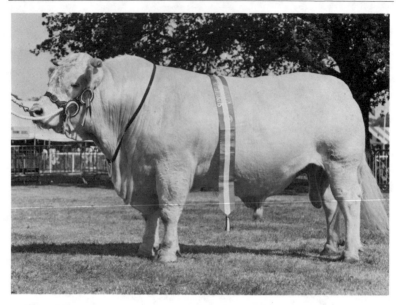

Charolais.

Hereford.

Aberdeen Angus.

Limousin.

Dairy Cattle. The Friesian breed is dominant in the dairy herds of England, Wales and Northern Ireland, and the Ayrshire breed in Scotland. Dairy cow numbers in the U.K. in 1979 totalled about 3.3 million (see Table 1), 63% of which were in herds of 60 and over, although average herd size was 49 cows, the largest in the E.E.C.

MILK production has been stimulated by advances in grassland management (◊ GRAZING SYSTEMS), new methods of grass CONSERVATION and programmes of controlled use of CONCENTRATES, while widespread use of progeny-tested sires has improved the genetic capabilities of the dairy herd. (Sources: Annual Review of Agriculture, H.M.S.O. 1980; and 'Agriculture in Britain', Central Office of Information, R5961/1977.)

**The main breeds of dairy cattle
now kept in the U.K.**

Guernsey.

Dairy Shorthorn.

Ayrshire.

Friesian.

Jersey.

Cattle Grid. A type of grate comprising parallel bars in a frame covering a pit in the road. Usually used to replace gates. The spaces between the bars are wide enough to discourage stock from crossing but are not a hindrance to humans or vehicular traffic.

Cattle Lifter. A stealer of cattle.

Cattle Plague. A viral disease of cattle characterised by inflammation and ulcerations of mucous membranes. Eradicated from Britain in 1877. Still seriously affects cattle in Asia and Africa. Also called Rinderpest or Bovine Typhus.

Caul. A membranous sac, part of the amnion, enclosing the head of an animal at birth.

Cauld. A Scottish term for a weir or dam in a stream.

Cauliflower. A variety of the cabbage species (*Brassica oleracea*) with an enlarged terminal bud in the early flowering stage. The head comprises short swollen flowering branches, tightly packed, surrounded by the upper leaves. (◗ BRASSICA, BROCCOLI)

Cavie. A hen-coop or cage.

Cavings. The small pieces of straw and leaves broken up by a THRESHING MACHINE and delivered from it after shaking out the grain and chaff.

Celery. An edible umbelliferous plant (*Apium graveolens*). The blanched leaf stalks are eaten cooked or raw. It is grown as a vegetable crop. Commercial production is mostly restricted to the rich organic soils of the FENS and the Lancashire mosses.

Cell. 1. One of the units of which plants and animals are constructed, comprising a mass of PROTOPLASM containing a NUCLEUS, surrounded by a thin membrane and (in plants) a cellulose cell wall (chitin in fungi). The larger plants and animals contain millions of cells. BACTERIA, fungal spores, certain ALGAE, etc., are single-celled organisms.
2. One of the hexagonal chambers in a HONEYCOMB.

Cell Wall. The limiting layer of a plant CELL enclosing the protoplasm, comprising the plasma membrane and CELLULOSE laid down as a crystal lattice.

Cellulose. A fibrous polysaccharide, the fundamental constituent of the CELL WALLS in plants. (♦ CARBOHYDRATES)

Central Council for Agricultural and Horticultural Cooperation. Established under the Agriculture Act 1967 with two functions; to promote and develop agricultural cooperation in the U.K. generally; and to administer, in conjunction with the Agricultural Ministers, the AGRICULTURAL AND HORTICULTURAL CO-OPERATION SCHEME. The Council also makes grants for desirable research.

Cereal(s). Cultivated members of the grass family whose seeds or grain are used to provide flour for breadmaking or as animal feed. The grain is rich in STARCH but also contains valuable PROTEINS and VITAMINS. The main cereals grown in Britain are BARLEY, MAIZE, OATS, RYE and WHEAT. Wheat is grown mainly in the eastern half of England. According to the 1980 ANNUAL REVIEW White Paper, over 6.6 million tonnes were produced in 1978, with average yields estimated at 5.26 tonnes per hectare (2.10 tons per acre). About 2.1 million tonnes of the 1978 wheat crop was used for flour milling, the remainder going mainly for animal feed. Since 1960 the area of barley has increased by about 75% and in general production has risen faster because yields have also increased. In 1978 production was almost 10 million tonnes, with average yields estimated at 4.19 tonnes per hectare (1.67 tons per acre). The malting (♦ MALTING BARLEY) and distilling market absorbs about one-fifth of the crop, most of the remainder being retained by farms or sold for animal feed. Barley is the main cereal crop in N. Ireland, where most of it is used for animal feed. The now almost universal use of the COMBINE HARVESTER has led to the installation of drying (♦ GRAIN DRIER) and storage facilities on most cereal-growing farms. Such equipment is also often used on a cooperative basis.

Cereal Seed ♦ SEED CERTIFICATION.

Cereal Substitutes. Substances used by animal feed compounders to provide the CARBOHYDRATE content of the feed instead of traditional cereals, e.g. MANIOC.

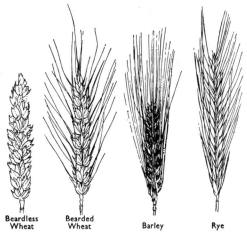

| Beardless Wheat | Bearded Wheat | Barley | Rye |

Ears of the Cereals.

Spreading Panicle of Oat.

r, rachis   p. pedicel   s. spikelet

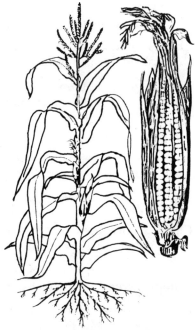

Maize Plant and Cob.

Certificate of Bad Husbandry. A certificate issued by an AGRICULTURAL LAND TRIBUNAL on receipt of an application from a landowner, and following a hearing. It is issued on the grounds that the tenant has failed to maintain a reasonable standard of efficient production (in respect of produce type, quality and quantity) or to keep the holding in such a condition to enable that standard to be maintained in the future. An incontestable notice to quit may be served on a tenant within 6 months of a successful application for a certificate, which deprives him of any entitlement to disturbance compensation.

Certification Centre ◗ DEADWEIGHT CERTIFICATION CENTRE, LIVEWEIGHT CERTIFICATION CENTRE.

Certified Seed ◗ SEED CERTIFICATION, SEED POTATOES.

Cesspit, Cesspool. A pit, well or pool into which liquid wastes and manure are directed. The solids are left behind as the liquid drains away.

Chafers. Various beetles, the grubs of which are pests in old grasslands, usually those which are in slightly dry situations.

Chaff. In general terms any worthless matter. The husks of corn separated from the grain during THRESHING or WINNOWING. Also short lengths of cut hay or straw. Also known as dowse or chop.

Chaffing. The chopping up of straw and hay for feeding to stock. Chaff is commonly mixed with CONCENTRATE rations for horses. It discourages them from bolting the ration and compels them to chew it.

Chain. A measure of length equivalent to 22 yds (20.1 m) or 100 LINKS. (◗ GUNTER'S CHAIN)

Chain Harness. A type of harness used to draw ploughs or other implements having no shafts. Also known as sling gear.

Chain Harrow. A type of HARROW, lacking a rigid frame, comprising a series of flexible chain links bearing spikes or knife-tines, generally double-ended and reversible. This arrangement permits the harrow to follow uneven ground very closely.

Chalaza. 1. The spiral albuminous band anchoring the yolk at each end to the shell in a bird's egg.
2. The base of the OVULE in plants where the FUNICLE or ovule stalk is primarily attached.

Chalk. A soft white or greyish, very porous type of LIMESTONE, very rich in CALCIUM CARBONATE. It consists principally of the calcareous sedimentary remains of small marine organisms and shells laid down under the seas about 100 million years ago. Chalk dominates the DOWNLAND of South-East England. It is spread on the land to correct soil acidity and calcium deficiency, either as rough chalk or after crushing (ground chalk) when it is more effective. It is sometimes kiln dried, finely ground, and sold as ground Carbonate of Lime. (◖ GROUND LIMESTONE, LIME)

Challenge Feeding. The feeding of dairy cows in early LACTATION with CONCENTRATES at a rate in excess of standard recommendations, to provide extra energy which would otherwise be derived by the breakdown of body tissue. Also called lead feeding. (◖ STEAMING UP)

Chambers of Agriculture. Local bodies founded in 1866 and affiliated to the Central Chamber of Agriculture with the object 'to promote and advance the best interests of agriculture, and to monitor all agricultural matters dealt with by Parliament and take such action as seemed desirable'. They lobbied successfully for the statutory establishment of the BOARD OF AGRICULTURE in 1889. Essentially the forerunner of the NATIONAL FARMERS' UNION.

Chandler. A retail dealer, e.g. corn-chandler.

Channel Island Breeds. The GUERNSEY and JERSEY breeds of cattle producing milk rich in BUTTERFAT, which attracts a special price and is marketed in bottles with a gold top.

Character. 1. The nature and way of behaviour of an animal.
2. A specific recognisable feature transmitted from generation to generation, e.g. height, colour, etc. In plant and animal breeding attempts are made to BREED-IN or BREED-OUT characters considered desirable or otherwise.

Charlock. Wild mustard (*Sinapsis arvensis*), a common yellow-flowered weed of cornfields.

Charolais. A breed of large, heavy, rapidly growing, beef cattle, developed in central France and descended from an indigenous breed in the Charolles district. Imported into Britain in the 1960s. White or cream in colour with soft hair, which may be long if wintered out of doors. They have a long, deep and wide, heavily muscular body. Some animals, known as culards, show double muscling in the hindquarters, a cause of calving difficulties (*see p. 88*).

Chase. The hunting of wild animals by pursuit. Also an open area or preserve for game. Beasts of chase are wild animals generally hunted, e.g. buck, doe, fox, marten and roe.

Chats. The smallest potatoes in a crop, either too small or of inadequate quality to be classed as WARE or SEED POTATOES.

Checking. The tilting of a plough to the land. (◗ BOARDING)

Cheek. One of two corresponding sides of an implement. Often used for the side of a plough facing the unploughed part of a field.

Cheese. A nutritious foodstuff prepared from the CURD of milk, coagulated by RENNET or acid, separated from the WHEY, and pressed into a solid mass. The nature of the BUTTERFAT content of the milk is important. Small fat globules (characteristic of AYRSHIRE and FRIESIAN milk) which rise extremely slowly, are required to produce an even texture to the cheese.

Hard cheeses include the main English varieties; Gloucester, Stilton, Cheddar, Cheshire and Wensleydale, and also the Gruyère cheese from Holland, Canada and France. The best known soft cheese is Camembert. Ewe milk is used to make Roquefort. Goats' milk is used for some French and Swiss cheeses.

Chemical Analysis. The determination of the composition of a substance using chemicals to break it down into simpler substances. The analysis may be qualitative, showing only the constituents present, or quantitative, showing the actual amounts present. Used, for example, to determine the FERTILIZER requirement of soils, or the COMPOSITIONAL QUALITY of milk. (♦ PHYSICAL ANALYSIS)

Chemical Weed Control. The application of HERBICIDES to crops on bare land to kill weeds.

Chenopodium. A genus of plants including a number of weeds, e.g. Fat Hen or Goosefoot (*C. album*), Good King Henry (*C. bonus-henricus*). South American Goosefoot (*C. quinoa*) is cultivated in Chile and Peru, the seeds and leaves being used like rice and spinach respectively.

Cherry. Many cultivated varieties of Cherry trees have been bred from the Wild Cherry or Gean (*Prunus avium*) and the mainly southerly Wild Dwarf Cherry or Sour Cherry (*P. cerasus*). They grow well in Kent. The edible fruit borne on a long stalk is round and fleshy, and contains a hard stone. (♦ TOP FRUIT)

Chester White. A breed of pig, originating in Chester County, Pennsylvania, derived from crosses between the YORKSHIRE, LINCOLN CURLEY COAT and the CUMBERLAND. The latter two are now extinct. A large white breed with blue freckles and semi-lop ears, producing high quality lean carcases and large litters.

Chestnut. 1. A horny knob on the inside of a horse's forelegs.
2. A reddish-brown colour, used descriptively of horses.
3. A hardwood tree (*Castanea sp.*). The Sweet Chestnut (*C. sativa*) is grown as a coppice crop, particularly in South-East England, much used for hop poles and fencing.

Cheviot. A hardy breed of pure white, short-woolled sheep, native to the Cheviot hills. A medium-sized sheep, clear of wool on its face and legs, with upright ears and an alert appearance. The North Country Cheviot (*see p. 392*) is a larger type which developed in Caithness, and the Sennybridge Cheviot is a third type developed in Glamorgan in the early nineteenth century.

Chewing the Cud. An activity of RUMINANTS, involving the mastication for the second time of food which has previously been swallowed and passed into the first stomach or RUMEN.

Chianina. A breed of beef cattle originating in the Chiana valley of Italy. Very muscular, heavy, long-legged cattle, standing up to 1.8 m (*c.* 6 ft) tall at the shoulders, formerly used for draught purposes. White, with black hooves, muzzle and horn tips. The long head bears short horns which curl forwards. There is much loose skin around the throat and a well developed dewlap.

Chick. A young bird about to be hatched or newly hatched.

Chicken. The young of the domestic fowl. Also a term for the flesh of a domestic fowl, not always very young.

Chifox. A breed of long-haired rabbit, variously coloured.

Chilver. A ewe lamb.

Chine. 1. The spine or backbone. A piece of the same with the adjoining parts used for cooking.
2. A ravine.

Chip. A fruit basket made from interwoven thin strips of wood.

Chinese Goose. A noisy, white or fawn breed of goose. Also known as the Swan Goose due to its small swan-like appearance.

Chisel Plough. A type of very heavy CULTIVATOR, with large section tines, the points of which incline forward, and are drawn through the soil at a depth greater than in normal ploughing. The underlying layers are burst without subsoil being brought to the surface.

A chisel plough.

Chit. A shoot or sprout.

Chitin. A nitrogenous polysaccharide (♦ CARBOHYDRATES) found in the exoskeleton of insects and in the cell walls of many fungi. It provides mechanical strength and resistance to chemicals.

Chitterlings. The smaller intestines of a pig or other animals sometimes used for food.

Chitting House. A building in which trays of potatoes are stored in stacks during the winter and stimulated to sprout before planting, by the provision of controlled heating to prevent frost damage and regulate the rate of growth, and movable fluorescent lighting.

Chlordane. A persistent, highly toxic, organochlorine insecticide, used to control earthworms in turf.

Chlorinated Hydrocarbons. A group of insecticides which mainly act on the central nervous system. They tend to be persistent in the environment and can build up to toxic levels in the body fats of other animals higher up food chains. They include D.D.T., ALDRIN, DIELDRIN and ENDRIN amongst others. (♦ ORGANO-PHOSPHORUS COMPOUNDS)

Chlorophyll. The green pigment found in plants essential to the process of PHOTOSYNTHESIS.

Chloroplast. A chlorophyll containing organelle in a plant CELL, the site of PHOTOSYNTHESIS.

Chlorosis. A condition of plants characterised by normally green parts becoming pale green or yellow (chlorotic). It is due to the prevention of CHLOROPHYLL formation, caused by various factors including lack of light, magnesium or iron deficiency, and excess calcium (when called Lime Chlorosis). Also called green sickness.

Chocolate Spot. A fungal disease (*Botrytis cinerea* and *B. fabae*) of beans, particularly when winter sown. It causes the development of chocolate coloured spots on the leaves, often quite rapidly, in warm wet conditions following late spring frosts. The shoots become black and die.

Choke. A fungal disease (*Epichloe typhina*) of COCKSFOOT, TIMOTHY and other grasses, characterised by fungal cylinders around the stems and leaves, affecting seed production.

Chop ♦ CHAFF.

Chopper Spreader. A mechanism sometimes attached to the straw outlet of a COMBINE HARVESTER which chops the straw into short lengths and spreads it over the field.

Chopping out. The removal of unwanted plants in a crop.

Chorion. The outermost membrane around a mammalian foetus (contiguous with the uterus surface), and around a developing chick in an egg.

Chromosome. A thread-shaped body, comprising mainly D.N.A. and protein, carrying the GENES, and present in the nucleus of a CELL. (♦ DIPLOID, HAPLOID, MEIOSIS, MITOSIS)

Chrysalis. A PUPA.

Chunk Honey. Honey containing at least one piece of COMB HONEY.

Churn. 1. A machine which agitates cream or whole milk turning it into butter.
2. A large metal milk container or can, commonly of 10 gallons capacity (occasionally 12 gallons), with a close-fitting, mushroom type lid. (♦ CHURN COLLECTION)

Churn Collection. Originally most milk was transported to dairies in CHURNS. Since the war bulk tankers have gradually replaced churns. In Britain milk collection is now 100% by bulk tanker. In January 1980 about 8% of milk collected in N. Ireland was still in churns, but the Milk Marketing Board of N. Ireland hope to phase out churn collection by the end of 1981.

Chute Parlour ♦ MILKING PARLOUR.

Clags. A type of wool oddment purchased by the BRITISH WOOL MARKETING BOARD comprising pieces of wool containing lumps of dried mud picked up by sheep during winter, usually when kept in a fold or run on arable land. A claggy fleece is one containing such lumps.

Clamp. 1. A traditional form of storage for potatoes or other roots, in which they are neatly piled and covered with straw and earth. Now mostly replaced by indoor storage. Also known by a variety of other names including bury, grave, hog, pie and pit. 2. A surface heap of SILAGE, either unwalled or walled using railway sleepers, wooden boarding or concrete to reduce side wastage. Some silage clamps are constructed below ground and are known as pits. Clamps are often enclosed in polythene and the air evacuated, the resultant carbon dioxide accumulation minimising respiration losses and facilitating fermentation.

A silage clamp covered with polythene sheeting held down by old tyres.

Clart. Another word for DAG, used mainly in North-East England.

Clasper. One of a pair of appendages at the tail end of a caterpillar by which it clings to leaves, twigs and other surfaces.

Clat. To remove wool from around the udder and the inside of the thighs of a ewe prior to lambing.

Clay. A constituent of SOIL comprising extremely fine-grained particles, less than 0.002 mm in diameter, characterised by high moisture retention. Sticky and plastic when wet, hardening and shrinking on drying. (♦ CLAY MINERALS, CLAY SOILS, LOAM, SAND, SILT)

Clay Minerals. The particles comprising CLAY consist of layers of hydrated aluminium and magnesium silicates forming a crystal lattice. Various types of clay mineral exist, e.g. kaolinite and montmorillonite, having different layer structures, and differing abilities to absorb and retain water, and to absorb and exchange cations (♦ CATION EXCHANGE). The agricultural productivity of a soil is related to the nature and content of the clay minerals in it and to its ability to retain (and subsequently release for crop use) cations, particularly those applied in the form of FERTILIZERS.

Clay Soils. Soils containing large amounts of CLAY, often called heavy soils because they require much more effort (i.e. greater tractor power) to plough and cultivate than sandy or LIGHT SOILS. They are sticky and swell when wet and are liable to POACHING, particularly in winter. They retain moisture well but become hard, shrink and crack in dry weather. Artificial drainage of clay soils is often carried out, particularly for arable cropping. (♦ LAND DRAINAGE, LOAM)

Clean Cattle. Cattle which have not been used for breeding, i.e. maiden HEIFERS, STEERS and young BULLS.

Clean Corn. A sample of wheat, barley, etc., not mixed with the seeds of other plants.

Clean Land. Land free from weeds.

Clean Legged. A descriptive term for a horse with few or no long hairs (known as feather) on its legs.

'Clean' Pasture. The sward after an arable crop, following reseeding, ungrazed by livestock and therefore presumed to be uncontaminated by parasitic worm larvae.

Cleaning Crop ♦ FALLOW CROP.

Cleansing ♦ AFTERBIRTH.

Cleck. A Scottish term meaning to HATCH. A 'clecking' is a BROOD.

Cleft Graft. A GRAFT involving the insertion of the SCION into a wedge-shaped cut in the centre of the STOCK.

Cleg. One of various species of blood-sucking, two-winged, tabanid flies (*Haematopota sp.*). They are ashy-grey with finely mottled wings and large bright eyes and are troublesome to horses. Also called horse-fly.

Cleveland Bay. A breed of horse, bay (reddish-brown to chestnut) coloured, once used for all agricultural purposes and for coaching. Native to the Cleveland Hills of Yorkshire.

Clevis. A U-shaped iron loop with a removable cross bolt used to couple or hitch an implement to a tractor drawbar.

Click Beetle. A beetle of the family Elateridae, characterised by their ability to leap into the air when they fall on their backs, making a clicking sound, as they try to land right side up. Their larvae are WIREWORMS. Also known as Skipjacks.

Cling. Diarrhoea in animals, particularly sheep.

Clip. 1. To remove or SHEAR the wool from a sheep.
2. The wool removed from either a single sheep or a whole flock.
3. One of several projections which prevent a horseshoe from shifting on a horse's foot.

Cloche. A temporary, portable covering providing protection to seedbeds from cold, wind and heavy rain, etc., and used to encourage early growth. Cloches, either glass or plastic, may be either simple single frames covering a few plants, or long continuous tunnels over rows of plants.

Clock. A Scottish term meaning to cluck or to BROOD. A clocker is a brooding hen.

Clone. A group of plants or animals all having an identical genetical make-up, being the descendants of a single parent. In plant breeding a clonal stock of plants can be vegetatively raised by taking CUTTINGS from a single parent plant.

Close-textured Soil ▶ SOIL TEXTURE.

Clostridium. A genus of ANAEROBIC bacteria responsible for various diseases such as botulism (*C. botulini*), tetanus (*C. tetani*) and clostridial enteritis (*C. Welchii*).

Clove. An old measure of weight for wool or cheese; 7, 8, or 10 lbs.

Cloven Hoofed. A descriptive term of animals with a divided hoof such as cattle and sheep.

Clover. A large genus (*Trifolium*) of leguminous plants characterised by trefoil leaves, having small flowers with tight roundish heads, and producing one- to four-seeded pods. Root nodules contain bacteria capable of NITROGEN FIXATION. A number of species are grown for FODDER or are sown with a mixture of grass seeds to produce LEYS, e.g. ALSIKE CLOVER (*T. hybridum*), WILD WHITE CLOVER or Kentish Clover (*T. repens*), Purple or RED CLOVER (*T. pratense*). Clovers usefully knit the sward together and help keep out weeds, especially in longer leys.

Clover Hay. HAY derived from a clover crop. Also hay from a temporary LEY as opposed to that derived from permanent grassland (MEADOW HAY). Also known as seeds hay.

Clover Rot. A fungal disease (*Sclerotinia trifoliorum*) of clovers and related plants, characterised in its early stages by leaves being covered with small brown spots. Plants later wither and die. Different species of clover vary in their susceptibility. BROAD RED CLOVER can suffer badly, ALSIKE and WILD WHITE CLOVER are less severely attacked.

Clover Sickness. A condition of a field on which clover has been grown for too long and which is heavily infected with EELWORMS and fungus diseases (e.g. CLOVER ROT).

Club Root. A fungal disease (*Plasmodiophora brassicae*) of cruciferous plants, particularly swedes, turnip, rape, cabbage and other BRASSICAS, which causes swelling and distortion of the roots. Growth is stunted and leaves become pale green. The SPORES can remain alive in the soil for several years and are particularly active in acid and wet soil conditions. Also called Finger-and-Toe.

Clun Forest. A hardy grassland breed of sheep derived from various local types in the West Midlands and developed as a compromise between Welsh Hill sheep and Down breeds from Shropshire. Characterised by its dark brown or tawny face, with wool extending over the forehead, and short upright ears.

Cluster. 1. The four teat cups of a MILKING MACHINE applied to a cow's udder.

2. A swarm of bees after settling.

3. The leaves, flowers or fruits growing from a common point on the stem or shoot of a plant.

Clutch. A batch of eggs laid by a bird. Also the brood of chicks hatched from such eggs.

Clydesdale. A breed of draught horse originating in Lanarkshire, developed by interbreeding with Flemish and English breeds, including the SHIRE. Similar to, but smaller than the Shire, with a narrower, less deep body and longer legs, feathered only on the back. Generally bay, brown or black in colour, rarely grey or chestnut, commonly with a white face and legs.

Coarse Fodder ⊕ DRY ROUGHAGE.

Coarse Grains. Cereal grains used to feed livestock.

Cob. 1. A short-legged, strong horse, often used for prolonged, slow saddle work.

2. Cubes manufactured from unmilled dried grass (⊕ MILL) or other CONCENTRATES.

3. A male swan.

4. The core of a head of MAIZE supporting the rows of grain.

5. A mixture of clay and straw, sometimes including gravel, used in building.

Cob Nut. A large hazel nut of the cultivated variety *Corylus avellana.*

Cobalt. (Co). A metallic TRACE ELEMENT present in small amounts in soil and herbage, required for the assimilation of iron into haemoglobin in red blood corpuscles. When deficient it causes 'pining' in sheep.

Cobweb Theorem. An economic theorem that producers of a commodity expect the price in the next period of production to be the same as current prices and thus plan production accordingly. It is most appropriate in the case of non-storable commodities.

Coccidiosis. An intestinal disease of livestock and poultry caused by various microscopic protozoan parasites of two genera, *Eimeria sp.* and *Isospora sp.*, belonging to the Order Coccidia. Characterised by diarrhoea and emaciation.

Coccus. 1. A spherical bacterium, as distinct from rod-shaped or vibrio (spiral-shaped) bacteria.
2. A portion of dry fruit containing one seed when it breaks up.

Cochin. A large feathery-legged domestic hen originating in China.

Cock. 1. A general term for a male bird. For poultry, usually applied to males over 18 months old.
2. Slang for woodcock.
3. A small pile of dung or of a cut crop drying in the field.

Cock Hop. A freak hop CONE bearing a single or several small leaves between the bracts.

Cock's Head ♦ SAINFOIN.

Cockchafer. One of two species of large beetles (*Melolontha sp.*), the larvae of which are destructive pests of crops. Also called Maybug, White Grubs and Rookworm.

Cockerel. A young male chicken, usually applied to one less than 18 months old. Also a male turkey less than 12 months old.

Cocking. The forming of heaps or COCKS of a loose cut crop for drying in the field.

Cockle. A shrunken seed, badly formed; crinkled or puckered.

Cocksfoot. A perennial grass (*Dactylis glomerata*), with characteristic one-sided flower-heads of rounded or oval individual flower spikelets on wiry branches, and dull green to deep blue leaves. Deep rooting, very high yielding, used extensively for pasture (*see p. 211*).

Coconut Cake. A CONCENTRATE feedingstuff reasonably rich in PROTEIN, used frequently in rations for dairy cows, fattening cattle and pigs. It produces a harder fat than most other oil seed products. Not as readily available as it used to be. (♦ CAKE)

Cocoon. A protective silky sheath spun by many insect larvae (e.g. the CATERPILLARS of butterflies and moths) in which they pupate (♦ PUPA). Also a similar sheath or capsule in which spiders and earthworms lay their eggs.

Cod ♦ SCROTUM.

Codlin(g). A type of elongated apple.

Codlin(g) Moth. A small moth (*Ernarmonia pomonella*), the caterpillars of which eat the inside of growing apples and cause them to fall prematurely.

Cod-liver Oil. The oil obtained from the livers of codfish, rich in VITAMINS A and D, valued for feeding to young animals to improve condition. It is particularly helpful in preventing RICKETS developing.

Coggins Test. A test used to detect anaemia in horses.

Coir Yarn. String or twine made of coconut fibre on which BINES are trained in hop gardens.

Colbred. A hybrid breed of longwool sheep, bred in Gloucestershire since the Second World War by Oscar Colburn by crossing the EAST FRIESLAND, DORSET HORN, BORDER LEICESTER and CLUN FOREST breeds. It is prolific, yields plentiful milk, and is mainly crossed with hill ewes producing crossbred breeding ewes. Colbred sheep are long, with a clean white face, and long legs.

Cold Bagged Hops. Pressed hops, not having been sufficiently well dried before packing, so that their preservation is endangered.

Cold Frame. A low protective structure mainly of glass or hard plastic, sometimes with brick or wood sides, under which seedlings are encouraged to grow faster than if left in the open. The frame provides protection from cold, wind and heavy rain etc. (♦ CLOCHE)

Cold Storage. The preservation of perishable produce (e.g. apples, meat, etc.), usually in bulk, in refrigerated chambers. The low temperature conditions inhibit the multiplication of bacteria and fungal spores etc.

Cole. A term for RAPE. Also a general name sometimes applied to any BRASSICA.

Cole Garth. A cabbage garden.

Cole Seed. The seed of RAPE.

Colewort. An old term for any BRASSICA, used nowadays mainly for the heartless kinds such as RAPE and KALE.

Coli Septicaemia. A bacterial infection (*Escherichia coli*) of poultry. It is usually secondary to other respiratory diseases and is characterised by inflammation of the air sacs which often contain purulent material. Young birds between 6 and 10 weeks of age are particularly susceptible, mainly in winter in conditions of poor ventilation or overcrowding.

Colic. A vague term used, particularly for horses, to describe severe abdominal pain. Also called gripes.

Coliform Bacteria. Those bacteria commonly present in animal intestines and passed out in faeces (e.g. *Escherichia coli*). The presence of such bacteria in milk indicates faecal contamination.

Collagen. A proteinous substance, the principal fibrous constituent of skin, tendon, ligament and bone. When boiled it produces GELATIN.

Collar. A device fitted round the neck of a horse by which a plough or cart, etc., is pulled. Normally of a closed design but sometimes open at the top with the two sides connected by a strap. The sides consist of metal or wooden bars called hames to which the TRACES are fixed, and each bears a ring through which the control reins pass.

Collar Work. Hard work done by a horse involving straining against the horse-collar, such as ploughing, as opposed to less strenuous work such as carting.

Collard. A variety of cabbage not forming a heart. Sometimes used to mean KALE or RAPE.

Collie. A long-haired, intelligent, Scottish breed of sheep-dog.

Colloid. A substance present in solution in a colloidal state, i.e. particles of 1–100 μm diameter dispersed in a continuous medium, having a high surface to volume ratio with pronounced absorptive properties. Clay particles and HUMUS in the soil are colloidal in that they can become dispersed in absorbed water, the soil swelling and becoming sticky. On drying, the particles release their water and the soil may shrink and clayey soils may crack. Clay and humus also act colloidally by absorbing CATIONS. (◊ CATION EXCHANGE)

Collop. A stock unit used in the letting of grassland, equal to 3 yearling cattle, to 1 yearling and 1 two-year-old bullock, or to 4 sheep.

Colluvium. A heterogeneous mixture of weathered material transported by gravity to be deposited at the base of a slope. (◊ ALLUVIUM)

Colon. The large intestine of an animal from the CAECUM to the RECTUM.

Colorado Beetle. An American beetle (*Leptinotarsa decemlineata*), yellow with black stripes, a pest of potato crops, feeding on the leaves.

Colostrum. The milk given by cows and other mammals during the first few days after giving birth. It is particularly rich in proteins including ANTIBODIES, and has a higher fat content than normal milk. Also called beestings or beistings.

Coloury. A descriptive term applied to dried hops with a distinctive green tint.

Colt. A young male horse less than four years old (five years if thoroughbred). (◊ FOAL)

Colter ◊ COULTER.

Colza. The oil yielded by RAPE seeds (also called cole seeds).

Comb. 1. A fleshy crest on the head of some birds.
2. The cellular structure constructed of beeswax in hives in which bees store surplus honey. In commercial beekeeping hives contain frames holding a thin sheet of wax on which bees construct the six-sided cells which form the comb. Also called honeycomb.
3. The lower blade of the cutting section of a powered sheep-shearing handpiece.
4. The edge at the top of a FURROW slice. Also called arris.

Comb Honey. Honey stored by bees in the cells of freshly built broodless (♦ BROOD) combs. Usually sold in sealed whole combs or in parts of such combs.

Combine Drill. A DRILL which sows seeds and fertilizer simultaneously, usually down the same seed-tube, although there are separate hoppers for seed and fertilizer. Combine drills are difficult to clean out effectively. (♦ SEED DRILL)

A self-propelled combine harvester.

Combine Harvester. Usually shortened to combine. A machine used to harvest all types of CEREALS, which cuts or picks up the crop from a WINDROW, threshes it (♦ THRESH), separates SEED and CHAFF from the STRAW, and cleans the seed. There are three types, viz, tractor drawn – p.t.o driven, tractor drawn – engine driven, and self-propelled. The last are the most widely used nowadays.

Harvesting rates depend on the size of machine and the condition of the crop. With the most modern machines, they can range from 4 tonnes/hour to 18 tonnes/hour of separated grain. Annual through-put of crop per machine depends on the type and range of crops to be harvested, but on average is between 40 ha (100 ac.) a year to 400 ha (1000 ac.) a year.

Various attachments are used with combines including straw

CHOPPER SPREADERS, ear lifters (GRAIN LIFTERS), STRAW SPREADERS, TRUSSERS, WINDROW PICK-UPS and maize headers. Recent developments include electronic monitoring of the functional efficiency of key components (e.g. shaft-speeds) and grain loss.

Commercial Recording. A service operated by the MEAT AND LIVESTOCK COMMISSION providing regular information on LIVEWEIGHT gains, mainly of cattle, but also sheep, either individually or in groups.

Commercial Seed. Seed of certain species of fodder, oil and fibre plants which is not inspected to determine varietal purity but which is marketed as being 'true to kind'. If seed is marketed as 'commercial' under a varietal description, it is on the supplier's declaration that the seed is of that variety. Seed standards for analytical purity and germination as prescribed must be met by official sampling and testing. (♦ SEED CERTIFICATION)

Commission. The executive and administrative body of the EUROPEAN ECONOMIC COMMUNITY, initiating proposals for policies and monitoring the implementation of regulations.

Common Agricultural Policy. Often abbreviated to C.A.P. The principal policy of the EUROPEAN ECONOMIC COMMUNITY with five main objectives set out in the TREATY OF ROME; viz, (*a*) increased agricultural productivity, (*b*) assurance of a fair standard of living for farmers, (*c*) stabilisation of agricultural markets, (*d*) guarantee of regular supplies of food, and (*e*) maintenance of a reasonable price for food. These objectives are pursued on the basis of four broad principles; viz, (*i*) the establishment of E.E.C. regimes for the main farm products, usually based on a system of COMMON PRICES with, in some instances, provision for INTERVENTION BUYING as the main method of internal market support, (*ii*) free trade between member states, (*iii*) a single trading system with non-member countries which, through the mechanism of common variable levies and/or common customs duties, ensures a preference for E.E.C.-produced supplies in internal E.E.C. trade, and (*iv*) joint responsibility by the member states for financing the C.A.P. through the EUROPEAN AGRICULTURAL GUIDANCE AND GUARANTEE FUND, which is an integral part of the E.E.C. budget. (♦ BASIC PRICE, TARGET PRICE)

Common Land. A tract of land in public ownership, open to common use. In early times common land was attached to a village or manor, on which the villagers had a right to graze animals. About 2 million ha (5 million ac.) of common land was enclosed by the Lords of the Manor in the fifteenth and sixteenth centuries and by Acts of Parliament in the eighteenth century, and passed into private ownership. By 1876 all ENCLOSURES were forbidden. The 600 000 ha (1.5 million ac.) of common land still existing in England and Wales is now administered by local authorities. The Commons Register, begun in 1965 and closed in 1970, contains a definitive record of all land held in common, and all common rights.

Common Market. A term for the EUROPEAN ECONOMIC COMMUNITY.

Common Prices. E.E.C. regulations for a range of products (including beef, cereals, milk products, pigmeat, oilseeds and sugar) which seek to provide all E.E.C. farmers with common guaranteed prices in terms of UNITS OF ACCOUNT. The regulations involve controls on imports from THIRD COUNTRIES and internal market support including INTERVENTION BUYING, operated on a similar basis in all member states. Regulations for potatoes and sheepmeat are being planned. Also called regimes (e.g. beef regime). Reviewed annually by the COUNCIL OF MINISTERS. (◆ ANNUAL REVIEW, BASIC PRICE, TARGET PRICE, COMMON AGRICULTURAL POLICY)

Common Scab ◆ SCAB.

Community. 1. A term applied to any assemblage of plants making up a distinct vegetation type, e.g. deciduous woodland.
2. A short term for the EUROPEAN ECONOMIC COMMUNITY.

Companion Crops. Two crops grown together in the same field, one of the crops, and often both, benefiting from the presence of the other. The term is particularly applied to crops grown together for seed production which have similar cultivation and management needs and whose seed can be separated after harvesting (e.g. clover and ryegrass, the clover facilitating soil NITROGEN FIXATION, beneficial to ryegrass growth, whilst the ryegrass assists in the harvesting of the clover).

Compensatory Amounts ◗ MONETARY COMPENSATORY AMOUNT.

Compensatory Growth. The phenomenon of animals turned out to pasture after a period of energy restriction exhibiting greater growth rate than animals which have not been so restricted. It is due to the greater feed intake following the period of energy restriction.

Complement. A constituent of blood SERUM, the presence of which is required for an ANTIBODY to kill an ANTIGEN.

Complement Fixation Test. A blood test used to diagnose certain bacterial diseases, e.g. BRUCELLOSIS.

Complete Diet. Winter feed for livestock comprising a mixture of CONCENTRATES and bulk forage fed as a complete mixture, the animals usually receiving no other feed.

Complete Food Wagon. A large mobile container, similar to a FORAGE WAGON, which mixes CONCENTRATES and bulk FORAGE (e.g. SILAGE) by internal augers to form a complete feed (◗ COMPLETE DIET), which is discharged to MANGERS via a side delivery mechanism (e.g. a moving floor) in an even steady flow.

Compositional Quality Scheme ◗ MILK COMPOSITIONAL QUALITY SCHEME.

Compost. A manure derived from decomposed plant remains, usually made by fermenting waste plant material (e.g. straw, grass mowings, etc.) in heaps, usually in alternate layers with added lime, nitrogen and water. Compost is usefully used in greenhouses, to enrich soil, either dug in or as a surface MULCH. Straw compost is valuable in building up poor soils for intensive vegetable cropping and assists in maintaining soil structure on heavier land.

Compost Activator, Compost Accelerator. A chemical substance which promotes decomposition and fermentation of decaying plant remains in a compost heap.

Compound Feed. Animal feed composed of several different FEEDINGSTUFFS (and including major minerals, TRACE ELEMENTS, VITAMINS and other additives) in proportions providing a balanced diet. The compound manufacturers provide a range of feeds suitably balanced for all types of stock at all growth and development stages. (◗ HOME MIXING)

Compound Fertilizer. A FERTILIZER containing a mixture of two or three of the major plant nutrients (i.e. NITROGEN, PHOSPHORUS and POTASSIUM). Many proprietary compound fertilizers are produced with differing nutrient ratios to suit specific crops and soils. Mostly produced in easily stored granulated form, but powders and less concentrated liquids are available. Also called mixed fertilizer. (◗ STRAIGHT FERTILIZER, UNIT OF FERTILIZER)

Compound Leaf. One divided into separate leaflets, e.g. CLOVER. (◗ SIMPLE LEAF)

Compressed Pockets. Hop POCKETS compressed to a high density for storage or export.

Compressibility – of Soil. The susceptibility of a soil to compact under a heavy load (i.e. a tractor).

Conacre. An Irish custom of letting land, usually already prepared, in small portions to grow a single crop. Also called cornacre.

Concave. A semi-circular grating in a THRESHING MACHINE, encircling the drum, against which the grain is dislodged by the rotating BEATERS on the drum.

Concentrates. A term for a variety of animal FEEDINGSTUFFS with a high food value relative to volume, with low FIBRE content, some rich in PROTEIN, others rich in CARBOHYDRATES or FAT. Concentrates include cereal grains and their by-products, leguminous seeds, oil seeds, oil CAKES and MEALS, and various animal by-products. Mainly supplied by compound feedingstuff manufacturers as expertly balanced formulations, often computer assisted, for all types of farm stock. Ingredients may vary in relation to current prices and availability, and TRACE ELEMENTS, VITAMINS and MINERALS are usually included to produce complete foods. Some farms mix their own concentrates using home-grown cereals. (◗ HOME MIXING)

116

Condensed Milk. Enriched milk produced by evaporating much of its water content, and by the addition of sugar. (♦ EVAPORATED MILK)

Condition Scoring. An animal can be brought into condition (good state of health) by careful feeding and good management. Condition scoring is the recording of the state of muscling and fatness (body condition) of breeding stock. Body condition scoring provides a convenient method, based on a standard handling technique, of subjectively assessing the overall nutritional status of breeding stock, and identifies individuals needing special feeding treatments. Score ranges vary, being 0–5 for cows, 1–5 for ewes and 1–9 for sows. The lowest figure indicates extreme thinness; the highest excessive fatness.

Cone. 1. The fruit of the female hop-plant, growing as a cluster of hops on a BINE.
2. The fruit of a coniferous tree, consisting of compact overlapping scales, arranged spirally around a central axis.

Conformation. The shape of an animal. Conformation is an important element in the BEEF CARCASE CLASSIFICATION SCHEME, and is concerned with carcase thickness and blockiness, and fullness of the round. There are five conformation classes with 5 the best and 1 the worst.

Conifers. A group of trees and shrubs with needle-like leaves and typically bearing CONES, e.g. pine, fir, spruce. Coniferous trees or softwoods grow quickly and provide early profit where grown commercially. They are especially suited to exposed hill and acid conditions. (♦ BROADLEAF)

Conservation. 1. The optimum rational use of natural resources and the environment, having regard to the various demands made upon them and the need to safeguard and maintain them for the future.
2. The protection of the soil against erosion or loss of fertility.
3. The preservation of grass as HAY and SILAGE and of fodder crops for winter feed. Also applied to the preservation of certain growing crops *in situ* for later use, e.g. frost-hardy THOUSAND HEADED KALE kept through the winter for folding (♦ FOLD) in spring.

Contact Animal. An animal which has been in contact with a diseased animal and may need to be isolated if the disease is infectious.

Contact Herbicides. HERBICIDES which kill only those parts of a plant with which they come into contact and used mostly to control annual weeds when seedlings. They have very little residual effect and are normally applied just before crop sowing. (♦ TRANSLOCATED HERBICIDES, TOTAL HERBICIDES, SELECTIVE HERBICIDE)

Contagious Abortion ♦ BRUCELLOSIS.

Contagious Ecthyma, Pustular Dermatitis ♦ ORF.

Continuous Flow Drier ♦ GRAIN DRIER.

Contour Ploughing. Ploughing on the level following the contour, as distinct from up and down the slope. This practice is an effective method of soil erosion control.

Control Area. An area declared by the Ministry of Agriculture, Fisheries and Food to be subject to various controls aimed at containing the spread of disease within the area and preventing animal to animal contact. Such areas are similar to Infected Areas (♦ INFECTED PLACE) but are usually much larger and cover a wider range of animals. Movements out of the area may be allowed at the discretion of a Ministry Veterinary Inspector and under a MOVEMENT LICENCE.

Controlled Environment Housing. Livestock buildings in which various conditions such as temperature, humidity, ventilation and lighting are strictly controlled. In poultry houses, for example, a steady temperature regime can minimise rearing losses, increase egg production and decrease feed requirements, thereby reducing production costs.

Controlled Grazing. The regulation of the grazing of pasture by adjusting livestock numbers according to the amount of grass and to its growth rate, or by alternate grazing and resting of the pasture. Livestock may be restricted to particular parts of a pasture by means of temporary fences.

Conversion Premium Scheme ♦ DAIRY HERD CONVERSION.

Cony. A rabbit. A cony burrow is a rabbit warren.

Cooler ◗ MILK COOLER.

Cooling Floor. An unheated part of an OAST where hops are placed to cool between drying and pressing. Fragility is reduced as moisture is absorbed from the atmosphere and is redistributed in the hop CONES.

Coomb. 1. A deep wooded valley or hollow in a hillside.
2. A measure of capacity equal to 4 BUSHELS. A sack containing such a quantity.

Coop. A box or cage in which animals, particularly poultry, are confined.

C.O.P.A. Abbreviation for Comité des Organisations Professionnelles Agricoles des Pays de la Communauté Economique Européenne. An umbrella body for the various farmers' unions in the E.E.C., based in Brussels, and officially recognised for negotiations between farmers and E.E.C. officials.

Coping. The stitching up of the open end of a filled hop POCKET.

Copper. (Cu). A metallic chemical element, essential to animals in trace amounts, but toxic if consumed in quantity. Also required as a TRACE ELEMENT by plants for various metabolic processes. Deficiency diseases may occur in plants and animals if it is not adequately available. (◗ SWAYBACK)

Copper Sulphate. ($CuSO_4.5H_2O$). A blue crystalline salt used as a fungicide in BORDEAUX MIXTURE. Also called bluestone or blue vitriol.

Coppice. A small wood of small trees and underwood cut periodically, essentially comprising sprouts from cut stumps.

Coppice with Standards. A system of commercial management of oakwoods in which only an open network of oaks or standards are left unfelled. Hazel, chestnut and other underwood grow back in the open areas and are cut or coppiced at 10 to 15 year intervals to provide hurdles, stakes, poles and fencing.

Copse. A small wood, COPPICE or clump of trees. Also called a spinney.

Corcass. An Irish term for a salt marsh or area of land liable to flood by a river.

Cord. 1. A measure of cut wood. A stack containing 128 cubic feet, measuring 8 × 4 × 4 feet. Originally determined by the use of a cord or string.

2. A hopper for carrying seeds during hand-sowing.

3. Another name for LIVER FLUKE disease.

4. A loose term for the umbilical cord of an animal.

Cordon. A fruit tree trained and closely pruned to grow as a single stem, supported by wire frames or walls.

Co-responsibility Levy. A levy introduced in 1977 throughout the E.E.C. on all milk deliveries to dairies to provide funds for expanding the market for milk and milk products with the intention of decreasing surpluses. The levy introduced the principle of shared responsibility amongst E.E.C. milk producers for the production of surpluses.

Corm. A short, swollen underground stem, carrying buds, acting as an organ for storage and VEGETATIVE REPRODUCTION. Similar to a BULB but the food is stored in the swollen stem as distinct from the scales of a bulb. Examples include crocus and gladiolus.

Corn. A general term for cereal crops (barley, maize, oats, rye and wheat) grown for their edible seeds or grain. The grain itself is also called corn. In America the term is restricted to maize. In Scotland and Ireland the term is generally restricted to oats, whilst in England it is mainly applied to wheat.

Corn Chandler. A retailer of grain.

Corn Cockle. A tall weed (*Agrostemma githago*) of cornfields. Hairy, long-stalked, with much divided leaves and single pale purple flower, with woolly sepals.

Corn Laws. Various statutes regulating trade in grain, particularly restricting importation to Britain by duties. Abolished in 1846 due to public pressure for cheaper bread.

Corn on the Cob ◗ SWEET CORN.

Corn Thrips. A small insect with four wings bearing fringes of long hairs and which sucks sap from cereal crops. (◗ THRIPS)

1 000 Corn Weight ◗ 1 000 GRAIN WEIGHT.

Cornacre ◗ CONACRE.

Cornage. A service or rent applied in feudal times and fixed according to the number of horned cattle owned.

Cornish ◆ INDIAN GAME.

Corolla. The inner ring of the parts of a flower consisting of the petals, usually conspicuously coloured. (◆ CALYX, PERIANTH)

Corpuscles. Red and white blood cells circulating in the blood of animals.

Corral. An enclosure in which cattle or horses are kept.

Cortex. 1. The bark or outer layer of a plant.
2. The outer layer of the various organs of plants and animals.

Cos. A type of lettuce with long upright leaves as distinct from the standard butter-headed, cabbage-like kind.

Cosset Lamb. A hand-reared lamb. Also called cade lamb, pet lamb or sock lamb.

Cossettes. Dried slices of potato used as a feedingstuff for stock.

Costard. A large kind of apple.

Cot House. A cottage occupied by a COTTAR.

Cote. A small house or shelter for animals, particularly for sheep (a sheep-cote or sheepfold) and some birds (pigeon-cote, dove-cote).

Cotland. Land held by a COTTAR.

Cotswold. A breed of sheep originating in the Cotswold hills but continually crossed with various longwool breeds. Very popular in the mid-nineteenth century, now very limited in number. White- or grey-faced with a well developed forelock.

Cott. A fleece, or part of a fleece, that has become cotted or matted usually due to insufficient wool oils being produced, often due to sickness or injury.

Cottage Pie ◆ GLOUCESTER OLD SPOT.

Cottage Piggery ◆ PIGGERY.

Cottar. An old Scottish term for a peasant occupying a cottage belonging to a farm, and paying rent in the form of labour. (◆ BONDAGER)

Cottier. An Irish term for a peasant holding a piece of ground under Cottier-tenure, now illegal, by which portions of land were let annually to the highest bidder.

Cottonseed Cake, Cottonseed Meal. An animal FEEDINGSTUFF made from the seeds of the cotton plant after the extraction of oil, mainly used for adult RUMINANTS. It has a high FIBRE content, is poor in calcium, with low PROTEIN quality. It is yellow-coloured due to the presence of gossypol which is toxic to young pigs and poultry, and is available either in DECORTICATED or undecorticated forms. (◗ CAKE)

Cotyledon. A seed leaf, forming part of the embryo of seeds. The first leaf to develop when a seed germinates, initially lacking CHLOROPHYLL. Monocotyledonous plants have one cotyledon in each seed whilst dicotyledonous plants have two. The two types of plants are also distinguished by their leaves. The former possess long, narrow leaves with parallel veins (e.g. grasses) whilst the latter have broad leaves with branching veins (e.g. broadleaf weeds such as dock, FAT HEN, etc.)

Couch Grass. A group of five grasses (*Agropyron sp.*) related to wheat, which they closely resemble. Common COUCH GRASS (*A. repens*) is a pestilent weed owing to its creeping RHIZOMES. Also called quick or twitch and by many local names.

Coulter. The part of a PLOUGH which makes the vertical cut in the soil, usually mounted in front of the SHARE, attached to the BEAM. There are three main types; knife, disc and skim coulters. Knife coulters were a feature of horse ploughs. Flat, circular, freely rotating disc coulters are used on most tractor-drawn ploughs. They rotate as the plough moves, separating a slice of soil from unploughed land, which the share and mouldboard then turn to produce a clean furrow slice. Skim coulters are attached to the beam, either in front of knife coulters or behind disc coulters. Each acts like a small DIGGER and turns a small furrow slice from the corner of the main furrow into the furrow bottom, thus completely burying crop residues or manures (*see p. 338*). Also called colter.

Coulter Face. The side of a SLICE cut by a COULTER in ploughing, and then turned over by the MOULDBOARD to become the front of the slice. (◗ SHARE FACE)

Council of Ministers. The ultimate decision and policy-making body of the E.E.C., comprising Ministers of the member states responsible for a particular subject which may be under review or discussion at any time. The Council of Ministers of Agriculture are responsible for the ANNUAL REVIEW determinations.

Count. 1. An estimate of numbers of bacteria or insects in soil, milk, etc.
2. ◗ BRADFORD COUNT.

Country Code. A series of ten reminders based on common sense concerning public conduct in the countryside, covering such matters as fastening gates and keeping dogs under control. Issued by the COUNTRYSIDE COMMISSION.

Country Landowners' Association. A national organisation of owners of rural land. Established as the Central Land Association in 1907 with the object of uniting all concerned with agriculture. The name changed to Central Landowners' Association in 1918, when it became an Association representing landowners' interests with the object being the prosperity of agriculture within the framework of private landownership. In 1949 the name was again changed to the Country Landowners' Association. Regional Secretaries administer County Branches throughout England and Wales, with a head office in London.

Countryside Commission. A statutory public body established under the Countryside Act 1968, charged with keeping under review various matters relating to the countryside, including facilities for its enjoyment, conservation, and enhancement of its natural beauty and amenity, and public access for recreation. The Commission is responsible to the Secretary of State for the Environment. It took over all the functions of the National Parks Commission.

County Agricultural Executive Committees. County-based Committees set up under the AGRICULTURE ACT, 1947 to 'promote agricultural development and efficiency' and to exercise various functions delegated by the Minister, and largely continuing and extending, on a statutory basis, the work of the WAR AGRICULTURAL EXECUTIVE COMMITTEES. They comprised members nominated by the Minister and others representing farmers, owners and workers, and acted through sub-

and district committees. They initially had the power to ensure the good management and husbandry of farmland with the ability to authorise and ultimately to dismiss a farmer or landowner from the land. The Agriculture Act, 1958 removed this power and Committees thenceforth had a general advisory role to promote general farming development. Abolished in 1972 when REGIONAL PANELS were established with similar functions.

Couple. A ewe and lamb.

Course. The period in a cropping sequence or ROTATION when the land is growing a particular crop in the sequence. Thus in the traditional four-course rotation a particular crop would only be grown on the same land in one year out of four.

Covent Garden Market. The main London wholesale market for fruit, vegetables and flowers, opened in 1661, reorganised in 1974 and transferred to a new site at Nine Elms Lane in Battersea, and now known as New Covent Garden Market.

Cover. 1. To BROOD or sit on a clutch of eggs.
2. To copulate with, as when a stallion covers a mare. Also called serve.
3. Undergrowth or thicket concealing game.

Cover Crop. 1. A crop which provides protection to a second crop sown beneath it. Temporary LEYS are sown usually using a cereal as the cover crop, the grass and clover seeds mixture being sown immediately following the sowing of the cereal seeds. Also called a nurse crop.
2. A crop grown to provide COVER for game birds, e.g. KALE.

Cow. The female of the bovine animals. In cattle terminology the female is termed a cow on bearing a second calf. (♦ HEIFER)

Cow Beef. Beef from dairy cows discarded due to age, or culled because of infertility, poor milk yield or udder disease.

Cow Calf ♦ CALF.

Cow-gait, Cow-gate. Pasture on which cows are grazed.

Cow-grass ♦ RED CLOVER.

Cow Heifer. A cow having borne only one calf.

Cowhouse. A building in which dairy cows are both housed and milked, being tied in pairs in stalls, as distinct from being kept in loose yards or CUBICLE sheds and milked in a separate MILKING PARLOUR. Cowhouses may be single or double range. In the latter the cows either face away from a central passage used for milking and dung removal, or towards it for central feeding. Also called a byre, shippen or shippon.

Cowman. A person in charge of the day-to-day supervision and management of a dairy herd, including milking.

Cox's Orange Pippin. A well-known variety of apple.

Crabtree Effect. The continued improvement in the feed-use efficiency of an animal under feed restriction, so that the feed supply has to be repeatedly reduced further in order to maintain a flat growth rate.

Cracked Wool. A fleece which felts together and cracks open, especially in wet weather.

Cradle. A set of fingers or light rods in a light frame set into a SCYTHE (a cradle-scythe), used to cut corn, laying it evenly with each stroke.

Craft. A class of work in agriculture or horticulture such as milk production, machinery operation and maintenance, or plant nursery practice. (♦ CRAFTSMAN)

Craft Certificate. A certificate issued by an AGRICULTURAL WAGES COMMITTEE certifying that a particular agricultural worker is proficient in one or more CRAFTS. Proficiency is determined by tests set and held under the authority of the National Proficiency Tests Council. (♦ PROFICIENCY TESTS)

Craftsman. A category of agricultural worker defined in the Agricultural Wages Order (♦ AGRICULTURAL WAGES BOARD) as (*a*) a qualified former Agricultural Apprenticeship Council apprentice or (*b*) a worker who is proficient (♦ CRAFT CERTIFICATE) in one or more CRAFTS and either is a qualified former AGRICULTURAL TRAINING BOARD apprentice or is 20 years of age and over.

Cramming ♦ GAVAGE.

Crane Fly. One of a large family of two-winged, two-legged flies (*Tipula sp.*), commonly called daddy-long-legs, the LARVAE of which, called LEATHERJACKETS, are abundant in grassland, and are pests, particularly of cereal crops following leys.

Cratch. A crib holding hay for cattle.

Craw. 1. The CROP, throat, or first stomach of fowls. Sometimes called maw.
2. A general term for an animal's stomach.

Crawler Tractor. A caterpillar or TRACKLAYER, usually with a large horse-power, capable of very heavy work such as pulling heavy 6-furrow ploughs.

Crazy Chick Disease. A nutritional disorder of CHICKS associated with fat-rich diets and possibly Vitamin E inadequacy. Characterised by falling over and paralysis.

Cream. An oily substance containing fats which rise on MILK left to stand. In manufacturing, milk is centrifuged to divide it into cream and SKIM MILK. The cream is sold as liquid, clotted or sterilised, or is churned into BUTTER.

Cream Cheese. Soft CHEESE made from CREAM strained through muslin.

Cream Separator. A machine which uses centrifugal action to separate milk into lighter CREAM for buttermaking and SKIM MILK or separated milk containing the heavier particles. Either hand-operated or power-driven. Farmhouse buttermaking is now uneconomic except for home consumption, and cream separators are now rarely seen.

Creamery. An establishment where butter and cheese and other commodities are made from milk supplied by producers. (♦ DAIRY)

Creep. 1. A fence or hurdle through which a young animal can pass but which restricts its mother.
2. A part of a farm building where young animals are fed but which is inaccessible to the mother(s). In a FARROWING house, for instance, the creep for the unweaned piglets is usually provided with a heat source.

Creep Feeding. The feeding of young animals, usually with proprietary or home-mixed MEAL, in the same building as the mother(s), the latter having no access to the feed. Unweaned piglets, and in-wintering lambs and calves are often creep fed.

Creep Grazing. A method of managing pasture in which lambs have access to certain areas of pastures before their mothers.

Crest. 1. The COMB or tuft on the head of a cock or other bird.
2. The mane of a horse.
3. The ridge on the neck of a horse or bull.

Crested Dogstail. A very common perennial grass (*Cynosorus cristatus*) of meadows and pastures, characterised by a one-sided flower-spike resembling a dog's tail.

Crew Yard. A cattle yard.

Crib. 1. A MANGER, or framed rack made of wood or iron, containing FODDER, to which animals help themselves.
2. A cattle stall.

Cribble. A coarse screen or sieve used for sifting corn.

Crimp. 1. In general terms the waviness of wool. One of the waves in wool fibres. Measured as the number of crimps per unit length and an important quality in assessing manufacturing value.
2. To crush grass with a CRIMPER.

Crimper. A machine with ribbed rollers between which grass is crushed during HAYMAKING, allowing sap to be removed by the action of sun and air. Similar to a ROLLER CRUSHER.

Crimson Clover ♦ TRIFOLIUM.

Crock. An old ewe or horse.

Croft. A small farm-holding, comprising a small area of land, mainly arable, usually adjoining a house. Characteristic of the Highlands and Islands of Scotland, although now becoming rarer. Often combined with fishing and other, often craft, occupations, the croft is regarded as being mostly self-sufficient. Their average size, excluding common hill pastures on which are kept a few sheep and cattle, is about 2 ha (5 acres).

Crone. An old ewe, often BROKEN-MOUTHED.

Crook. A staff with a long handle and hook-shaped end used by SHEPHERDS to catch sheep by trapping their legs.

Crop. 1. Plants, carefully selected and developed over many years, sown on cultivated land to produce food for man and animals or raw materials, e.g. barley, potatoes. The term is also applied to plants which are not sown but come up naturally in cultivated land from wild seed, e.g. a crop of thistles.
2. To mow, cut, reap or gather a cultivated crop.
3. The total quantity produced, cut or harvested from cultivated land.
4. To bite off in eating, e.g. grazing sheep will crop a LEY.
5. The expanded part of a bird's oesophagus or gullet, in which food is stored before being digested. Also called the craw.
6. A hunting whip with a loop as distinct from a lash.
7. An entire hide.

Crop Bound. A bird with its CROP obstructed by material which cannot pass on to the stomach and GIZZARD, e.g. feather, straw, wool, etc.

Crop Certification Schemes. Various schemes operated by the Ministry of Agriculture, Fisheries and Food covering herbage and cereal plants, potatoes, hops and certain horticultural crops, designed to provide for the production of seeds and stocks, involving registered growing, inspection and certification in respect of health and purity. (◗ SEED CERTIFICATION, SEED POTATOES)

Crop Husbandry. The practice of growing and harvesting crops. The main object is the production of crops economically without detriment to the land.

Crop Rotation ◗ ROTATION.

Crop Spraying ◗ SPRAY.

Crop Variety Testing. Varieties of all the more important agricultural and field vegetable crops are tested by the NATIONAL INSTITUTE OF AGRICULTURAL BOTANY. The results are summarised in Farmers' and Growers' leaflets published by the Institute.

Cropper. A cultivated plant which yields a good CROP.

Cropping. 1. The raising of crops.

2. The cutting or harvesting of crops.

Cross. A crossbred animal or a HYBRID plant. (♦ CROSS-BREEDING)

Cross Crop. To sow a crop out of its normal ROTATION order.

Cross Drill. To sow seed by drilling (♦ DRILL) in two directions, i.e. up or down the field, and across the field.

Cross Fertilization. FERTILIZATION by the transference of pollen from one flower or plant to another flower or plant, as distinct from SELF FERTILIZATION.

Cross Plough. To plough a field in both directions, side to side and top to bottom.

Cross Pollination ♦ POLLINATION.

Cross-block. To thin out a root crop by hoeing or cultivating across the crop rows as distinct from along the lines. The crop is thinned by hand by removing some of the clumps of plants thus produced.

Cross-breeding. The mating of animals (or plants) of different breeds, but both of which are purebred, in order to combine the best characteristics of the two breeds. The progeny, known as crossbreds, possess HYBRID VIGOUR, often expressed in females as increased mothering ability. They are sometimes crossed with a male of a third breed to introduce further qualities. (♦ IN-BREEDING, LINE-BREEDING, OUT-CROSSING)

Crosskill Roller. A type of ROLLER, with tooth-edged rings, used to crush clods. Similar to the CAMBRIDGE ROLL.

Crosslands. Diagonal strips of unploughed land left for ploughing out the corners when PLOUGHING a field 'round and round'.

Croup. 1. A horse's rump.

2. A chronic respiratory disease of poultry. Also called roup.

Crown. 1. The top part of a plant, such as the top branches of a fruit tree or the leaves of certain root crops, e.g. sugar beet.

2. The junction of the root and stem of a plant.

3. The perennial rootstock of certain plants, e.g. rhubarb.

Crown Furrows ▶ SCRATCH.

Crown Graft. A type of GRAFT in which the SCION is fixed into a vertical slit in the bark and sap wood of the STOCK. Also called a rind graft.

Crucifer. A plant belonging to the cabbage family (*Cruciferae*) characterised by four petals and sepals, both arranged in the form of a cross.

Crude Fibre. A constituent of animal FEEDINGSTUFFS measured by PROXIMATE ANALYSIS, comprising mainly CELLULOSE, LIGNIN and related compounds. (▶ FIBRE)

Crude Protein. An approximate assessment of the PROTEIN content of animal food derived by PROXIMATE ANALYSIS, and based on the nitrogen content. The formula CP = %N ×(100/16) is used and assumes that protein contains all the nitrogen in food and that all proteins contain 16% nitrogen. (▶ DIGESTIBLE CRUDE PROTEIN, TRUE PROTEIN, PROTEIN EQUIVALENT)

Crue. A sheep fold.

Cruels ▶ ACTINOBACILLOSIS.

Crumb. Particles of clay, sand and silt in soil loosely held together or aggregated by positive and negative charges or forces on the surfaces of the particles.

Crupper. A strap attaching to the back of a saddle, passing under the horse's tail.

Crush. A narrow passage, usually comprising a steel or wooden framework on each side which can be sealed in front of and behind an animal. Used to closely confine an individual animal for close inspection, administering injections, taking blood samples, etc.

Crushed Grain. Cereal grains crushed in a CRUSHING MILL before feeding to livestock to assist in their digestion. Also called rolled grain.

Crushing Mill. A machine which crushes cereal grains before feeding to livestock, by rolling it between two iron cylinders rotating close together. Also called ROLLER MILL. (▶ HAMMER MILL, PLATE MILL)

Crutch. A pole with a cross-bar at the end, used in DIPPING sheep to push the shoulders down, immersing the head and shoulders (*see p. 145*).

Cryogenic Flask. A flask used in ARTIFICIAL INSEMINATION to transport semen at very low temperatures.

Cub. 1. The young of certain animals, e.g. fox.
2. A cattle pen or small enclosure such as a chicken COOP or rabbit hutch.
3. A CRIB or fodder rack.

Cuber. A machine which forces MEAL, usually binding it with molasses at the same time, through circular holes to produce extruded pencils of meal which are chopped into cubes or pellets of various lengths ranging from 3.25 mm for poultry to 13 mm for cattle. (♦ CAKE)

Cubicles. Rows of partitioned compartments in a building, similar to the stalls in a COWHOUSE. Each cubicle is usually covered with BEDDING, retained by a raised kerb, and will permit one untied dairy cow to lie down comfortably. The cubicles are backed by a dunging passage with a solid concrete or slatted floor.

Winter housing of cows in cubicles. The end of the building is clad with Yorkshire boarding.

Cuckoo Corn. Spring cereals sown after mid-April and likely to produce poorer yields than earlier sown cereals due to the contracted growing period. The name is taken from the usual earliest recorded cuckoo song.

Cuckoo Lambs. Late-born lambs, born after mid-April.

Cucumber. A creeping plant (*Cucumis sativus*) cultivated under glass and in the open where there is no frost risk, both as a market garden and vegetable garden crop. Used as a salad ingredient and for pickling.

Cud. The partly digested food brought back from the first stomach or RUMEN or a RUMINANT to be chewed again.

Culard ♦ CHAROLAIS.

Cull. An animal separated and removed from the herd or flock, being unsuitable, of poor quality or too old.

Culm. The flowering stem of a grass. (♦ MALT CULMS)

Cultivar. A variety of a cultivated plant.

Cultivations. The various TILLAGE operations carried out to the land prior to and during the growing of crops.

A spring-tined cultivator.

Cultivator. An implement used to break down the soil into a tilth before sowing a crop, for ripping out weeds, raising unrotted TRASH to the surface to be killed by the sun, and (by attaching special blades) for row crop hoeing. Cultivators have heavy TINES and penetrate the soil deeply, as distinct from the lighter, shallower working HARROWS. (◗ CHISEL PLOUGH)

Culture. Bacteria growing on or in a substrate or substance which provides suitable feeding conditions.

Culvert. A channel for carrying water under a road, railway, etc.

Cumberland. A breed of pig, now extinct, from which, together with the YORKSHIRE and LINCOLN CURLY COAT, was derived the CHESTER WHITE in America.

Cup Drill, Cup Feed Drill. A type of DRILL comprising discs of tiny cups which rotate in the seed box and deliver the seed to the sowing tubes.

Curb. 1. A chain or strap attached to the BIT for restraining a horse.
2. A hard swelling on a horse's leg due to a strained ligament.

Curd. Milk thickened or coagulated by acid (◗ RENNET) in CHEESE MAKING, and distinguished from the more watery WHEY.

Cure. The preservation of meat by salting, smoking or pickling for storage.

Curry. To rub down, clean and groom a horse, usually using a special comb called a curry comb.

Custock ◗ CASTOCK.

Custom Hatching. A service provided by a HATCHERY involving the HATCHING of eggs for another hatchery on a fee-paying basis.

Cut. 1. A term used to describe the steepness of the crest of a FURROW slice.
2. A term used to describe the mowing of grassland for SILAGE.

Cutter. 1. ◗ PIG.
2. A term for the SHIN of a mouldboard.

Cutter Bar. A mechanism on a BINDER, COMBINE HARVESTER or MOWER which cuts the crop and consists of a fixed finger bar (◗ FINGERS) and reciprocating KNIVES. The knives of a mower operate at a faster speed than those on binders and combines since grass is more difficult to cut than the dry straw stems of CEREALS.

A cutter bar mower mounted on the three-point linkage of a tractor.

Cutter Blower ◗ FORAGE BLOWER.

Cutting. A cut section of material removed from a living plant which, when placed in a suitable rooting medium, will produce roots and give rise to a new plant. Cuttings are used extensively in the propagation of certain types of plant, e.g. Chrysanthemums, Geraniums, etc. (◗ GRAFT, BUDDING, LAYERING)

Cutworm. A CATERPILLAR of various noctuid moths, active at night only, attacking roots and stems and named after its habit of eating completely through young plant stems at ground level. Also called surface caterpillar.

Cyanamide. (Ca CN$_2$). Calcium cyanamide, an ARTIFICIAL FER-TILIZER, converted by soil water into AMMONIA. Also called nitrolime.

Cyst. 1. A membrane secreted around the resting stages of many animals during their development, e.g. that enclosing EELWORM eggs in the soil.

2. A swelling or hollow tumour containing soft or watery matter as distinct from pus.

D

2,4-D. An abbreviation for 2:4 di-chloro-phenoxy-acetic acid (also abbreviated to DCPA). A TRANSLOCATED HERBICIDE used to control many broad-leaved annual and perennial weeds post-emergence in cereals (except spring oats) and grass seed crops, in grassland and turf. Closely related chemically to 2,4,5-T.

D-value. The percentage of digestible organic matter in the dry matter of animal feeds such as hay, silage and dried grass. (◗ DIGESTIBILITY TRIAL, DIGESTIBLE ENERGY VALUE, TOTAL DIGESTIBLE NUTRIENT VALUE)

Dag, Dag-Lock. A tuft of wool clotted with dirt, mainly faecal, especially round the tail of a sheep. Also called tag or tag-lock. Such clots in a fleece are known as daggings. Dagging also means the removal of such clots or soiled wool in an effort to prevent STRIKE. They are purchased by the BRITISH WOOL MARKETING BOARD as a WOOL ODDMENT.

Dairy. 1. A farm building, often a single room, in which milk is cooled and temporarily stored prior to collection for transportation to a commercial dairy (see 2 below), and in which it may be treated and made into cream, butter and cheese – a practice now becoming rare on farms. Milking equipment is usually washed and kept in the dairy.
2. A commercial processing plant to which fresh farm milk is transported by tanker (◗ BULK MILK) for treatment and distribution as bottled milk, and for the production of cream, butter, cheese, yogurt and other products for retailing. Some dairies are purely milk bottling plants whilst others are mainly CREAMERIES. (◗ MILK MARKETING BOARD)

Dairy Cattle ◗ CATTLE.

Dairy Cows. Cows and heifers kept for producing milk or for rearing calves for a dairy herd, as distinct from BEEF COWS.

Dairy Farm. A farm on which the principal activity is milk production.

Dairy Followers. Young stock in a dairy herd not yet in-milk, growing up to replace their mothers as DAIRY COWS.

Dairy Herd Conversion Scheme ◗ MILK NON-MARKETING AND DAIRY HERD CONVERSION SCHEME.

Dairy Shorthorn. A dual-purpose breed of cattle, red, white, red and white, or roan in colour, with characteristic short forward-curling horns. Good beef steers are produced and cows yield milk containing 3.6% butterfat. The breed is now losing popularity (*see p. 91*). (◗ SHORTHORN, MILK YIELDS)

Dairy Trade Federation. An organisation representing the 220 first-hand buyers of milk in England and Wales, comprising public companies, independent buyers and co-operative retail societies. The Federation has four constituent associations: the National Dairymen's Association, the Creamery Proprietors' Association, the Co-operative Milk Trade Association and the Amalgamated Master Dairymen, each with representation on the Federation's Council. The Federation, with 4500 members, is an employers' association, negotiating wages and conditions for the employees in the dairy industry. It also negotiates prices with the MILK MARKETING BOARD.

Dairymaid, Dairyman. A person employed to work with dairy cattle. The terms are variously used in different regions. In some areas the work may relate purely to the milking of a dairy herd and/or to the production of butter or cheese in a DAIRY, whilst elsewhere the terms can be synonymous with COWMAN or HERDSMAN. Also a person employed by a commercial dairy.

Dale, Dalesmen. A wide valley, usually open and unfenced, often with a river flowing through it. Mainly in the North of England and in Southern Scotland. Farmers from the dales are known as dalesmen.

Dam. The mother of an animal. Usually used when describing the PEDIGREE of a farm animal. (◗ SIRE)

Damping-off. A disease of seedlings caused by *Pythium* and various other fungi in excessively moist conditions. It causes wilting and roots to die.

137

Damson. A small black oval plum (*Prunus domestica* var *damascena*).

Dandy Brush. A stiff bristled brush used to groom horses.

Dandy Cock, Dandy Hen. A BANTAM.

Danish Piggery ♦ PIGGERY.

Danish Red. A breed of dual purpose cattle predominant in Denmark and used in the U.K. in recent years to improve the productivity of the RED POLL.

Darnel. A common weed (*Lolium temulentum*) of cereal fields and pastures.

Dartmoor. A breed of hardy, long curly-woolled, white-faced, sheep. Similar in appearance to the DEVON LONGWOOL and SOUTH DEVON breeds, all having a common origin. Large, heavily-boned sheep, generally polled, although rams are occasionally horned, and producing high quality fleeces.

Day-old-chick. One that is despatched to the buyer within 24 hours of hatching.

Day-workers ♦ CASUAL WORKERS.

DCPA ♦ 2,4-D.

D.D.T. An abbreviation for dichlor-diphenyl-trichlor-ethane. A persistent organochloride used as a contact insecticide.

D.E. Value ♦ DIGESTIBLE ENERGY VALUE.

Dead-in-shell. In general terms a fertile egg in which the developing embryo has died. In specific terms, embryo death occurring between the 14th and last day of incubation. (♦ ADDLED)

Deadman ♦ ANCHOR WIRE.

Deadstock. The equipment required to operate a farming enterprise, such as tractors, implements, hay racks, and hurdles, etc.

Deadweight. The weight of a DRESSED CARCASE. Under the BEEF PREMIUM SCHEME and the FATSTOCK GUARANTEE SCHEME there are a number of Ministry of Agriculture specifications for the dressing of carcases presented for certification, which either include or exclude the kidney knob and channel fats. Deadweight is an indication to a butcher of the meat available on a carcase. (♦ LIVEWEIGHT)

Deadweight Certification Centre. A slaughterhouse or other place, approved by the MEAT AND LIVESTOCK COMMISSION, where fat sheep or clean cattle are slaughtered and their carcases presented for certification by an M.L.C. fatstock officer as eligible, respectively, for a fatstock guarantee payment (◗ FATSTOCK GUARANTEE SCHEME) or a beef premium payment (◗ BEEF PREMIUM SCHEME). (◗ LIVEWEIGHT CERTIFICATION CENTRE)

Deciduous. A descriptive term of plants, particularly trees, which shed their leaves annually, usually in the autumn. (◗ EVERGREEN)

Deciduous Teeth. The first teeth to appear in a young animal, later shed and replaced by permanent teeth. Also called milk teeth.

Decortication. The removal of husks from seeds. Oil seeds are frequently decorticated before oil extraction. Cattle CAKE derived from oil seeds (e.g. COTTONSEED CAKE) may be either decorticated or undecorticated (the husks having been left on).

Deep Litter. A system of BEDDING for cattle or poultry using straw, shavings or sawdust. (*a*) For cattle, bedded areas are usually adjacent to open or covered feeding areas. Warmth is released as the faeces ferments in the litter. Fresh litter is added on top periodically and depths of 2 to 3 ft may be attained before

Cattle on deep litter in a covered yard.

the cattle are turned out in spring when the litter is removed. (*b*) For poultry, a layer of compost, top soil or decomposed horse manure is added beneath the litter and then mixed in, providing bacteria which break down the faeces to a dry powder. Fresh litter is periodically added. The litter is removed after each crop of birds, although it may be returned after heaping, which generates heat and kills parasites.

Deep Storage Drying ◗ STORAGE DRYING.

Deficiency Diseases. Diseases of animals and plants due to an inadequacy of one or more essential food substances, such as a VITAMIN, mineral ELEMENT, PROTEIN or AMINO ACID, etc., e.g. GREY LEAF, BROWN HEART. (◗ MINERAL MIXTURES)

Deficiency Payments ◗ GUARANTEED PRICES.

Definitive Host ◗ HOST.

Deflocculation. The disassociation of large particles into smaller ones. Particularly applied to soils and the disintegration of CRUMBS under wet conditions. (◗ FLOCCULATION)

Defoliation. The removal of the leaves of a plant. A defoliant is a chemical which causes this.

Deforestation. The process of clearing forests. (◗ AFFORESTATION, REFORESTATION)

Degras. A FAT derived from sheepskins.

Dehiscent. A descriptive term for fruits which split open to release the seeds, e.g. peas.

Dehorning. The removal of an animal's horns to prevent it wounding others. The best method is to DISBUD the young animal, but horns can also be removed under anaesthetic by electric saw. Also called dishorning.

Dehydrated Fodder ◗ DRIED GRASS.

Dell. A small valley, usually wooded.

Denaturing. The treatment of wheat grain so as to make it unusable for human consumption either by staining with a suitable dye or by the addition of fish oil to give a smell to it. Wheat, when denatured, can be used in animal feed, but not for flour-milling. Surplus E.E.C. milk powder sold as animal feed (other than for young calves) is also denatured in a similar way to prevent its resale for human consumption.

Dene. A small valley, usually steep-sided.

Denitrification. The breakdown of nitrates and nitrites by denitrifying bacteria in the soil in anaerobic conditions with the release of gaseous nitrogen. (♦ NITROGEN CYCLE)

Dental Formula. A formula indicating the number of each kind of teeth for a given animal species. The teeth are shown in the order incisors (i), canines (c), premolars (p) and molars (m), with teeth numbers for one side of the upper jaw shown above those for the lower jaw. The formula for a pig for example is $i \frac{3}{3}$, $c \frac{1}{1}$, $p \frac{4}{4}$, $m \frac{3}{3}$.

Dentition. The cutting or growing of teeth. Also the arrangement and number of teeth. (♦ DENTAL FORMULA)

Depasture. To graze.

Depreciation. A figure included in the annual farm accounts as an expense indicating the amount by which various assets such as machinery and implements, etc., have lost value with ageing.

Depth Wheel. A wheel on an implement used to control the depth of working.

Derbyshire Gritstone. A hardy breed of sheep, largest of the hill and mountain breeds, originating in the Peak District. Hornless, with a mottled black and white face and legs, producing a fine, soft fleece of a high quality (*see p. 391*).

Derris. An insecticide powder derived from the root of the tropical plant, *Derris elliptica*. Used against fleas, lice and WARBLES.

Desiccant. A chemical applied to a crop causing the foliage to dry up so that harvesting can commence.

Devon. A breed of hardy, docile, deep cherry-red cattle, popularly known as Red Rubies, with fairly long, often curly hair, medium horns, and a short head and legs. They are small in size but neat in appearance and are found mainly in the South-West of England. Now mainly regarded as a beef breed but at one time kept as dual-purpose cattle, cows yielding rich milk with more than 4% BUTTERFAT. An early maturing breed able to use both rough and good grass. Sometimes called North Devon Cattle as a distinction from SOUTH DEVON cattle.

Devon Closewool. A breed of white-faced, medium-sized, polled sheep, used as a grassland breed and found in North Devon. The fleece produced is very thick and medium length. The breed was derived from crosses between the DEVON LONGWOOL and EXMOOR HORN.

Devon and Cornwall Longwool. A breed of sheep established in 1977 with the amalgamation of the DEVON LONGWOOL and SOUTH DEVON breeds, combining the characteristics of both breeds.

Devon Longwool. Until 1977 a breed of sheep with a long heavy fleece and wool on the cheeks. It was similar to but smaller than the LINCOLN and originated from crosses between the ancient BAMPTON NOTT and the LEICESTER. Crossed with the EXMOOR HORN it gave rise to the DEVON CLOSEWOOL. The breed was important locally in the South-West of England, and is now amalgamated with the SOUTH DEVON to form the DEVON AND CORNWALL LONGWOOL.

Dew-claw. The rudimentary fifth digit protruding from the heel of cattle, pigs and dogs.

Dewlap. 1. The loose flesh which hangs from the throat of oxen and cows.
2. The fleshy wattle of the turkey.

Dewponds. Shallow, artificial ponds, lined with clay or cement, found on chalk DOWNS. They retain water for long periods, especially in droughts, and were once believed to be filled by moisture condensing from the air. They are used for watering cattle.

Dexter. A relatively rare breed of cattle of uncertain origin imported from Ireland in the eighteenth century. A dwarf breed, horned, with a large head, short legs, and sometimes with slightly deformed feet and leg bones. Usually black, occasionally red. They are dual-purpose cattle, fattening easily, with a lower feed requirement than other dairy herds, and able to use poor quality grazing. (◆ MILK YIELDS)

Dextrose. A term for GLUCOSE.

Dey. A DAIRYMAID.

Diamond Tup. A shearling male lamb. Also called a dinmont, particularly in the Borders. (◆ SHEEP NAMES)

Diastase. An ENZYME produced in germinating seeds and by the pancreas which converts STARCH to SUGAR. (◆ MALTING)

Dib. A small bone in a sheep's leg.

Dibber, Dibble. A pointed hand tool used to make holes in the ground for seeds or plants.

Dicotyledon ◆ COTYLEDON.

Dieldrin. An insecticide ten times more powerful than D.D.T. A CHLORINATED HYDROCARBON, highly poisonous to birds and fish, banned from use in sheep DIPS in 1966 and as a winter wheat dressing in 1975. Its use is now limited under the PESTICIDES SAFETY PRECAUTIONS SCHEME to specified circumstances where no suitable alternative exists.

Digestibility Trial. The determination of the digestibility of a food using live animals held in special crates. Measurements are made of the weights of food given and excreted (ignoring urine) on a daily basis, the composition of both being analysed. Usually carried out over a 7–10 day period after acclimatisation for about 10–14 days. Various figures relating to the food samples can be derived from a trial including the DIGESTIBLE ENERGY VALUE (expressed in MJ/kg dry matter), D-VALUE, TOTAL DIGESTIBLE NUTRIENT VALUE, DIGESTIBLE CRUDE PROTEIN, etc.

Digestible Crude Protein. A measure of the PROTEIN value of a FEEDINGSTUFF, determined by a DIGESTIBILITY TRIAL. It is more useful than CRUDE PROTEIN or TRUE PROTEIN content in formulating the diets of RUMINANTS since they can vary considerably in digestibility and composition. (♦ PROTEIN EQUIVALENT)

Digestible Energy Value. The amount of utilisable energy in an animal food after deduction of the energy value of undigested faecal residues. (♦ DIGESTIBILITY TRIAL)

Digestion Coefficient. The percentage digestibility of a constituent of a foodstuff (e.g. CRUDE FIBRE, PROTEIN) calculated from the results of a DIGESTIBILITY TRIAL.

Digestive Juices. Various juices secreted into the digestive tract of animals which cause chemical changes to food and facilitate its absorption into the body, e.g. saliva, BILE, GASTRIC JUICE, PANCREATIC JUICE and INTESTINAL JUICE.

Digger. A type of MOULDBOARD, abruptly concave and twisted, that produces a rough well-broken furrow slice, as distinct from the continuous smooth slice of the LEA PLOUGH. Used for deep ploughing and popular for general work. (♦ SEMI-DIGGER, GENERAL PURPOSE MOULDBOARD)

Dike ♦ DYKE.

Dinmont ♦ DIAMOND TUP.

Dioecious. A term applied to plants having the male and female flowers borne on different individual plants of the species, e.g. willow. (♦ MONOECIOUS)

Dip. 1. A proprietary chemical which is diluted in a recommended volume of water in a DIPPING BATH for DIPPING animals, mainly sheep. Dips containing DIELDRIN were more advantageous than arsenic and sulphur based dips as they persisted in the fleece. They significantly reduced STRIKE in sheep. Dieldrin was banned from use in sheep dips in 1966, being highly poisonous, and a possible meat contaminant. Nowadays dips mainly contain organo-phosphorus chemicals.
2. A DIPPING BATH.

Dip Slope. The long gentle slope of a line of hills following the inclination of the rock strata, as distinct from the shorter, steeper SCARP SLOPE.

Diploid. A term applied to living cells having two sets of CHROMOSOMES in the nucleus, present in pairs. Characteristic of most animal and plant cells. In REPRODUCTION, the pairs separate in the process of MEIOSIS producing GAMETES which are HAPLOID, having only one set of chromosomes.

Dipped Sheep Market. A market, or a separate and distinct part of a market, at which, for a particular period, only dipped sheep are sold. (◗ DIPPING)

Dipper. A DIPPING BATH.

Sheep being immersed in a dipping bath with the aid of a crutch.

Dipping. The temporary total immersion of an animal in a DIP contained in a DIPPING BATH. Dipping is carried out mainly with sheep to kill common parasites (e.g. KEDS, LICE, TICKS), to arrest the spread of or to cure SHEEP SCAB, and as a preventive measure against BLOWFLY attack. Beef cattle may also be dipped, particularly in the tropics, to control ticks. Dipping practice and timing varies from area to area according to climate, breeds kept and custom. The Ministry of Agriculture can issue Dipping Orders requiring that sheep in an area be dipped within specific periods to control diseases such as Sheep Scab. In 1979 the whole U.K. was declared a Sheep Scab infected area after a series of outbreaks and a universal Dipping Order issued.

Dipping Bath, Dipping Tank. Also called a dip. A bath in which a chemical DIP is diluted in water, and in which animals, mainly sheep, are immersed, to control parasites. The commonest type is the short swim bath. The sheep are lowered into the bath, pushed under with a CRUTCH, usually kept in the bath for about a minute, and then allowed to swim to and walk up the exit ramp to a DRAINING PEN (*see p. 145*).

Dipping Notice. A notice served under the 1977 Sheep Scab Order by a veterinary inspector to the owner or person in charge of sheep requiring that they be dipped in the presence of, and to the satisfaction of a local authority inspector, on or before a given date, wherever required. Such a notice is issued to control the spread of SHEEP SCAB. After DIPPING a MOVEMENT LICENCE may be issued.

Dipping Order ◗ DIPPING.

Diptera. Two-winged insects or flies.

Direct Drilling. The sowing, by DRILL, of seeds direct into a field, without previous cultivation and usually following the application of weed control sprays.

Directive. A regulation issued by the E.E.C. COUNCIL OF MINISTERS, or the COMMISSION, binding on the member states in respect of its objective, but each member able to determine the method of achievement.

Direct Reseeding. Sowing grass seeds (usually a SEEDS MIXTURE) without a COVER CROP in a field in which the immediate past crop was grass. (◖ UNDERSOWING)

Direct Seeding, Direct Sowing. Sowing grass seed (usually a SEEDS MIXTURE) on bare ground without a COVER CROP. (◖ UNDERSOWING)

Dirty Land. Land full of weeds.

Disaccharide ◖ CARBOHYDRATES.

Disbud. 1. To remove BUDS from a plant.
2. To remove or prevent the growth of the horn buds of calves by painting them with a caustic compound or by use of a hot iron, usually in the first 7–10 days after birth. KIDS and lambs are also sometimes disbudded.

Disc Coulter. A flat circular kind of COULTER which rotates as the plough moves and makes a vertical cut in the soil, separating a slice of soil from unploughed land, which the SHARE and MOULDBOARD then turn to produce a clean furrow slice (*see p. 338*).

Disc Harrow, Disc Cultivator. A type of HARROW with a number (normally four) of sets or gangs of sharp-edged concave rotating discs, the angle of which can be varied in relation to the

A heavy duty disc harrow.

147

direction of travel, altering the severity of CULTIVATION according to requirements. The gangs can be positioned so that disc faces in different gangs face in opposite directions. Often used on light land, and on freshly ploughed grassland or stubble since unrotted trash is not brought to the surface.

Disc Plough. A type of plough with heavy dish-faced wheels, individually mounted, and usually adjusted at an angle (between 35 and 45°) to the line of travel, which cut and turn the soil to the side, partly inverting it. The discs are usually between 50–80 cm (20–32 in.) in diameter and are sometimes fitted with scraping devices to remove accumulating mud. The discs may be plain or cut-away. The latter give more bite and cut into hard soils. Disc ploughs are used for deep, rapid cultivations on rough land but are not very common. MOULDBOARD ploughs produce better quality work.

Diseases of Animals Act 1950. An Act consolidating earlier legislation providing measures for the control of specific animal diseases including notification (◗ NOTIFIABLE DISEASE), isolation of affected, suspected, or CONTACT animals, declaration of infected places, restrictions on animal movement, disinfection of premises, and procedures when disease is identified in markets, fairs, etc. The instruments of control are contained in various Orders.

Diseases of Animal Inspectors. Inspectors employed by local authorities since 1974, with responsibility for a wide range of animal health matters including, for example, assisting the Ministry of Agriculture, licensing and overseeing livestock markets, monitoring animal movements and issuing MOVEMENT LICENCES, and disease prevention (e.g. RABIES control, monitoring the use of waste food, etc.). Many of these duties were formerly carried out by the Police, as they still are in parts of Scotland.

Dished Face. A face having a distinct concavity, e.g. the MIDDLE WHITE PIG.

Dishley Leicester. A breed of improved LEICESTER sheep produced by Robert BAKEWELL in the late eighteenth century. The name was derived from Dishley Grange, where Bakewell lived.

Dishorning ◗ DEHORNING.

Disinfection. 1. The cleansing of appropriate buildings, animals, implements and utensils, etc., which may be harbouring harmful bacteria or viruses, after an infection has existed on a farm. The process can involve physical cleaning, the application of approved chemical solutions, steam cleaning and fumigation. 2. The cleaning of farm equipment and utensils (e.g. milking plant) using suitable detergents and disinfecting agents (e.g. sodium hypochlorite) to reduce bacterial contamination. 3. The chemical treatment of seed before sowing to control fungal disease in crops, particularly cereals.

Distillers' Grains. A by-product of whisky manufacturing similar to BREWERS' GRAINS consisting of the remains of malted barley. Used as an animal FEEDINGSTUFF either wet or dry, but mainly preferred dry as wet grain retains raw alcohol which may intoxicate the animals.

Ditch. A trench dug in the ground to drain water from farmland, usually discharging to a stream or other watercourse. The water reaching a ditch may be by surface run-off, by natural underground flow, or via an artificial underfield drainage system, the DRAINS of which discharge into the ditch.

Ditcher. A machine designed to excavate, clean or repair a DITCH.

Ditch Water. Stagnant, foul-smelling water such as that often found in ditches.

Diviner. One who divines or detects the presence of water beneath the ground. (♦ DOWSER)

DNA. Deoxyribonucleic acid. The main constituent of CHROMOSOMES in cell nuclei. (♦ GENE)

D.N.O.C. Di-nitro-ortho-cresol. An insecticide, ACARICIDE, and powerful weedkiller applied as a foliar spray.

Dock. 1. A weed (*Rumex sp.*) with large leaves and a long root. 2. To remove the whole or part of an animal's tail. Lowland breeds of sheep are customarily docked as lambs to avoid the accumulation of dirt and faeces in the tail.

Dodded. Hornless or POLLED. A doddy is a Scottish term for a hornless cow.

Doddle. Loose flesh hanging from the throat of a goat. Also called tassel or wattle.

Doe. A female deer, hare or rabbit.

Dole ♦ NIB.

Dolly. An ornament made of straw usually placed on top of a stack. Also called corn dolly.

Dolphin ♦ BEAN APHIS.

Dominant Gene. The stronger GENE in a gene pair which asserts itself over the other recessive gene in determining a specific character in the offspring of a plant or animal (e.g. eye colour, height).

Dominant Plant. The major plant in a vegetation community, usually the most numerous or the tallest, e.g. beech trees in a beech wood.

Dorking. A table breed of fowl, square-bodied, variously coloured (including red, white, silver-grey and barred), with a single comb and five-clawed white legs. Hens produce white eggs.

Dormancy. A state of low metabolic activity in living organisms, particularly in seeds and spores, usually associated with unfavourable environmental conditions (e.g. low temperature, inadequate moisture, short day length) for growth and thought to be controlled by HORMONES. Stored seeds can remain dormant for years and will germinate after sowing if conditions are right. Deciduous trees lie dormant in the winter after shedding their leaves.

Dormant Spray. A spray applied to a crop during a dormant period.

Dorset Down. A breed of hill sheep originally bred from mainly polled local Dorset sheep and crossed with the SOUTHDOWN and the developing HAMPSHIRE breed, combining the carcase quality and early maturity of the former with the good growth rate of the latter. The breed is brown-faced with wool growing over the forehead down to the eyes, and over the cheeks. It produces a good quality fleece and has been extensively exported, particularly to Australia.

Dorset Horn. A breed of sheep derived mainly from crosses between MERINO-type sheep and tan-faced sheep (which gave rise to the PORTLAND) in the South-West of England. It is characterised by an exceptionally white fleece, a white face with pink nostrils and lips, and long curly horns in both sexes. The breed can lamb at any time of the year, ewes sometimes producing two crops a year. They have been exported throughout the world and a polled variety has been developed in Australia.

Dorset Wedge Silage. SILAGE produced by filling a SILO in wedge-shaped layers with cut grass or other greenstuff, covering with plastic sheeting and sealing the silo to ferment.

Double-breasted Plough ◗ RIDGER.

Double Chop Harvester. A type of offset-trailed or semi-mounted FORAGE HARVESTER, consisting of a cutting or pick-up mechanism of the flail-type, with an auger delivering the cut crop sideways to a flywheel chopper mechanism for further chopping before it is loaded by chute into the trailer. This type of machine produces a non-uniform chop length. Also called a Flywheel Chopper.

Double-cut Red Clover ◗ RED CLOVER.

Doubles. Twins, especially lambs.

Double Scalp ◗ CAPPIE.

Double-suckling. A method of feeding beef cattle in which a second calf is introduced and allowed to suckle alongside a cow's own newborn calf. The second calf is sometimes rubbed with part of the PLACENTA to encourage the cow to accept it as her own. The system is only used with cows producing sufficient milk to feed two calves. (◗ SINGLE-SUCKLING, MULTIPLE-SUCKLING)

Double-work. A method of ensuring a successful fruit-tree GRAFT between a SCION and STOCK which otherwise would be incompatible, by grafting between the two a scion of a variety compatible with both, called an intermediate stem piece.

Down. The soft plumage covering young birds, or that found under the feathers of certain birds (e.g. ducks, geese).

151

Down Breeds. Breeds of hornless, short-woolled sheep, with coloured faces and legs, producing good dense fleeces and high quality meat. Developed in hilly areas (♦ DOWNS) and including the DORSET DOWN, HAMPSHIRE DOWN, OXFORD DOWN, SHROPSHIRE, SOUTHDOWN and SUFFOLK.

Down-calver. A cow or heifer about ready to give birth to a calf.

Downs. A term derived from the Old English word dun, meaning a hill. Open, mostly treeless, rolling hill land, especially the chalk hills of Southern England (e.g. North Downs, South Downs, Hampshire Downs) with a relatively thin covering of soil, mainly used for grazing sheep and cereal growing.

Dowse ♦ CHAFF.

Dowser. A water diviner – a person able to locate underground water using a dowsing-rod (a forked twig) which quivers when over a water source.

Draff ♦ BREWERS' GRAINS.

Draft. The force exerted on a trailer or other trailed implement by a tractor, measured in newtons (N). Also called draught and drawbar pull. (♦ DYNAMOMETER)

Drafted Animal. An animal either added to (drafted in) or removed from (drafted out) a herd or flock. Draft ewes are those sold from a breeding flock of sheep whilst they still are able to produce lambs. Usually they are sold from a hill flock and sent to a lowland farm when they are considered to lack hardiness for continued hill breeding, normally after producing three crops of lambs.

Drag. A kind of fork with curved prongs at right-angles to the handle, used to draw manure off a load or from a heap.

Drag Harrow. A kind of heavy HARROW, with TINES having curved ends, and often with wheels, capable of quite deep work. The action is similar to that of a CULTIVATOR which has fewer tines and can work deeper. Drag harrows are preferred on difficult land for seed bed preparation, bringing fewer unweathered clods to the surface. Also called a duck-foot harrow.

Drail. The iron bow of a horse-drawn plough from which the TRACES draw.

Drain. 1. An artificially dug channel or DITCH to carry surplus water away from farmland. Common in the FENS. Usually maintained by an INTERNAL DRAINAGE BOARD.
2. An underground channel in a field drainage system usually comprising a piped duct of clay, porous concrete or plastic tube. (◗ MOLE DRAIN, TILE DRAIN, UNDERDRAIN)

Drain Gauge. A tank placed in a ditch beneath a land UNDERDRAIN outlet to catch and measure the volume of water passing through the drain.

Drainage ◗ LAND DRAINAGE.

Drainage Area ◗ CATCHMENT.

Drained Honey. Honey obtained by draining uncapped broodless honeycombs.

Draining Pen. A pen into which sheep emerging from a DIPPING BATH are directed to allow surplus DIP to drain off the fleece and return via a sieving device to the bath.

Drake. A male DUCK.

Draught ◗ DRAFT.

Draught Horse. A horse used to pull carts, trailers and farm implements (e.g. the SHIRE horse).

Draw. 1. ◗ YELM.
2. To help in the delivery of lambs. Also to select ewes for the formation of groups.

Draw Hoe. A type of hand HOE with a curved blade which is drawn towards the user. (◗ DUTCH HOE)

Drawbar. A metal bar at the back of a tractor to which implements are attached.

Drawbar Pull ◗ DRAFT.

Dray. A low strong cart with no fixed sides, used for carrying heavy loads.

Dredge Corn. A mixture of cereals sown together (e.g. oats and barley) producing a yield of grain better than if either crop were grown separately. (◗ MASHLUM)

Drench. A liquid medicine poured down an animal's throat.

Dress. In general terms to clean, prepare or repair.

1. To apply fertilizer to land (e.g. TOP DRESSING).

2. To chemically treat seeds before sowing to control fungal disease in growing crops, particularly cereals.

3. To clean corn seeds by removing CHAFF, weeds and other unwanted matter.

4. To remove the tough fibres from flax.

5. To sort potatoes into various sizes.

6. To prune a hop plant by cutting the previous year's growth from the crown.

7. To recut grooves in a millstone.

(and ♦ CARCASE)

Dressed Carcase ♦ CARCASE.

Dried Blood. An ORGANIC FERTILIZER, derived from waste blood from slaughterhouses, dried by evaporation. It has a nitrogen content of about 13%.

Dried Grass. Grass which, after cutting, has been thoroughly dried by artificial means, and is used as an ingredient in animal feedingstuffs. It is either milled and mixed with molasses to produce cubes or pellets, or pressed directly into cobs or wafers. Used for many years for poultry and pigs, and now becoming popular as a dairy cow ration combined with a CONCENTRATE and roughage. Sometimes called dehydrated fodder.

Drier ♦ GRAIN DRIER.

Drift, Drift-way ♦ DROVE.

Drillage ♦ HAYLAGE.

Drill. 1. An implement used to sow seed in rows (♦ SEED DRILL). Seed and fertilizer may also be sown simultaneously using a COMBINE DRILL.

2. A ridge with seed or growing plants on it (e.g. turnips, potatoes), or the plants in such a row.

Drill Barrow. A hand drive SEED DRILL.

Drill Harrow. 1. A light flexible CHAIN HARROW trailed behind a SEED DRILL to cover the seed with extra soil.

2. A drill mounted on a SPRING TINED HARROW. Also called a tine harrow.

Drill Plough. An old type of trailed plough with an apparatus attached to the beam which drilled seed directly into the newly produced furrow.

Drilled to a Stand. Seeds sown by a SEED DRILL at predetermined intervals so as to produce a crop with a specific plant population which requires no thinning. (◊ THIN).

Drone. A male bee.

Drought. An extended period without rain. (◊ ABSOLUTE DROUGHT, PARTIAL DROUGHT, DRY SPELL)

Drove. 1. A number of cattle or sheep being driven from place to place.
2. A track along which cattle or sheep are driven. Also called drift, drift-way.

Drum. The cylinder of a THRESHING MACHINE or COMBINE HARVESTER bearing BEATERS, which revolves and dislodges the grain from the ears of cereals. Also called a threshing drum.

Drumhead. A kind of CABBAGE having a flattish round head.

Drupe. A kind of fruit comprising a hard kernel containing a single seed enclosed within a fleshy pulp, e.g. plum, cherry. Also called a stone-fruit.

Dry Bible. A disease of horned cattle in which the third stomach, or bible, is very dry.

Dry Cow. A cow not producing milk.

Dry Cured Bacon. Bacon preserved by being rubbed with dry salt as opposed to being soaked in brine.

Dry Feeding. A method of feeding MEAL to animals in a dry state without the addition of water. Used with pigs and poultry.

Dry Flock ◊ FLYING FLOCK.

Dry Mash. A mixture of feedingstuffs fed to animals in a dry state.

Dry Matter. The various mineral and organic components (e.g. ASH, CRUDE PROTEIN, etc.) in feedingstuff derived by PROXIMATE ANALYSIS.

Dry Measure. A system of measuring by bulk. (◊ BUSHEL, PECK)

Dry Period. A period between LACTATIONS when a cow is allowed to rest from the strain of producing milk, usually of 6–8 weeks duration.

Dry Rot. 1. The decay of timber caused by fungi (particularly *Merulius lacrymans*). The wood is destroyed and reduced to a dry, brittle, powdery mass.
2. A storage disease of potatoes caused mainly by the fungus, *Fusarium caeruleum*. Spores infect the tubers via wounds and bruises caused by rough handling during harvesting. Tubers are caused to wither and wrinkle.

Dry Roughage. Bulky, dry, non-succulent animal foods (e.g. hay and straw), rich in CRUDE FIBRE, used mainly in MAINTENANCE RATIONS for cattle and sheep. Also called coarse fodder. (♦ GREEN FORAGES, CONCENTRATES, SUCCULENT FOODS)

Dry Spell. A period (in the U.K.) of at least 15 consecutive days, during which less than 10.16 mm (0.4 in.) of rain has fallen on any day. Not an internationally accepted definition. (♦ ABSOLUTE DROUGHT, PARTIAL DROUGHT and WET SPELL)

Dry Stone Wall. A wall built of stones without mortar cementing them together. Often constructed to mark boundaries, particularly characteristic of certain hill areas (e.g. the Lake District), and often used by sheep for shelter in the winter.

Drying-off. Gradual reduction of the volume of milk taken from a cow so as to cause it to cease lactating.

Dual-purpose Breeds. Breeds of (*a*) cattle considered useful for both beef and milk production, (*b*) poultry considered good both for egg laying and as table birds, and, (*c*) pigs, the females of which are kept generally for crossing with either bacon or pork type boars.

Dubbing. The removal of the COMB of a cock. Also the trimming of part of the comb in DAY-OLD-CHICKS. The practice reduces frost bite, and is advantageous if serious CANNIBALISM or pecking occurs.

Duck. A smaller bird of the duck family (*Anatidae*). The female adult duck as distinct from the male adult duck or drake. (♦ GOOSE)

156

Duckfoot Harrow ♦ DRAG HARROW.

Duckling. A young duck, usually less than 12 weeks old.

Duff. To change the brand on stolen cattle.

Dug. A nipple or udder, particularly of a cow.

Dump Box. A large portable container similar to a FORAGE WAGON, placed adjacent to a SILO, into which trailer loads of FORAGE from the field are tipped, and from which the forage is delivered in an even steady flow, usually by a moving floor mechanism or elevator belt, for loading into the silo. It may also be used to deliver forage directly into a MANGER. (♦ FORAGE BOX)

Dump Rake. An expression for a hay rake.

Dung. The faeces of animals. Also to excrete it while grazing, or to spread it on the land.

Dung Liquor. The dark coloured liquid which drains from a manure heap, containing soluble compounds of ammonia and organic matter.

Dunging Channel. The passage(s) in a building housing livestock (e.g. COWHOUSE, PIGGERY) into which the animals void faeces whilst kept in stalls or pens, and from which it can be removed. They may be central or along outside walls according to the design (*see p. 131*). Also called a grip in a cowhouse.

Dungmere. A manure pit.

Dunt ♦ STURDY.

Durham. A breed of large-framed, high yielding, shorthorn cattle developed in the Tees Valley, and improved by interbreeding in the late eighteenth century after which it spread throughout Britain and gave rise to the main SHORTHORN varieties (♦ DAIRY SHORTHORN, NORTHERN DAIRY SHORTHORN, BEEF SHORTHORN, WHITEBRED SHORTHORN and LINCOLN RED). Also known as Teeswater, Holderness, Yorkshire.

Duroc. A breed of pig imported from North America and used in some CROSS-BREEDING programmes.

Durum. A variety of wheat used for the production of spaghetti and other pasta products. Sometimes called 'spaghetti' wheat.

Dusting. The application of insecticide or fungicide to crops in a dry, powdery state. It is less efficient than spraying, usually requiring about twice the amount of chemical per unit area.

Dutch. A breed of rabbit with distinct coloured cheeks and hindquarters (mainly black, blue or brown), but otherwise white.

Dutch Barn. A storage building mainly for hay and straw, but with many other uses (e.g. storing implements, housing livestock, covering a grain silo or silage pit). Construction may be of timber, concrete or steel. The latter, with a curved roof, is very popular, although concrete barns with asbestos-cement roofs are common. The sides may be completely open or some may be partly or wholly covered for protection. The design incorporates a number of standard bays which can be used for different purposes simultaneously.

A Dutch barn.

Dutch Belted. A rare breed of dairy cattle which originated in Holland. It is characterised by a white 'belt' round the otherwise black body between the shoulders and hips.

Dutch Harrow. A type of HARROW with a heavy wooden frame fitted with spikes, used to break soil clods and to level SEED BEDS.

Dutch Hoe. A hand HOE with a blade attached like a spade, pushed away from the user. (♦ DRAW HOE)

Dutch White Clover. A variety of white clover (*Trifolium repens*) used for FODDER in temporary LEYS. It is shorter lived, larger and roots less at the nodes than WILD WHITE CLOVER. (♦ CLOVER)

Dwarf French Beans ♦ BEANS.

Dyke. 1. A ditch or minor lowland watercourse.
2. A raised bank of earth, stones, etc., constructed to keep a river or sea water from flooding adjoining, usually low-lying, land. Also called Dike.

Dynamometer. An instrument which measures power. Used in agriculture to measure the resistance or DRAFT of any implement drawn by a tractor.

E

E. Coli ▶ ESCHERICHIA COLI.

Ea. A term used in the FENS for a drainage canal, probably derived from the French word *eau*, meaning water.

Eadish ▶ AFTERMATH.

Ean. To give birth to a lamb. Also called yean.

Eanling. A young lamb.

Ear. 1. The spike or flowering head of a grass, usually applied to cereals. Also to form such ears.
2. An obsolete term meaning to plough or till.

Early Bite. A sudden growth of grass which can be grazed before the main crop of grass has really begun to grow.

Early Flowering Red Clover ▶ RED CLOVER.

Early Potatoes ▶ POTATO.

Early Soil, Early District. A soil which warms up rapidly in spring and is able to grow a crop before other soils. An early district is an area with such soils.

Earmarking. The marking of an animal's ear for identification, by attaching a marked metal or plastic tag, tattooing, or cutting distinctive notches.

Earth. 1. A fox's burrow.
2. To cover with soil.

Earth-board. A MOULDBOARD.

Earth Nut. Another name for GROUNDNUT.

Earthing Up. The building up of a ridge with soil removed from adjacent furrows, particularly used to cover potatoes during planting.

Earthworm. One of a number of species of segmented, bristled worms present in the soil, sometimes in millions per acre, which aerate the soil and assist drainage and the upward movement of nutrient salts. The largest and commonest is *Lumbricus terrestris*, which can be a foot long.

Ease. To milk a cow a little in order to relieve pressure on its udder.

Easement. A legal term for a right to use something, particularly land, belonging to another (e.g. for access, pipelines, etc.) or restricting the owner's use of it.

East Friesland. A breed of sheep originating in Holland, renowned for its prolificacy and high milk yield rich in BUTTERFAT. Used in the development of the COLBRED and in many current breed improvement schemes. A large, long-bodied, slim breed, with a white face, long ears, bald tail, and producing fine white wool.

East Lothian Rotation. A six-course system of crop ROTATION which comprises rootcrops – barley – seeds – oats – potatoes – wheat. This sequence is suited to mixed cropping in East Lothian and other areas of Scotland where growth conditions favour oats which follow seeds and benefit from increased fertility provided by clover. (◆ NODULES)

Eastrip Special Blend. A hybrid sheep, the product of a cross between the BLUEFACED LEICESTER and POLL DORSET, said to be prolific.

Easy Feed. A method of livestock feeding in which the stock are allowed easy access to feed. Usually the feed is transferred mechanically to large hoppers or feeding passages.

Eatage. An obsolete term for grass used for grazing as distinct from for hay, particularly AFTERMATH. Also the right to use aftermath for grazing.

Eatche. A Scottish term for ADZE.

Ecology. The study of the relationship between living organisms and their environment.

Ectoparasite. A PARASITE living on the outside of its host, e.g. a LOUSE. Also called exoparasite. (◆ ENDOPARASITE)

Edaphic. Of or pertaining to the soil.

Edaphology. The study of the influence of soils on living organisms, particularly plants, including man's use of land for growing crops.

Eddish ◗ AFTERMATH.

E.E.C. ◗ EUROPEAN ECONOMIC COMMUNITY.

Eelworms. Tiny, generally microscopic, nematodes or roundworms. There are many species, some free-living in the soil or fresh water, others parasitic in plants and insects, e.g. stem eelworm, *Ditylenchus dipsaci*, which causes the stem base of oats to become swollen or 'tulip rooted', and potato cyst eelworm, *Heterodera rostockiensis*, which causes the roots to become disorganised and develop cysts containing swollen female eelworms.

Effluent. Liquid, solid or gaseous waste material, e.g. SLURRY, industrial waste, sewerage.

Egg. 1. An oval-shaped body laid by female birds (and some other animals) comprising a calcareous shell containing ALBUMEN and a YOLK, the latter developing to produce a new young bird which hatches as a CHICK.
2. An ovum or female GAMETE. Also called egg cell.

Egg Bound. A term for a bird unable to lay an egg.

Egg-cell ◗ EGG.

Egg Classes. There are three quality classes under current E.E.C. marketing regulations: Class A ('fresh eggs'); Class B ('second quality or preserved eggs'); and Class C ('non-graded eggs intended for the manufacture of foodstuffs for human consumption'). (◗ EGG WEIGHT GRADES)

Egg Tooth. A horny knob on the bill of a newborn bird used to crack the eggshell in hatching, and discarded soon after.

Egg Weight Grades. There are seven weight grades under E.E.C. regulations for Class A and B eggs (◗ EGG CLASSES). Grade 1 are 70 g or over, Grade 7 are under 45 g.

Egg White ◗ ALBUMEN.

Eggery. A place where eggs are laid.

Eggler. A dealer in eggs.

Eggs Authority. Established by the Agriculture Act 1970 to replace the British Egg Marketing Board and financed by a levy payable by hatcheries. It has five main functions, (*a*) to support the market by buying and disposing of eggs as thought fit (this right has not been exercised much, particularly since the U.K. joined the E.E.C.), (*b*) to collect and disseminate market intelligence, (*c*) to advertise and promote egg sales, (*d*) to encourage or undertake research into the demand for eggs and egg products, marketing, storage, distribution and processing, and (*e*) to initiate schemes for quality control and weight grading of eggs. (⬥ EGG CLASSES, EGG WEIGHT GRADES)

Eild. 1. Not yielding milk.
2. BARREN.

Eirack. A Scottish term for a young hen.

Electric Dog. A mechanism to encourage cows to enter a MILKING PARLOUR from a collecting yard, comprising an electric wire sandwiched between two parallel side wires which is drawn up behind the cows gradually as they go through the parlour, by means of a cord in the parlour. (⬥ ELECTRIC FENCE)

Electric Fence. A system of mobile fencing usually used to facilitate CONTROLLED GRAZING, consisting of easily moved insulated posts, usually about 3 ft high, supporting thin wires (usually one for cattle and horses, two for sheep and pigs) carrying an electric current, either battery generated or from the mains. The current is not continuous but in 'pulses' at the rate of about 60 per minute.

Element. A chemical substance, consisting entirely of like atoms, that cannot be split into simpler substances and from which all other substances are built up, e.g. oxygen, carbon.

Elevator. A machine for raising a cut crop, GRAIN, or BALES to the top of a SILO, storage bin, or bale stack, etc., usually for storage. Various designs exist. Grain elevators usually consist of a continuously moving belt bearing buckets, or an endless chain bearing flights or steps (operating like a moving staircase). Elevators for hay or straw bales often comprise a continuous moving chain bearing spikes or cross slats. (⬥ AUGER)

Elevator Digger. An implement for harvesting potatoes. There are three kinds; trailed, semi-mounted, and fully mounted. They lift potatoes from the field rows and separate the soil from the crop on a chain web and deposit them in a row on the field for collection. The Potato Harvester is essentially an improved elevator digger bearing a platform for the pickers to stand on, and a system of elevators delivering the potatoes to a sorting table, and conveying them to the side for collection and transport. Some complex modern machines incorporate electronic automatic sorting and require only one operator.

A semi-mounted elevator digger.

Elt. A young sow.

Embden. A heavy white breed of goose, with blue eyes.

Emblements. 1. Crops raised by cultivation (e.g. wheat, potatoes) as distinct from the fruits of trees or grass.
2. In law, such crops sown by a tenant which he has the right to return to and harvest when his tenancy has terminated before harvest. (♦ AWAY-GOING CROP)

Embryo. A young animal or plant in its earliest stages of development after fertilization, e.g. a young animal in its mother's womb, or the innermost part of a plant seed.

E.M.L.A. Scheme. A description applied to the best fruit tree material produced at East Malling and Long Ashton Research Stations and issued by them to those growers interested in the propogation of E.M.L.A. material. Special precautions are taken to ensure that fruit trees thus propagated retain the original E.M.L.A. health and identity when they reach the grower.

Emmer. A species of WHEAT, *Triticum dicoccum*.

Enclosure. 1. An area of land bounded by fences, hedges or walls. 2. COMMON LAND converted to private ownership in the Middle Ages by erecting fences and hedges, etc. Sometimes spelt inclosure.

Endemic. 1. A descriptive term for plants or animals naturally occurring in a particular area. Also called indigenous. (◆ EXOTIC)
2. A descriptive term for a disease constantly or generally found in an area, even though it may not be expressing itself at a particular time. (◆ EPIDEMIC)

Endoparasite. A PARASITE living on the inside of its host, e.g. TAPEWORM. (◆ ECTOPARASITE)

Endosperm. Food storage tissue surrounding and nourishing the EMBRYO in plant seeds. By the time the seed is fully developed the embryo may have completely absorbed the endosperm in some seeds, e.g. peas and beans, whilst in others part may remain and be absorbed after germination, e.g. wheat. Cereals are grown for the endosperm content of the grain, which is a valuable food source.

Endrin. A very toxic insecticide. A poly-CHLORINATED HYDROCARBON.

Energy of Combustion ◆ GROSS ENERGY VALUE.

English Horse Bean ◆ TICK BEAN.

English Leicester. A long-woolled breed of sheep improved by BAKEWELL with the introduction of LINCOLN and RYELAND blood, and subsequently used in the development of many other long-wool breeds. The modern Leicester is large and hornless, with a white face. It produces a curly glossy fleece. The breed is mainly restricted to parts of Yorkshire. Also called Leicester or Leicester Longwool.

Ensilage. To make SILAGE by the storage of green fodder in pits or silos.

Entire. An animal which has not been castrated. A stallion.

Entomology. The science of insects.

Entry ◗ TAKE.

Environment. Surrounding conditions, e.g. climate, landform, etc.

Enzootic Abortion. A virus infection of sheep affecting the placenta, causing abortion about 10–14 days prior to lambing time. Common in South-East Scotland and North-East England.

Enzootic Disease. An animal disease prevalent in particular areas or districts or during certain seasons (e.g. ENZOOTIC ABORTION) as distinct from epizootic diseases which spread rapidly and attack animals in large numbers almost simultaneously over large areas, even over entire continents (e.g. FOOT AND MOUTH DISEASE). (◗ ENDEMIC, EPIDEMIC)

Enzymes. PROTEINS present in living organisms which act catalytically to promote chemical changes whilst remaining unchanged themselves, e.g. ALPHA AMYLASE.

Epidemic. The outbreak of an epizootic disease affecting great numbers of animals in one area at the same time and transmitted from place to place. The disease may not necessarily be one that is normally present in an area. (◗ ENDEMIC, ENZOOTIC DISEASE)

Epidermis. The outermost layer(s) of cells of plants and animals, e.g. skin.

Epizootic ◗ ENZOOTIC DISEASE.

Epsom Salts. ($MgSO_4 \cdot 7H_2O$). Magnesium sulphate, a white crystalline soluble salt, used often as a saline purgative, and as a FERTILIZER, mainly as a foliar spray on fruit. It contains about 10% magnesium. (◗ KIESERITE)

Equine. Pertaining to a horse or horses.

Eradication Area. An area in which a programme designed to eradicate a particular animal disease is operating. Eradication schemes operating in such areas often involve the compulsory slaughter of animals or herds found, on testing, to have the specific disease. (♦ ATTESTED AREA)

Ergot. 1. A fungal disease (*Claviceps purpurea*) of grasses, particularly cereals and especially rye. It attacks the seed heads and destroys the grain. Animals can suffer serious poisoning (ergotism) by eating diseased grain or products derived from such grain (e.g. bread, maize meal). In man it causes the condition known as St Anthony's Fire.
2. A small patch of horn, hidden by a tuft of hair, and located on the back of a horse's fetlock.

Erosion. The wearing away of the land surface, particularly soil, by running water, ice and wind, etc. Ploughing up and down slopes can lead to gully formation, and light and friable soils can blow away in strong winds when exposed, usually in the spring. Wind erosion is a serious problem in the FENS.

Erysipelas. An infectious disease, mainly of pigs, in which hot, reddish inflammations of the skin and high fever occurs, and when severe lameness and difficult breathing follows. The disease commonly causes infertility and abortion.

Escherichia coli. A bacterium found in the alimentary canal of most mammals. Certain strains cause disease, e.g. COLI SEPTICAEMIA in poultry and diarrhoea in various animals. Usually abbreviated to *E. coli*.

Espalier. A lattice-work, usually of wood, to train trees on. Also fruit trees so-trained, usually so that the branches grow horizontally.

Essential Amino Acid ♦ AMINO ACID.

Essex, Essex Saddleback ♦ BRITISH SADDLEBACK.

Estate. The total property in single ownership, often applied to the property of large landowners, including farms, cultivated land, woodland, houses, etc.

Ether Extract. One of the fractions determined by the PROXIMATE ANALYSIS of animal feedingstuffs, containing mainly fats and oils together with various minor compounds, e.g. fat-soluble vitamins.

Etiolation. A condition of plants when grown in darkness, characterised by a general pale yellowness due to a lack of CHLOROPHYLL, and by long internodes and small leaves.

Euk ♦ YOLK.

European Agricultural Guidance and Guarantee Fund. An E.E.C. fund in two sections: (*a*) The guidance section finances measures to improve agricultural structure in the E.E.C., providing grants towards the cost of projects promoting the objectives of the COMMON AGRICULTURAL POLICY. The government of the member state concerned must approve each project and be prepared also to provide financial aid. Grants are only available for physical investment, e.g. buildings, drainage, equipment, etc. (*b*) The guarantee section finances market support arrangements. Also known as F.E.O.G.A. (Fonds Européen d'Orientation et de Garanties Agricoles). (♦ COMMON PRICES, GUARANTEED PRICE)

European Currency Unit ♦ UNIT OF ACCOUNT.

European Economic Community. Often called the Common Market, established in 1957 by the Treaty of Rome, with six member states including Belgium, France, West Germany, Holland, Italy and Luxembourg. In 1973 Denmark, Ireland and the U.K. joined the Community and Greece became a member in 1981. Negotiations are in hand for the further expansion of the E.E.C. to include Spain and Portugal. (♦ COMMON AGRICULTURAL POLICY)

Eutrophic. A descriptive term for nutrient or soil solutions with nutrient concentrations optimal or nearly optimal for animal or plant growth.

Evaporated Milk. Milk thickened and enriched by evaporating much of its water content, but not sweetened like CONDENSED MILK.

Evapotranspiration. The loss of water from an area (e.g. a field) by evaporation at the soil surface and by plant TRANSPIRATION. Potential Evapotranspiration is the maximum transpiration possible under given weather conditions with a low-growing crop which does not shade the soil completely and has an adequate water supply. (◊ IRRIGATION)

Evergreen. A plant which keeps its leaves throughout the year. Mainly coniferous trees and shrubs. (◊ DECIDUOUS)

Eviscerated. Having had the viscera or bowels removed.

Ewe. An adult female sheep. (◊ SHEEP NAMES)

Ewe Hogg ◊ SHEEP NAMES.

Ewe Lamb. A female lamb under six months old.

Ewe-necked. A term for horses with thin hollow necks.

Exmoor Horn. A hill breed of sheep found mainly on Exmoor, of rounded compact appearance, with a white face, curled horns (in both sexes), and thick soft fleece. The breed was crossed with the DEVON LONGWOOL to produce the DEVON CLOSEWOOL.

Exoparasite ◊ ECTOPARASITE.

Exotic. A descriptive term for plants or animals introduced into an area or country from outside or abroad, not naturally occurring there. (◊ ENDEMIC)

Expander ◊ MOLE PLOUGH.

Expeller Process. A process for removing the oil from oilseeds in which the seeds are forced along a tapering tube, squeezing out the oil. The residues, containing usually less than 5% oil, may be used as animal feeds. (◊ CAKE)

Extensive Farming. A method of farming in which a large amount of land is used to raise stock and produce crops, yields usually being about average, as distinct from INTENSIVE FARMING.

Extracted Honey. Honey obtained by centrifuging uncapped broodless honey-combs.

Extra-late-flowering Red Clover ◊ RED CLOVER.

169

Eye. 1. An obsolete term for a brood, particularly of pheasants. 2. The seed-bud of a potato or an axillary bud on a plant.

Eyespot. A fungal disease (*Cercosporella herpotrichoides*) which attacks all cereals, causing oval or eye-shaped areas of black dots on leaf sheaths as they emerge from the soil in spring, and causing straws to weaken and lodge.

F

F₁ and F₂ Generations. Genetical terms for the offspring generations produced by a parental generation of plants or animals. F_1 is the first filial generation, and F_2 is the second filial generation (i.e. the offspring of F_1).

Factor ♦ GENE.

Facultative Parasite ♦ PARASITE.

Faeces. Solid waste or undigested material expelled by animals. Also called dung or manure. (♦ FARMYARD MANURE)

Fagg ♦ TICK.

Fagging Hook ♦ BAGGING HOOK.

Fairy Ring. An arc or ring of lush, usually darker coloured grass in a pasture caused by fungi which grow in expanding circles in the form of a perennial MYCELIUM. Such rings may be many years old and many yards across. The ring of lush grass is due to nitrate production, which acts as a fertilizer.

Fall. 1. An American term for autumn.
2. The dropping of an animal from its mother's uterus to the ground during birth. Also the number of offspring born in a herd or flock, etc., in a season.

Fallen Wool. A type of wool oddment purchased by the BRITISH WOOL MARKETING BOARD comprising wool which is either clipped from a dead sheep, or is badly shorn and is in the form of torn or broken pieces as distinct from an intact fleece.

Falling Number, Falling Time ♦ HAGBERG TEST.

Fallow. Land left unsown, usually for a season, during which it is ploughed and cultivated to kill perennial weeds by desiccation. The practice of fallowing is now less common.

Fallow Crop. A crop grown in well spaced out rows, e.g. potatoes or turnips, so that the land between the rows can be hoed and cultivated to control weeds. Also called a cleaning crop.

171

False Gid, False Staggers. A 'dazed' condition of sheep due to inflammation of the head affecting the brain, caused by maggots of the SHEEP NOSTRIL FLY feeding on the mucous membranes of the sinuses.

Family Farm. A farm run by a farmer and his family, usually with no other employees, and generally small to moderate in size.

Fan. A basket used to WINNOW corn.

Fanged. A term for root crops having forked roots.

Fank. A Scottish term for a sheep FOLD.

Fardel-bound. Constipated. Especially used for cattle and sheep having food retained in the OMASUM.

Farina. Ground corn or cereal meal.

Farm. An area of land, usually with a house and necessary buildings, used for agricultural purposes (♦ AGRICULTURE). The term originally meant such land leased or rented by a landlord and worked by another person, the farmer. According to the 1980 ANNUAL REVIEW White Paper, 65% of farms in Great Britain were wholly or mainly owner-occupied in 1979 compared with 54% in 1960/61. In N. Ireland virtually all farmers are owner-occupiers. In recent years the number of farms in the U.K. has been declining and in 1979 the total number of holdings was 257 000. The average farm size is about 60 hectares (150 acres), although the size of individual enterprises has been expanding in recent years.

Farm and Horticultural Development Scheme. A scheme begun in 1974 and due to finish in 1982, partly financed by the EUROPEAN AGRICULTURAL GUIDANCE AND GUARANTEE FUND, offering financial assistance to eligible farmers and growers to develop their businesses, particularly by capital expenditure. The scheme, administered by the Ministry of Agriculture, Fisheries and Food, requires the submission of a development plan, to last for up to 6 years, which will result in the business being able to support at least one person putting in a full year's work and provide incomes for other employees comparable with those in other industries. Usually abbreviated to F.H.D.S.

Farm Animal Welfare Council. An advisory body established in 1979 to keep under review the welfare of farm animals on agricultural land, at markets, in transit and at places of slaughter, and to advise the Agriculture Ministers of any legislative or other changes that may be necessary.

Farm Bottled Milk ♦ UNTREATED MILK.

Farm Capital Grant Scheme. A scheme begun in 1970, and varied subsequently, administered by the Ministry of Agriculture, Fisheries and Food, providing grant aid towards capital expenditure on farms for specified developments, e.g. buildings, field drainage, roads, orchard grubbing, etc. Usually abbreviated to F.C.G.S.

Farm Fresh Eggs. A description which, under the E.E.C. egg marketing arrangements (♦ EGG CLASSES), can only be applied to class 'A' eggs.

Farmers' Goods Licence. A special rate of excise duty available to farmers licensing motor lorries to be used solely for conveying the produce of, or articles required for, his agricultural land.

Farmers' List. Those veterinary medicines, as prescribed under the Medicines Act 1968, which can be supplied to farmers, or other commercial keepers of animals, by agricultural merchants, without a prescription or supervision of a pharmacist or veterinarian. Also called the merchants' list.

Farmer's Lung. A distressing respiratory condition caused by the inhalation of dust particles, e.g. from mouldy feed or litter, containing fungal spores, which results in an allergic reaction in the air sacs of the lungs and causes breathing difficulties. Cows and horses can also suffer.

Farming and Wildlife Advisory Group. An independent body, with local committees in many counties, established in 1969 to encourage and stimulate understanding and liaison between farming and conservation interests. The Group undertakes promotional activities, conferences, publicity, and on-farm demonstrations, and acts as a forum for collecting, exchanging and disseminating information, research results and experience on wildlife conservation on farms, both nationally and locally. Usually abbreviated to F.W.A.G.

Farming – Types of ♦ LAND USE.

Farmstead. A farm house and the farm buildings associated with it.

Farmyard. A yard or enclosure surrounded by farm buildings and usually adjacent to the farmhouse.

Farmyard Manure. The faeces and urine of farm animals mixed with litter, mainly straw, to absorb the urine. This mixture is also called 'Long Dung'. Its composition is variable dependent on the animals contributing the dung, their diet and the kind of litter used. Farmyard manure is usually stored in manure heaps

A farrowing crate. The piglets are kept warm by the heat source suspended on chains.

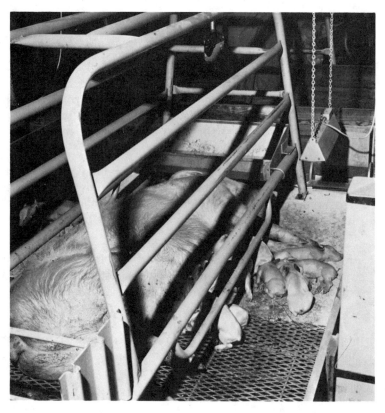

where bacterial activity releases ammonia, fermentation occurs, and the material generally degrades into simpler compounds. It becomes structureless and the liquids drain as DUNG LIQUOR. In this condition it is called 'Short Dung'. This material is spread on farmland to improve soil structure (being rich in ORGANIC MATTER) and also as a FERTILIZER, its main constituents being compounds of Nitrogen, Potassium and Phosphorus. Also called muck. (♦ LIQUID MANURE, SLURRY)

Farrand Test. A method for determining the ALPHA AMYLASE content of MILLING WHEAT. Used as an alternative to the HAGBERG TEST. An extract of ground wheat is reacted with a solution of 'limit' dextrin (a derivative of wheat starch). After a measured time it is mixed with iodine solution, and the colour produced compared with that of a suitably diluted iodine – 'limit' dextrin mixture. Loss of colour, due to Alpha Amylase in the wheat extract reacting with the 'limit' dextrin, indicates the activity of the enzyme, which is expressed as units. Sound wheat has an activity of 10 units or less. Wheat with an activity over 40 units is not normally accepted for milling. 40 Farrand Units is approximately equivalent to a Hagberg Falling Number of 160.

Farrier. A person who makes and fits horseshoes.

Farrow. 1. To give birth to a LITTER of pigs. Also the litter itself.
2. A descriptive term for a cow or heifer not in-calf, or barren.

Farrowing Crate, Farrowing Rail. A steel crate used in a pig pen to contain a sow and prevent it from lying on and killing the piglets, whilst allowing them access to her teats. They are normally in the form of a long narrow rectangle, but are sometimes circular (the Ruakura pen). A farrowing rail serves the same purpose and is usually fitted a short distance from the pen walls and floor.

Farrowing Index. The number of LITTERS produced per year by a breeding sow.

Fat. 1. A term applied to animals reared for their meat (e.g. fat cattle, fat sheep) which have been fattened up and are in a ready condition for sale to a butcher at a market.
2. A dry measure of nine BUSHELS.
3. ♦ FATS.

Fat Class. A visual appraisal of the degree of external fat development on a beef or sheep carcase, being one of the characteristics assessed in CARCASE CLASSIFICATION. The fat classes currently used by the MEAT AND LIVESTOCK COMMISSION are as follows:

CLASS	PERCENTAGE FAT	
	Beef Carcases	*Sheep Carcases*
1	<4.5	<6.0
2	4.5–7.4	6.0–9.9
3L	7.5–8.9	10.0–11.9
3H	9.0–10.4	12.0–13.9
4	10.5–13.4	14.0–17.9
5	>13.4	>18.0

In the case of pigs fatness is measured with an Intrascope which indicates fat depth to the nearest millimetre.

Fat Hen. A weed (*Chenopodium album*), particularly of arable land.

Fatling. A young animal fattened for slaughter.

Fats. Storage materials found in living organisms, mainly in liquid form (oils) in plants, and in solid form (also called adipose tissue) in animals, comprising mixtures of glycerides (condensation products of glycerol and FATTY ACIDS). Most animals deposit fat in the body as an energy store. Some plants, mostly tropical, produce seeds rich in oil which is extracted mainly for cooking, the residue being used for cattle CAKE. (♦ ANIMAL FATS, STEAPSIN)

Fatstock. Livestock fattened for sale in a market.

Fatstock Guarantee Scheme. A scheme administered by the Ministry of Agriculture, Fisheries and Food, providing a guaranteed price or payment for certain sheep and their carcases sold (*a*) by auction at an approved fatstock market, (*b*) privately, or (*c*) by grade and deadweight through organisations such as the FATSTOCK MARKETING CORPORATION, or other recognised schemes. (♦ DEADWEIGHT CERTIFICATION CENTRE, LIVEWEIGHT CERTIFICATION CENTRE)

Fatstock Marketing Corporation. A limited company established in 1954 by the Farmers' Unions of the U.K. (♦ NATIONAL FARMERS' UNION), becoming a public company in 1962. F.M.C. is a wholesale meat company and the biggest meat group in Europe. It is involved in the procurement of livestock on a national basis and in meat distribution. It operates a network of abbatoirs and meat processing factories and is also involved in the curing of bacon, and the canning and freezing of meat products. F.M.C. has plants for FELLMONGERING and wool production, and hide and skin markets throughout the British Isles, and deals in bone phosphates and animal proteins. It also functions in Eire and New Zealand.

Fatty Acids. Organic acids, thirteen of which occur in natural FATS. They are either (a) saturated (i.e. molecules to which no further atoms can be added), found mainly in solid form in ANIMALS FATS, or (b) unsaturated, found mainly in liquid form (oils) in plants. The most common fatty acids in natural fats are palmitic, oleic and stearic. (♦ STEAPSIN)

Fauna. The population of animal life in a place or area, or during a particular period. (♦ FLORA)

Faverolle. A heavy table breed of poultry, with a single comb and slightly feathered white legs bearing five toes, producing white eggs. Variously coloured including blue, buff, salmon and white.

Fawn. A young deer.

f.c.s. An abbreviation for 'filled, carted and spread'. An agricultural valuers' expression for the labour required to transport manure from the farmyard to the fields. Now rarely used, the expression 'labour to farmyard manure' being more common.

Feather. 1. One of the dermal outgrowths forming the plumage of a bird.
2. Long hairs found on the sides and on the back of the legs of some breeds of horse, e.g. the SHIRE.

Feathered Maiden. A MAIDEN TREE which has grown a number of side shoots.

Federation of Agricultural Co-operatives (U.K.) Ltd. The representative body in the E.E.C. for all U.K. co-operatives including the MARKETING BOARDS.

Fee. An obsolete term for livestock, particularly cattle.

Feed ◗ FEEDINGSTUFF.

Feed Block. A block of foodstuffs left on pastures, particularly in hill areas, for sheep to lick at will as an aid to maintaining CONDITION. Normally they contain a CARBOHYDRATE source (e.g. cereals, glucose, or molasses), PROTEIN and UREA (sometimes excluded), MINERALS, VITAMINS and TRACE ELEMENTS.

Feed Conversion Ratio ◗ FOOD CONVERSION RATIO.

Feed Off. The practice of allowing a crop to be eaten by animals whilst still in the ground.

Feed Ring. A railed circular container mainly used for hay and straw, from which livestock feed around its circumference.

Feedingstuffs. The various foods available for farm animals, loosely classified into four main groups, viz, (*a*) GREEN FORAGES, (*b*) succulent foods, mainly roots and tubers, e.g. turnips, potatoes, etc., (*c*) dry roughage or coarse fodder, e.g. HAY derived from grasses, cereals and legumes, or STRAW derived from cereals and the haulms of peas and beans, and (*d*) CONCENTRATES. Also called feeds.

Feedlot. An area of land in which animals (very commonly beef cattle) are accommodated at very high density. The feedlot does not contribute at all to the production of animal feed, all of which has, therefore, to be brought to the animals from outside the feedlot.

Feer. To plough the first furrow when starting ploughing. (◗ FINISH)

Fell. 1. The hide or skin of an animal, particularly when covered with hair. A fleece.
2. A term mainly used in Northern England for mountainous moorland and hill pastures.
3. To cut down a tree.

Fellmonger. A person who prepares animal skins for tanning.

Felloe, Felly. The circular rim of a wheel to which the spokes are attached.

Fen. A lowlying, marshy area of land, with a PEAT soil which may be alkaline to slightly acid, as distinct from the very acid soil of a BOG. Such areas in Eastern England have been drained to produce very fertile land and are known as the Fens.

Fenland Rotation. A three-course system of crop ROTATION developed on the deep, rich, silts and FENS of East Anglia, in the sequence potatoes – sugarbeet – wheat. The short sequence increases the likelihood of attack by pests and disease, and other high value crops are frequently included, such as rootcrops, brassicas, bulbs, peas and celery.

F.E.O.G.A. ◗ EUROPEAN AGRICULTURAL GUIDANCE AND GUARANTEE FUND.

Ferm. A farm.

Fermentation. The breakdown of organic substances by the action of ENZYMES, usually secreted by living organisms such as BACTERIA and YEAST. Heat and gas (e.g. carbon dioxide) are usually evolved. The principle of fermentation is used in the making of SILAGE, wine, vinegar, etc.

Ferret. An albino variety of the polecat, half-tamed and used to catch or unearth rabbits and rats.

Fertile. 1. A descriptive term for plants or animals able to reproduce successfully, as distinct from being STERILE. Also those able to produce offspring prolifically. Used also to describe eggs capable of producing a chick, or seeds capable of GERMINATION.
2. A descriptive term for land or soil, rich in nutrients, and capable of producing good crops.

Fertilization. 1. The enrichment of soil by the application of FERTILIZER.
2. The union of male and female gametes in REPRODUCTION to produce a ZYGOTE.

Fertilizer. In general terms anything added to the soil to increase the amount of plant nutrients available for crop growth. Normally applied to inorganic chemicals and non-bulky organic substances (i.e. excluding COMPOST, FARMYARD MANURE, etc., which are commonly called MANURES). Fertilizers basically provide the major plant nutrients (i.e. NITROGEN, PHOSPHORUS and POTASH), either individually (◆ STRAIGHT FERTILIZER) or as a combination of two or three (◆ COMPOUND FERTILIZER), sometimes together with TRACE ELEMENTS.

(*a*) The nitrogenous fertilizers include AMMONIUM NITRATE, NITRATE OF SODA, SULPHATE OF AMMONIA, ANHYDROUS AMMONIA, UREA, AMMONIUM PHOSPHATE and LIQUID NITROGEN FERTILIZERS.

(*b*) The phosphatic fertilizers may be quick acting (water soluble) or slow acting (water insoluble) and include AMMONIUM PHOSPHATE, SUPERPHOSPHATE, TRIPLE SUPERPHOSPHATE, BASIC SLAG, MINERAL PHOSPHATES, BONE FLOUR, BONE MEAL and FISH GUANO.

(*c*) The potash fertilizers include MURIATE OF POTASH, SULPHATE OF POTASH, KAINIT and NITRATE OF POTASH.

Fertilizers applied to the soil usually quickly dissolve in the SOIL WATER and are taken up by plant roots or absorbed by soil COLLOIDS (◆ CATION EXCHANGE). Excess fertilizer may be leached through the soil into the GROUNDWATER. Fertilizers nowadays are mostly produced in easily stored granulated form, but powders and crystals, and also less concentrated liquids are available. Inorganic chemicals such as LIMESTONE and GYPSUM applied to correct such soil conditions as acidity and salinity may also be considered as fertilizers. (◆ FERTILIZER DISTRIBUTOR, ORGANIC FERTILIZER)

Fertilizer Distributors. Machines which distribute FERTILIZER on the land usually by broadcasting if in a solid form (i.e. crystals, granules or powders), or by spraying if in liquid form, and sometimes by injection into the soil (e.g. ANHYDROUS AMMONIA). Broadcast types comprise a hopper which feeds the fertilizer to a spreading mechanism, the most popular being either a spinning disc or oscillating spout, which distribute fertilizer in swaths of variable width. These mechanisms are less accurate than the 'full width' distributors such as the once

A spinner broadcaster capable of handling prilled, semi-prilled, granular and crystalline fertilizers, and also cereal and grass seeds.

popular 'plate and flicker' type or the more modern agitator or pneumatic type spreaders. Fertilizers can also be applied by a COMBINE DRILL, PLACEMENT DRILL, and by aircraft as TOP DRESSING, particularly to cereals.

Fertilizer Requirement. The quantity of particular plant nutrients required, in addition to those already contained in the soil, to increase crop growth to a desired optimum.

Fertilizer Value ♦ MANURIAL VALUE.

Fescue. The common name for *Festuca*, a genus of grass, including many valuable pasture and fodder species, e.g. Sheep's Fescue (*F. ovina*) Red Fescue (*F. rubra*), Meadow Fescue (*F. pratensis*) and Tall Fescue (*F. arundinacea*).

181

Fetlock. The tuft of hair which grows above a horse's hoof or the part of the leg where it grows.

Fibre. A constituent of animal FEEDINGSTUFFS consisting mainly of CELLULOSE and LIGNIN, which aids digestion by stimulating muscular activity in the digestive tract. It also facilitates the digestion of CONCENTRATES by opening them up to the action of the digestive juices. All farm animals require a certain amount of fibre except when very young and on a liquid diet. RUMINANTS require the most fibre and derive energy from it by bacterial digestion in the RUMEN. Cows require adequate fibre to produce milk with a high percentage of BUTTERFAT. Sheep and horses require less fibre than cattle but more than pigs. Excess fibre in the diet can prevent an animal eating sufficient food to derive adequate essential nutrients. HAY and STRAW are characterised by a high fibre content.

Fiddle. A small portable device carried on the shoulders for broadcast sowing, especially grass and clover seeds in hilly areas. It comprises a hopper which slowly drops seed on to a ribbed disc which the operator causes to rotate and throw the seed by pushing and pulling an attached 'bow'. Very rarely used.

Field. An area of land, usually enclosed by a ditch, fence, hedge, etc., for cultivation or grazing purposes. Fields are usually sown to a single crop at one time. In recent years the trend has been the removal of field divisions (e.g. hedges) to facilitate cultivation and increased efficiency, and the larger fields created are sometimes termed management units. Such larger fields also allow cropping flexibility and sometimes several different crops are grown in the field at the same time.

Field Beans ♦ BEANS.

Field Capacity. The state of soil when all the soil moisture that is able to freely drain away has done so. The remaining moisture is held by the forces of surface tension around soil particles and in capillary pores (♦ CAPILLARY WATER). (♦ also PERMANENT WILTING PERCENTAGE, READILY AVAILABLE MOISTURE and SOIL MOISTURE DEFICIT)

Field Drainage ♦ LAND DRAINAGE.

Field Room. The time needed for a crop stooked (♦ STOOK) in a field to dry sufficiently for stacking.

Field Sports. Sports carried out in the countryside such as foxhunting, hare coursing, horse racing, etc. (♦ BLOOD SPORTS)

Fieldman. A farmworker, the major part of whose work is in the fields, as distinct from around and in the farm buildings.

Filament ♦ STAMEN.

Filbert. The nut of the cultivated HAZEL.

Fill. 1. The contents of the digestive tract of an animal.
2. The space between the shafts of a cart in which a horse stands when harnessed, the horse itself being the filler.

Filler. 1. A tree temporarily growing in between fruit trees in an orchard until they mature and require the space.
2. ♦ FILL.

Filly. A young female horse between 1 and 4 years of age. (♦ FILLY FOAL, MARE)

Filly Foal. A female horse up to 1 year of age. (♦ FILLY)

Finger-and-toe ♦ CLUB ROOT.

Fingers. Units with flat faces (known as ledgers) projecting from and bolted to the cutter bar of a BINDER, MOWER or COMBINE HARVESTER, across which the knives slide backwards and forwards and cut the grass blades or crop against the fingers.

Finish. 1. An OPEN FURROW being the last one ploughed in a field. (♦ FEER)
2. ♦ FINISHED.

Finished. A term applied to animals, particularly beef cattle, which have been carefully fattened and are ready for sale at a market, having an acceptable degree of fat cover. Finished animals are expected to produce good quality carcases.

Fire Blight. A NOTIFIABLE DISEASE caused by bacteria (*Erwinia amylovora*), mainly of pears and apples, giving a scorched appearance to the fruit. Spread by insects.

First-calf Heifer. A term used in some areas for a HEIFER after having borne its first calf but before it has borne a second calf.

183

First Cross. The progeny from the mating of a male and female from separate pure breeds of animals or varieties of plants.

First Early Potatoes ◗ POTATOES.

First-year-poling. A common expression for the provision, for the first time, of full-height support to the BINES of young hop plants.

Fish Farming. The breeding and rearing of fish in tanks, ponds, cages, etc., ultimately for harvesting and slaughter for the table.

Fish Guano. A phosphatic FERTILIZER made from unmarketable fish and fish waste products after the extraction of oil, of variable composition. The NITROGEN content is usually 7–8% and the PHOSPHORIC ACID content 4–8%. Usually only used for horticultural purposes. Also called fish meal. (◗ MANURE, GUANO)

Fish Manure. The unprocessed offal from gutted fish and waste or unsold rotting fish, spread on fields as MANURE. (◗ FISH GUANO)

Fish Meal. 1. An animal FEEDINGSTUFF consisting of dried and ground waste or unsold fish and fish filleting residues, by law containing not more than 6% oil and 4% salt. White fish meal is derived from non-oily 'white' fish (e.g. cod, haddock, etc.) and usually contains about 66% protein. Other fish meals are derived from oily fish (e.g. herring, pilchards, etc.) after most of the oil has been extracted and are usually richer in protein than white fish meal. Fish meal is mainly fed to pigs and poultry, but is also included in rations for dairy cows, calves and other farm animals.
2. ◗ FISH GUANO.

Fives. A term for samples taken from hop POCKETS (for assessing type) whose identity numbers end in 5 (i.e. 5th, 15th, 25th, etc.). Samples are thus taken from one pocket in ten.

Fixed ◗ BREED IN.

Fixed Equipment. In general terms anything on a farm, excluding growing crops, which cannot readily be moved. The term is used by agricultural valuers and others in situations involving farm sales, termination of farm tenancies, etc. It is defined by the Agriculture Act 1947 as including 'any building or structure affixed to land and any works on, in, over or under land, and also includes anything grown on land for a purpose other than use after severance from the land, consumption of the thing grown or of produce thereof, and amenity'. This definition is usually applied in situations involving OWNER-OCCUPIERS. In those involving TENANT FARMERS, improvements carried out by the tenant at his expense are not normally regarded as fixed equipment.

Flag. Cereal leaves.

Flail. 1. A hand implement used to thresh corn, comprising a wooden beating bar or swingle, hinged to or loosely attached (often by leather) to a handle.
2. One or more free swinging arms on certain machinery (e.g. FLAIL FORAGE HARVESTER, FLAIL HEDGER) which, when rotated at speed, sever vegetation.

Flail Forage Harvester. A type of FORAGE HARVESTER consisting mainly of one or more swinging arms or flails, each bearing a number of knives, and attached to a horizontal high-speed rotor. The flails are covered by a hood from which a chute delivers the cut forage to a trailer. Flail forage harvesters may be trailed or semi-mounted on a tractor, and offset versions of both types exist which avoid tractor wheel damage to the crop.

Flail Hedger. A type of HEDGE CUTTER commonly used nowadays for regular trimming. They are usually driven hydraulically or mechanically by p.t.o. from a tractor and consist of a high-speed cylinder bearing free swinging FLAILS (*see p. 186*).

Flaked Maize. A highly digestible, starch-rich FEEDINGSTUFF derived from MAIZE which has been steam treated, rolled and dried. It has a relatively low oil content and is often given to pigs.

Flank. The side of an animal between the ribs and the thigh.

Flare ♦ FLECK.

Flash Method ♦ PASTEURISED MILK.

Flat Deck Rearing Piggery ♦ PIGGERY.

Flat Rate Feeding. A system of feeding by which cows are fed a standard fixed quantity of CONCENTRATE per day, from calving to turnout, thereafter depending on grass alone. A variation commonly used with autumn calving cows is stepped feeding, in which the level of concentrate fed to groups of cows in the herd is reduced in one or two steps through the winter, with concentrate feeding being discontinued on turning out to grass. The success of these systems depends on the cows having unrestricted access to good quality forage, usually grass SILAGE. If necessary, the silage intake can be reduced by supplementation with other suitable feedingstuffs, e.g. BREWERS' GRAINS, DRIED GRASS.

Flat Roll. A type of ROLLER with 2 or 3 smooth surfaced cylinders mounted on an axle. (◆ CAMBRIDGE ROLL)

Flatworms. A phylum of worms (Platyhelminthes) with flat segmented bodies including parasitic TAPEWORMS and FLUKES, and free-living planarians.

A hydraulic flail hedge trimmer.

Flax. Varieties of the plant, *Linum usitatissimum*, selected and grown for their long stems rich in FIBRE, used in linen manufacture. Flax is rarely grown now in the U.K. Other varieties of the same species have been selected for their seeds which are rich in oil. (♦ LINSEED)

Flax Brake ♦ BRAKE.

Flea Beetle. A small dark beetle (*Phyllotretra sp.*) which causes extensive damage to BRASSICA seedlings, eating small holes in the cotyledonous leaves (♦ COTYLEDON), especially during hot dry weather between April and mid-May.

Fleam. A veterinary instrument for bleeding cattle.

Fleck. A fat layer surrounding the kidneys of pigs. Also called flare.

Fleece. A sheep's coat of wool. Also the wool shorn off a sheep and maintained in one piece. (♦ SHEARING, WOOL, WOOL ODDMENTS)

Flemish Giant. A very large breed of rabbit, dark grey in colour with white specks.

Flesh Fly ♦ BLOW FLY.

Flick. Rabbit fur. Also waste rabbit fur, used as a nitrogenous FERTILIZER.

Flies. A general term for winged insects, but particularly applied to the two-winged insects of the Order Diptera, the mouth parts of which form a PROBOSCIS for piercing and sucking, especially blood. Many are important in agriculture, causing or transmitting disease to plants and animals. Some are parasitic (e.g. CLEG, GADFLY, KED, etc.) whilst the LARVAE of others attack livestock and crops (e.g. BLOW FLY, CRANE FLY, FRIT FLY, WARBLE FLY, etc.). Clouds of flies, especially when biting, make cattle very restless.

Flight. Oat HUSKS.

Flinty. Used to describe cereals when very ripe and hardened like flint.

Flitch. A SIDE of bacon, salted and cured.

Float. 1. A low cart used to carry livestock or produce.
2. ♦ PLANKER.

Flocculation. The aggregation of small particles to form larger ones. Particularly applied to soils and the formation of CRUMBS. (♦ DEFLOCCULATION)

Flock. 1. A collection or company of animals. Particularly applied to sheep, goats and birds.
2. A lock or tuft of wool (and cotton or hair), especially when waste and used to stuff upholstery, mattresses, etc.

Flock Book. A record for a particular breed of sheep, maintained by the BREED SOCIETY, of registered sheep conforming to the breed type and their PEDIGREE relationships. (♦ HERD BOOK)

Flock Mating. Allowing several COCKS to mate at will with the hens in a flock of poultry, as distinct from using only one cock run with the hens in a pen (pen mating).

Flockmaster. The owner or person in charge of a flock of sheep.

Flora. The plant population of an area or country. Also a botanical ordering of those plants together with descriptions for identification. (♦ FAUNA)

Flour. A powder of finely ground cereal grains, especially wheat, used for breadmaking. Also any similar powder.

Flower. The blossom of a plant containing the reproductive organs, arranged on a central axis or RECEPTACLE in four whorls, viz, the outermost CALYX (the sepals), then the usually brightly coloured COROLLA (the petals), inside which are the STAMENS and the innermost CARPEL(S). (♦ PERIANTH)

Flue Dust. A FERTILIZER containing 5–15% POTASH derived from industry, mainly as waste products from iron and steel manufacturing processes.

Flukes. Parasitic flatworms found in the liver, guts, lung and blood vessels of animals. LIVER FLUKES are especially common in sheep and cattle.

Flukey Area. A badly drained pasture where mud-snails are found, which are the alternative HOST in the transmission of LIVER FLUKE disease.

Flush. 1. A sudden, generally rapid growth of foliage, usually applied to grass.
2. To give an animal a richer diet prior to mating in order to improve its CONDITION.

Fly Wool. Pieces of loose wool that fall from a fleece during shearing. (♦ WOOL ODDMENTS)

Flying Bent. A coarse tussock-forming grass (*Molinia caerulea*) found in wet habitats on MOORLAND, HEATH, FENS and MARSHES, with purple flowers, eaten only by sheep. Also called Purple Moor Grass or Purple Melick-Grass.

Flying Flock. A flock of sheep temporarily imported onto a farm, normally for less than a year, and then sold out. There are three kinds, viz, (*a*) STORE lambs for fattening, also called a dry flock, (*b*) EWE LAMBS to be reared for sale as young ewes for breeding, and (*c*) draft ewes for breeding (♦ DRAFTED ANIMAL), often brought in to graze the remaining grass on pastures vacated by dairy cattle moved to winter housing.

Flywheel Chopper ♦ DOUBLE CHOP HARVESTER.

Foal. 1. A young horse in its first year. A colt foal is an uncastrated male foal, and a filly foal is a female foal.
2. To give birth to a foal.

Fodder. Food supplied to livestock, particularly dry roughages such as HAY and STRAW. Sometimes used loosely to mean FORAGE.

Fodder Beet. A root crop of the genus *Beta* which also includes SUGAR BEET and MANGEL, usually grown following cereals and used to feed stock. It is intermediate between mangel and sugar beet, having a higher DRY MATTER and sugar content, but smaller root than mangel, but a lower dry matter and sugar content and larger root than sugar beet. Fodder beet may be classified into medium dry matter (14–18%) and high dry matter (19–20%) varieties.

Fodder Crop. A crop grown for use as animal feed, e.g. HAY, KALE, RAPE, LUCERNE, TRIFOLIUM, etc., and usually consumed in the green state. Also called green crop or forage crop. There has been a decline in the area of traditional fodder crops since 1960 as farmers have become able to provide winter feed more economically by increased grass production and improved methods of conservation. SILAGE production in particular more than trebled between 1964 and 1978. Apart from grass the main fodder crops now grown are field beans (♦ BEANS), MAIZE and kale, which are principally used to supplement conserved grass for winter feeding.

Fodder Unit ♦ FOOD UNIT.

Foetus. An embryo of an animal at that stage of development in the womb when the main adult features are recognisable.

Fog. Pasture grasses allowed to grow during late summer and autumn, providing winter grazing for sheep and cattle. Land left with such grass on it. Also to graze such grass in winter. (♦ YORKSHIRE FOG)

Foggage. Synonymous with FOG. Also used to describe rough coarse grasses of mountain and hill areas and ROUGH GRAZING.

Fold, Folding. An enclosure or pen for sheep, usually movable and made with HURDLES or wire. Usually used to enclose sheep (and also pigs) in a field with a growing arable crop (e.g. turnips, kale), moved each day to another part of the field or enlarged once the crop in that area has been consumed. This practice is termed folding.

Fold Units. Mobile units for poultry consisting of a house comprising a sleeping area and nest box attachment together with a RUN. Usually designed to accommodate about 20 laying hens, and moved regularly to provide clean land for the hens to spread the faeces evenly over the land.

Followers. 1. Young cows in a dairy herd not yet in milk and still maturing to replace their mothers.
2. Cattle put out to graze a pasture more thoroughly after a dairy herd has taken the best grass.

Following Crop ♦ AWAY-GOING CROP.

Food ♦ FEEDINGSTUFFS.

Food and Agricultural Organisation. An autonomous agency of the United Nations which aims to raise nutritional levels and living standards, particularly in poor countries, to improve the production and distribution of food and agricultural products, and to better the condition of rural populations, thereby ensuring freedom from hunger in the world and contributing to the expansion of the world economy. F.A.O. employs over 3 000 professional planners and technicians working on a wide variety of projects in many parts of the world.

Food Conversion Ratio. The number of kgs of food consumed by an animal required to produce a liveweight gain of 1 kg. A measure of the efficiency of the animal in converting food into flesh, a small ratio indicating high efficiency. Also called feed conversion ratio.

Food Unit. 1 kg of average barley, equivalent to 0.71 kg STARCH EQUIVALENT. Also called a fodder unit.

Foot and Mouth Disease. A highly infectious viral NOTIFIABLE DISEASE of cattle, pigs, sheep and goats, characterised by the development of blisters in the mouth, causing considerable salivation, and on the feet resulting in lameness. Death is not usual, but animals cease gaining weight and milk production in dairy cows drops. Epidemics occur from time to time and the disease has been ENDEMIC in many parts of the world. The disease can be spread by infected bones, offals, birds, etc., and by the wind. Diseased animals and their contacts are slaughtered in the U.K.

Foot-bath. 1. A shallow concrete or wooden trough containing disinfectant solution through which sheep are driven as a preventative or curative measure for various diseases (e.g. FOOT ROT).
2. A shallow bath containing water, with a base of narrowly spaced horizontal pipes, through which cattle are driven, to remove mud from their hooves, and usually containing copper sulphate to harden the hooves.

Foot Rot. 1. A highly infectious bacterial disease (*Fusiformis sp.*) of sheep, affecting the horny parts of the feet and adjacent soft parts. It produces painful swellings with foul-smelling discharges, causing lameness, and can lead to hooves being shed. Sheep are unable to feed properly and lose CONDITION. The disease is common on wet land.
2. Erosion of the horny parts of pigs' feet mainly caused by bruising, injury and wear in damp, dirty conditions, or where

housed on rough concreting, leading to cracks and infection.
3. A fungal disease (*Fusarium solani*) of peas, causing yellowing and wilting of plants.

Forage. Certain crops consumed in the green state by livestock, particularly cattle and horses (e.g. KALE, MAIZE, LUCERNE, etc.) or made into SILAGE. Also to rummage in search of food. Sometimes used loosely to mean FODDER.

Forage Blower. A machine used to fill a SILO, particularly of the tower type, with finely chopped material for making SILAGE, by blowing it up a duct, often to a height of 15 m (50 ft) into the silo. Some kinds of forage blower also chop the crop, these being called cutter blowers.

Forage Box. A large mobile container, usually with wheels, similar to a DUMP BOX, used to transport FORAGE from a SILO to a MANGER. The forage is discharged in a steady even flow, usually by a transverse moving floor mechanism or elevator belt, driven by p.t.o. from a tractor. Also sometimes used to transport forage from the field to mangers for herds under ZERO-GRAZING.

Forage Harvester. A machine which cuts, chops and loads (by chute) green crops (e.g. grass, lucerne, etc.) into an adjacent or

A metred (precision) chop forage harvester picking up cut grass from a windrow. The tractor is fitted with a safety cab.

towed trailer, for subsequent SILAGE making. Sometimes the crop is pre-cut and allowed to wilt in WINDROWS before being collected by a pick-up reel substituted for the cutter mechanism. There are three main types, viz, DOUBLE CHOP HARVESTER, FLAIL FORAGE HARVESTER and METERED CHOP HARVESTERS. They are mainly trailed or semi-mounted but self-propelled machines are available.

Forage Rape ♦ RAPE.

Forage Wagon. A large mobile container, similar to a FORAGE BOX, having a pick-up mechanism at the front and used to collect cut FORAGE from WINDROWS and to transport it to a SILO or to a discharge point for a herd kept under ZERO-GRAZING. There is no specialised unloading mechanism as in a forage box, and the load is discharged from the back of the wagon in the mass.

Force. To provide artificial heat and light and plentiful nutrients to plants, usually in a GLASSHOUSE, so that they flower or produce fruit out of season. Also to cause a plant to grow and ripen rapidly.

Force Feed ♦ SEED DRILL.

Force Moulting. A practice of poultry breeders whereby prolific egg layers selected for breeding are moved to strange quarters in early August and/or their food supply is limited and sometimes restricted to oats only. Within 7–10 days heavy moulting commences and feeding is increased steadily and artificial light provided. This practice ensures the birds are brought into lay early in the season to provide hatching eggs when required. Normally prolific layers do not start moulting naturally until October/November and are not back in lay until later in the season.

Forced Draught. Hot air blown through a GRAIN DRIER as distinct from a natural draught resulting from ventilation.

Foremilk. The first few squirts of milk drawn from a cow's udder by hand into a STRIP CUP before starting milking. This milk has a high bacterial content.

Forest. A large area of uncultivated land covered by trees and underwood. Blocks of woodland in excess of 1 000 acres are classed as forest by the FORESTRY COMMISSION. The original meaning was unenclosed woodland or open, mainly treeless, areas reserved for hunting, usually belonging to the Crown, e.g. the New Forest.

Forestry Commission. Established in 1919. Under the Forestry Act 1967 it now promotes forestry interests, the development of AFFORESTATION and the production and supply of timber in the U.K., with powers to collect and disseminate forestry information, to develop education and training, to conduct research, to provide advice and financial aid to private forestry, to regulate felling, and to control tree pests and diseases. It is also responsible for establishing and managing forests, to provide new material for industry and employment in rural areas. Under the Countryside Acts it provides recreational facilities and is responsible for conservation. It thus has a dual role as both an 'Authority' and an 'enterprise'.

Formalin. An aqueous solution of formaldehyde used as a sterilant, mainly of glasshouse and potting soils and implements, and occasionally used as a disinfectant.

Forracre ◗ HEADLAND.

Fors. Hairs in a fleece, mixed with the wool. The removal of such hairs from wool is called forsing.

Forward. A descriptive term applied to both livestock and crops that are more advanced in their development at a particular time than normal.

Foul Brood. A bacterial disease (*Bacillus sp.*) of bees which kills the LARVAE and causes their decomposition.

Foul-in-the-foot. A bacterial infection (usually *Actinomyces necrophorus*) of the hooves, particularly of the hind feet, of cattle, normally around wounds or cuts. Swelling, heat, pus production and lameness occur. Muddy fields and gateways are particular sources of infection.

Four-tooth Sheep. Sheep of about 18–21 months of age. (◗ TWO-TOOTH, SIX-TOOTH, FULL-MOUTH)

Fowl ♦ POULTRY.

Fowl Cholera. A contagious bacterial disease (*Pasteurella avisepticus*) of poultry, characterised by sudden high fever and profuse green diarrhoea.

Fowl Paralysis ♦ MAREK'S DISEASE.

Fowl Pest. A term mainly used for NEWCASTLE DISEASE. Also another name for fowl plague, a similar viral disease of poultry which is rare in the British Isles.

Fowl Pox. A viral disease of fowls causing warts on the comb, wattles and other parts of the head.

Fowl-sick ♦ SICK.

Fowl Typhoid. An infectious bacterial disease (*Salmonella gallinarum*) of poultry, mainly affecting PULLETS, causing drowsiness and loss of appetite.

Fox. A carnivorous canine predator (*Vulpes vulpes*) which is known to take young lambs and poultry and is a vector of RABIES.

Foxy Hops. Over-mature dried hops with a reddish-brown colour. The colour may also be caused by disease or decay.

Frame. 1. ♦ CLOCHE.
  2. ♦ BEEHIVE and COMB.

Frame Harrow. A type of HARROW with an articulated sectional frame and no wheels, providing flexibility to follow the contours of the land.

Free-grazing. The most common method of grazing stock with few, if any, controls. STOCKING RATE is usually adjusted to the supply of herbage but over- and under-grazing often occurs. Fencing costs are kept to a minimum and intensive stock supervision is avoided. (♦ GRAZING SYSTEMS)

Free-martin. A HEIFER CALF born as a twin with a BULL-CALF, usually sterile.

Free-range. A system of keeping poultry in which the birds are provided with small houses and are allowed to run free over a field or large enclosed area. Each house is usually designed to accommodate 50–150 birds and is raised slightly above the ground. Most poultry are nowadays kept under more intensive systems such as battery cages. (♦ BATTERY HENS)

Pigs may also be kept under a free range system, with breeding sows provided with cheap mobile housing or arks for protection, and having the run of an area of grass. The majority of pigs are also kept under more intensive systems.

French Bean ♦ BEANS.

French Rice ♦ AMEL CORN.

Fresh Eggs ♦ EGG CLASSES.

Freshen. A term applied to a cow as it approaches calving.

Friesian. A breed of black and white cattle originating in Holland and exported across the world, so that some variations now exist. The British Friesian is intermediate between the well-muscled, short-legged, dual-purpose Dutch Friesian, and the Holstein-Friesian (♦ HOLSTEIN) of Canada which produces a high yield of milk and BUTTERFAT. It is characterised by longer legs than the Dutch Friesian, with a deep body, long and wide hindquarters and a large udder. The average annual milk yield in 1978/79 (♦ MILK YIELDS) was 5 511 kg (5 368 litres; 1 179 gal.) with a 3.79% butterfat content. Red and white types are also known. The British Friesian is the most popular dairy breed in England and Wales (*see p. 92*).

Frit Fly. A small shiny black fly (*Oscinella frit*) that lays its eggs on oat seedlings and TILLERS, which are then attacked by the LARVAE. The next generation of flies lay eggs in the GLUMES and the larvae attack the developing grain. Wheat and maize may also suffer from Frit Fly attack.

Frog. 1. A V-shaped wedge of elastic horn on the base of a horse's hoof.
2. A metal casting on a plough, bolted onto the BEAM, and to which are attached the SHARE, MOULDBOARD and LANDSIDE.

Front Loader Arms. Hydraulically operated arms fitted on either side of the front end of a tractor to which various implements may be attached.

Fructose. A monosaccharide sugar (♦ CARBOHYDRATES) present in fruit, nectar and honey. Also called fruit sugar or laevulose.

Fruit. The ripened OVARY of a plant, containing the seeds, classified into true fruits (those derived from the carpel(s)) and false fruits (those also containing parts of other organs).

True fruits may be fleshy or dry. Fleshy fruits include DRUPES such as plums and cherries, and Berries (♦ BERRY) such as tomatoes, gooseberries and currants. Dry fruits are numerous and varied and include the CARYOPSIS of grasses and the pods of peas.

False fruits include apples and strawberries, the flesh of which are actually swollen RECEPTACLES. (♦ SOFT FRUIT, TOP FRUIT)

Fruit and Vegetable Quality Classes. E.E.C. regulations prescribing grade classes for specified fruits and vegetables each including standards for colour, shape, cleanliness, blemish, size, etc., together with packing, presentation and marking. The lowest class (Class III) is only applicable to some commodities when supplies in higher classes are inadequate to meet consumer requirements (♦ HORTICULTURAL MARKETING INSPECTORATE)

Fruit Bud. A BUD which develops into a fruit as distinct from a shoot or leaf.

Fruit Fly. 1. Small yellow-brown flies (*Drosophila sp.*) which feed on fruit.
2. Large flies, e.g. celery fly, gall flies, etc., whose larvae feed on fruits.

Fruit Sugar ♦ FRUCTOSE.

Full Mouthed. A descriptive term for livestock in which a complete set of permanent teeth have grown, e.g. after 3 years in sheep.

Fumigate. To destroy BACTERIA, insects and other pests by exposing the place they are infecting, e.g. a farmbuilding, to poisonous gas or smoke. (♦ DISINFECTION)

Fungicide. A chemical used to destroy fungi and thus control fungal diseases.

Fungus. One of a large group of plants including moulds, mushrooms and yeasts, characterised by lack of CHLOROPHYLL. They are either PARASITES, many of which cause crop diseases (e.g. POTATO BLIGHT, MILDEW, ERGOT, RUST) or SAPROPHYTES which are important in the release of nutrients to the soil from dead plants and animals. They are mainly constructed of thread-like hyphae and are classified into three groups, viz, ASCOMYCETES, BASIDIOMYCETES and PHYCOMYCETES.

Funicle. The stalk attaching an OVULE to the OVARY wall.

Furlong. A measure of distance equal to 220 yards.

Furrow. The groove or trench in the soil made by a plough as the MOULDBOARD turns over the soil in a more or less continuous strip (the furrow slice). A completely ploughed field, before other cultivations are undertaken, is said to be 'in the furrow'.

Furrow Opener. Another name for a COULTER.

Furrow Press. A heavy type of multi-ring ROLLER usually trailed behind a plough, with wedge-shaped rims to some of the wheels (the press wheels) which compress the lower parts of the furrow slices. Sometimes used when preparing light land for wheat following a LEY. Also called press.

Furrow Wheel. A wheel on a trailed PLOUGH which runs in the previously turned furrow.

fym. An abbreviation for FARMYARD MANURE.

G

Gadfly. A large, dark, blood-sucking fly (*Tabanus sp.*) with clear wings and large brilliant eyes. Gadflies attack livestock between May and September, normally on warm sunny days, and their biting can cause distress.

Gadsman. A Scottish term for a person who drives horse-drawn ploughs.

Gait. 1. A term used in some cases for a sheaf of corn set on its end.
2. The manner of walking, running, trotting, etc., of an animal.

Gall. 1. An abnormal growth of plant tissue, usually in response to parasitic infection, mainly by a variety of INSECTS (e.g. certain wasps, saw-flies, midges, APHIDS, and some moth and beetle LARVAE) and MITES, the offspring of which are provided with food and shelter by the swollen tissue. EELWORMS and FUNGI can also cause galls to be produced.
2. ♦ BILE.
3. A painful swelling, particularly on a horse, or a sore caused by rubbing.

Gall Bladder. A small sac in which BILE produced by the liver is stored until required for digestion.

Gallon. A measure of volume or capacity, equivalent to 8 pints or 4.55 litres, used mainly for liquids but also as a dry measure for grain. The British Imperial gallon is equal to 277.3 cubic inches. The American gallon is equivalent to 0.83 British gallons.

Galloway. A breed of usually black (but sometimes greyish-brown), hornless, beef cattle, originating in South-West Scotland, with a short, broad head. Under the thick, long, rough, wavy hair they also have a woolly undercoat providing winter insulation. It is a hardy breed that does well on the rough grazing of infertile, exposed uplands, and can survive outside wintering. They mature slowly but produce good quality carcases without excess fat (*see p. 87*). (♦ BELTED GALLOWAY)

Gambrel. 1. A bent stick once used to hang carcases.
2. A horse's HOCK.

Game. In general terms, any animal particularly valued for sport or its meat. In specific terms, those animals protected by Game Laws. A number of definitions of game are contained in various Acts of Parliament. The traditional meaning is covered by the Game Act, 1831, which lists 'game' as including 'hares, pheasants, partridges, grouse, heath and moor game and black game', and under the Act other species are regarded as game in certain circumstances. Other distinctions in law are made between live and dead game, and wild and tame game. Rabbits are classified as game in two Acts. Deer are included under the Game Licences Act, 1860, and for compensation purposes under the Agricultural Holdings Act, 1948. With few exceptions, game may only be killed by persons possessing a Game Licence. A licence is also needed to sell game. It is illegal to shoot grouse, ptarmigan, black game, partridges and pheasants on Sundays or Christmas Day, and game must not be shot during 'close seasons'. Various quarry species (e.g. snipe, woodcock, wild duck, geese, etc.) are protected by the Protection of Birds Acts, 1954 and 1967, and may only be shot during specified open seasons. Game is regarded as a valuable asset on many farms and is carefully managed for sport shooting.

The Game Conservancy. An independent organisation devoted to wildlife conservation, particularly of game species, and to assisting landowners, farmers and shooting men, to make the best use of their sporting assets in the U.K. and also abroad. A team of scientists carry out research into practical problems concerning game, and wildfowl habitat. Experienced field advisers are stationed throughout the country to give advice on shoot and woodland deer management, and a series of booklets on related subjects is published. Courses, lectures and symposia are also organised.

Game Fowl. Breeds of FOWL descended from ancient fighting cocks, usually of low fertility, but carrying plentiful breast meat.
(♦ INDIAN FOWL, OLD ENGLISH FOWL)

Game Licence ♦ GAME.

Game Tenant. A person who rents the right to shoot or fish in an area.

Gamete. A reproductive cell produced by an animal or plant, usually HAPLOID. A female gamete (e.g. an OVUM in animals, an egg cell within an OVULE in plants) unites in the process of FERTILIZATION with a male gamete (a SPERM in animals, a POLLEN grain in plants) to produce a ZYGOTE which is DIPLOID, and which develops into a new animal or plant. Some animals and plants (e.g. APHIDS, dandelions) produce ova which develop without fertilization into new individuals. These gametes are mainly diploid.

Gammon. The thigh and adjacent parts of a SIDE of bacon.

Gander. An adult male goose.

Gang. 1. A number of employees working together on a particular job, harvesting crops, sometimes called gang labour and operating under a foreman (or ganger).
2. An area of pasture on which cattle are allowed to graze.
3. A set of tools used together, such as a set of DISCS mounted on a single axle in a DISC HARROW.

Gantry. An annex to an OAST, normally with a slatted floor for ventilation, in which green hops are stored in POKES while the kilns are reloaded. Also called green stage.

Gapes. A disease mainly of young chickens and turkeys due to small worms in the windpipe, causing birds to gasp for breath or 'gape'.

Garden. An area of land used for growing vegetables, fruit or flowers. (♦ HOP GARDEN, MARKET GARDEN)

Garden of England. A term applied generally to South-East England, and particularly to Kent, because of its favourable environmental conditions, enabling the growing of a more extensive range of horticultural produce than can generally be grown in other regions. Historically, this produce has been mainly sold through the London markets (e.g. COVENT GARDEN market).

Garget. 1. ♦ MASTITIS.
2. A swelling in the throat of cattle and pigs.

Garner. A GRANARY.

Garron. A small Scottish horse, slightly larger than a Highland pony, cream or greyish-brown coloured, with a black tail and backbone, and used for work on some hill-farms.

Garthman. An old term for a person in charge of cattle in a yard.

Gas Liquor ♦ AMMONIA.

Gas Store. 1. A sealed store in which fruit, particularly apples, are kept under controlled atmospheric conditions. Temperature is kept low, and the atmospheric levels of nitrogen and oxygen increased and decreased, respectively, in order to reduce cellular RESPIRATION so as to retard the rate of maturing and to preserve the fruit.
2. Sometimes applied to a gas-tight cold store in which fruit, particularly apples, are preserved by low temperature and accumulating carbon dioxide released by the fruit during respiration.

Gaskin ♦ Gean.

Gastric Juice. A DIGESTIVE JUICE secreted into the stomach containing various ENZYMES (i.e. LIPASE, PEPSIN and RENNIN) and hydrochloric acid necessary to provide the acid conditions for their action. Also called stomach juice.

'Gate' Money ♦ KEY MONEY.

Gathering. A method of SYSTEMATIC PLOUGHING in which the tractor and plough are turned to the right each time they come out or go into work, so that already ploughed land is circled in a clockwise direction. Usually carried out alternatively with CASTING when ploughing marked LANDS in a field (*see p. 339*).

Gavage. The force feeding of poultry, usually with the aid of a stomach tube. Also called cramming.

Gean. The wild cherry (*Prunus avium*). Also called gaskin or mazzard. (♦ CHERRY)

Gebur. A historical term for a tenant-farmer.

Geese ♦ GOOSE.

Gelbvieh. A breed of cattle developed in Germany and first imported into the U.K. in 1973 for use in beef production. A large-framed breed, golden coloured, with a gentle disposition.

Cows have excellent milking qualities and produce strong, rapidly growing calves. The breed is able to withstand low temperatures and bulls are becoming increasingly popular for CROSS-BREEDING.

Geld. 1. To castrate (♦ CASTRATION), especially a horse.
2. A tax paid by landowners to the Crown under the Saxon and early Norman kings.

Gelding. A castrated male horse.

Gelt. 1. ♦ GILT.
2. An obsolete alternative for GELD.

Gene. A hereditary factor transmitted from generation to generation of plant or animal, responsible for the determination of a particular characteristic (e.g. colour, height, sex, etc.). Genes are arranged in pairs on homologous CHROMOSOMES, each parent contributing one gene to the pair. Sometimes one gene in the pair is DOMINANT over the other recessive gene. Genes are thought to consist essentially of DNA.

Genera. Plural of GENUS.

General Purpose Mouldboard. An almost flat MOULDBOARD, twisted along its length, producing an almost unbroken furrow slice. It works well to a depth of about 20 cm (8 in.), and is particularly used for winter ploughing (allowing the alternate wet and dry conditions and frost action to produce a tilth), and for ploughing shallow soils or preparing seed beds for cereals or LEYS. (♦ DIGGER, SEMI-DIGGER, LEA PLOUGH)

Genetics. The study of heredity.

Genotype. The total genetic constitution of an individual plant or organism including both DOMINANT GENES and recessive genes present. (♦ GENOTYPIC SELECTION, PHENOTYPE)

Genotypic Selection. The selection of animals in stock breeding on account of their GENOTYPE, revealed by their performance following close study of their breeding behaviour and the results of PROGENY TESTING. (♦ PEDIGREE SELECTION, PHENOTYPIC SELECTION)

Gentles. A term for MAGGOTS.

Genus. A taxonomic group in the classification of living organisms, consisting of one or more closely related SPECIES bearing the same generic name, e.g. *Agropyron caninum*, and *A. repens*, bearded and common Couch Grass, respectively.

Geoponics. The study of AGRICULTURE.

Georgic. Relating to agriculture.

Gerber Test. A test used by the MILK MARKETING BOARD to determine the BUTTERFAT content of milk. It involves the use of sulphuric acid and amyl alcohol to break down the PROTEINS which surround the fat globules in milk, releasing the fat for measurement. (♦ BABCOCK TEST)

Germ. 1. A micro-organism causing disease, particularly BACTERIA.
2. A loose term for seed or the nucleus of a seed.

Germicide. A chemical substance that kills GERMS.

Germination. The commencement of growth of a SEED (or SPORE) with the production of roots and shoots (♦ PLUMULE, RADICLE) usually requiring favourable environmental conditions (e.g. adequate moisture, temperature and daylength, etc.). Most crops grown in Britain can germinate at fairly low temperatures but seeds and crops which are sensitive to frost are not usually planted in cold soils. (♦ DORMANCY, GERMINATION CAPACITY, GERMINATION ENERGY)

Germination Capacity. A term used in SEED TESTING for the number of seeds in a sample which germinate in a given time.

Germination Energy. The rate of GERMINATION of a seed.

Gestation Period. The length of time between conception and birth, during which a developing young animal is carried in its mother's womb. For cows about 283 days, for sheep 144–150 days, and for pigs 116–120 days.

Ghyll ♦ GILL.

Gid, Gig ♦ STURDY.

Gigot. A leg of mutton.

Gill. 1. The WATTLE below the bill of a FOWL.
2. A steep-sided valley, wooded glen or brook, in Northern England. Also called ghyll.

Gilt. A young female pig not having produced a LITTER. Also called hilt, yelt or yilt. (♦ SOW)

Gimmer ♦ SHEEP NAMES.

Gin. A spring trap or snare used to catch rabbits. Now illegal.

Girnel. A Scottish term for a granary.

Gizzard. The muscular stomach of a bird in which food is ground up, assisted by swallowed grit and small stones.

Glasshouse. A building constructed almost entirely of glass providing a sheltered environment, and warmer conditions than the open air, for the growing of flowers and vegetables. Artificial heating may also be provided. Also called a greenhouse.

Glasshouse Crops. Much of the glasshouse sector of the horticultural industry has been re-equipped since the mid-1960s. Widespread use is made of units with automatic control of heating and ventilation, semi-automatic control of watering, and carbon dioxide enrichment of the atmosphere. Tomatoes are the most important glasshouse crop and, together with lettuce and cucumbers, represent some 95% of the total value of glasshouse vegetable output. The area under sweet peppers and celery has increased significantly.

Glat. A gap in a hedge. Repairing such gaps is known as glatting.

Glean. To gather corn left by a reaper or by those hand-harvesting a grain crop.

Glebe. Land attached to a church. Also an archaic term for the soil.

Glen. A long, narrow, steep-sided valley in Scotland, usually carrying a river or stream and often wooded.

Gley. A type of soil, usually clayey, developed in a waterlogged site where ANAEROBIC CONDITIONS exist. Gley soils are characteristically grey or blue-grey in colour with rusty mottles in the subsoil.

Gloucester. 1. A type of soft quick-ripening cheese. Double Gloucester is rich orange-red in colour, crumbly and ripens more slowly.

2. A minor breed of dark brown cattle, with a distinctive white stripe down the back, hindquarters and tail and running between the hind legs to cover the udder and belly, and with relatively short horns. The milk contains small fat globules and was once used in the manufacturing of Double Gloucester cheese. The breed is sometimes called Old Gloucester.

Gloucester Old Spot. A comparatively rare, hardy breed of pig mainly restricted to the Severn Valley where it originated. It is characterised by lop ears and a white coat bearing black spots, although breeders have recently produced almost entirely white types. It is a grazing breed able to survive on a poor diet of fallen fruit in orchards and hence once called the 'orchard pig'. Also sometimes called the 'cottage pig'.

Glucose. A monosaccharide sugar (♦ CARBOHYDRATES) found in honey, sweet fruits, grapes, etc. Complex storage carbohydrates (e.g. STARCH, GLYCOGEN) are converted into glucose, the form of sugar used by plants and animals to provide energy (♦ RESPIRATION). Glucose is the primary product of PHOTOSYNTHESIS. Also called dextrose or grape sugar.

Glume. One of two small BRACTS enclosing a grass SPIKELET. The glumes enclosing seeds of cereals and grasses are usually removed by THRESHING but may remain attached as HUSK.

Gluten. A mixture of insoluble PROTEINS present in wheat (and other cereal grains) giving wheat flour its distinctive character. It is a sticky substance remaining after the STARCH has been extracted.

Glycogen. A soluble polysaccharide (♦ CARBOHYDRATES) consisting of chains of GLUCOSE molecules. An energy-rich compound stored by animals mainly in the liver and muscles and converted to glucose when the energy is required (♦ RESPIRATION). Also called animal starch.

Goad. A pointed stick used to urge cattle to move faster.

Goaf. A term used in some parts for a stack of hay, straw, etc. in a barn.

Goat. A RUMINANT (*Capra hircus*) allied to the sheep, mainly kept in Britain in small numbers for milk production although some large herds exist. British milk breeds have been mainly derived from types originating in Switzerland (e.g. BRITISH ALPINE, SAANEN, TOGGENBERG). Goat milk has a higher PROTEIN and BUTTERFAT content than cow's milk and is used in medicine due to its small fat globules and softer curds which are more easily digested. In many countries goats are reared for meat. They browse a wide variety of herbage including shrubs and coarse vegetation, and are therefore sometimes called the 'poor man's cow'. Owing to their browsing ability goats are sometimes used for scrubland reclamation, although in certain parts of the world excessive goat grazing has caused widespread destruction of vegetation. Some wild herds are found on the hills of Scotland, Wales and Northern England.

Goatling. A female goat between one and two years of age.

Gobbler. An adult male turkey. Also called stag turkey or turkey-cock.

Go-down. A cutting in the bank of a stream or river allowing animals to get to the water to drink.

Goggles ♦ STURDY.

Gold Top Milk. Milk produced by CHANNEL ISLAND BREEDS of dairy cattle which must contain a minimum of 4% BUTTERFAT, for which a premium is paid, and which is marketed in bottles with a gold coloured aluminium foil top.

Gonads. The organs or glands in animals which produce GAMETES, i.e. OVARY, TESTIS.

Goose. A larger bird of the duck family (Anatidae). Domestic geese are descended from the Greylag Goose (*Anser anser*), and are mostly kept in small flocks under FREE-RANGE, providing table birds mainly at Christmas. Some large commercial units exist. The term goose is usually applied to the adult female. (♦ GANDER, GOSLING, DUCK)

Gore. A triangular piece of land.

Gosling. A young GOOSE.

Gout Fly. A small yellow and black fly (*Chlorops pumilionis*) related to the FRIT FLY, the LARVAE of which feed, in consecutive generations, on the shoots and the developing EARS of cereals, particularly BARLEY. The leaf sheaths become swoollen (or 'gouty') and twisted. Also called the Ribbon-footed Corn Fly.

Graded Seed ◗ RUBBED SEED.

Grading Up. The establishing of a pedigree herd of cattle by mating pedigree bulls with a core of non-pedigree cows, and subsequently mating each generation of females with further pedigree bulls of the same breed, thereby increasing the percentage of pedigree 'blood' with each generation.

Graft. 1. A shoot of a desired plant variety (a scion) bearing a BUD, joined on to the stem of another variety (the stock) with a weaker or stronger rooting system, as required, so as to develop into a whole plant, usually a tree or shrub, of the same variety as the scion. Grafting is a common method of PROPAGATION (◗ BUDDING). Cherry varieties are often grafted onto Wild Cherry stocks.
2. A ditching spade with a long, narrow, concave blade.
3. A spade's depth. Also called spit.

Grain. 1. Cereal seed, either in bulk after harvesting or an individual seed. Also a growing cereal crop.
2. The smallest British unit of weight, 1/7000 of a pound, 0.0648 gm, equivalent to the average weight of a seed of corn.
3. A tine or prong on a fork.
4. ◗ BREWERS' GRAINS, DISTILLERS' GRAINS.

Grain Drier. A machine used to dry grain after it has been harvested while too moist for safe storage with the possibility of FERMENTATION occurring. There are three main types.

Batch Driers: The grain is held in batches in special containers and dried using a hot air flow, followed by cooling with a cold air flow. The batches may be in bags laid on a mesh or grid floor (platform type – now less popular being labour intensive), or more commonly in bulk form on trays with perforated floors (tray type) or alternatively contained in perforated chambers.

Continuous Flow Driers: The grain is conveyed through the drier in a continuous flow in a shallow bed, either at right-angles to the flow of high temperature air (cross-flow type) or in the opposite direction (counter-flow type)

Storage Driers: These are the most popular type in Britain, involving slow drying with large quantities of warm air until equilibrium is reached. They include bins or silos with perforated floors (in-bin type), or buildings with flat concrete floors overlaid by ventilating air ducts (on-floor type).

The rate of drying varies from 6% per day in silos to 6% per hour in small driers. The safe, long-term storage of grain requires a moisture content of 14%.

Grain Elevator ⧫ ELEVATOR, AUGER.

Grain Lifters. Projecting devices fitted to the CUTTER-BAR fingers of a COMBINE HARVESTER to lift cereals which have been flattened or laid-down by bad weather, so that the crop can be cut and gathered by the combine's reel.

Grain Moisture Content. The percentage moisture in harvested cereal grain. For long-term safe storage grain needs to be dried (⧫ GRAIN DRIER) to 14% moisture content. Above this level there is a possibility that the grain will ferment and heat up, totally spoiling the grain if not well ventilated and occasionally turned. Above 20% moisture content, turning will not prevent heating. Grain to be used as stock feed can be stored in a more moist condition (i.e. 18–24%) in airtight stores, in chilled stores, or by the addition of a preservative, e.g. propionic acid. MILLING WHEAT is normally purchased from producers by millers at about 16% moisture content (for milling at 15.0–15.5% moisture content). Deductions are usually made or premia paid, if moisture content is above or below this level.

Grains ⧫ BREWERS' GRAINS, DISTILLERS' GRAINS.

1 000 Grain Weight. The weight of 1 000 grains of a random sample of a cereal or grass variety. A value quoted in recommended variety lists published by the NATIONAL INSTITUTE OF AGRICULTURAL BOTANY and of particular use to farmers in determining the weight of cereal seed required for precision drilling. (⧫ DRILL)

Graip. A mainly Scottish term for a fork with three or four prongs used for digging potatoes or lifting manure.

Graminae. The botanical term for the grass family, flowering plants with long, narrow leaves and tubular stems, including pasture grasses, CEREALS, reeds (but not SEDGES), bamboo and sugar-cane.

Granary. 1. A farm building for storing GRAIN and other feeding-stuffs (e.g. CAKE, MEAL) and for processing them (i.e. grinding and mixing) before feeding to livestock. Also called a grange.
2. A figurative term for a rich grain-growing area or region.

Grange. 1. A farmhouse with accompanying buildings and stables. Also a country house.
2. A GRANARY.

Grant Aid. Financial aid provided to farmers by the Government or E.E.C. for the development and improvement of farms, usually under specific schemes, e.g. FARM AND HORTICUL-TURAL DEVELOPMENT SCHEME, FARM CAPITAL GRANT SCHEME. Also called improvement grants.

Grape Sugar ◖ GLUCOSE.

Grass. 1. A plant of the GRAMINAE family, but usually excluding the CEREALS.
2. A term used in the general sense for herbage used for grazing or making into HAY, and for MEADOWS, PASTURES and GRASSLAND. (◖ GRASSLAND SPECIES)

Grass Drying. The artificial drying of green FODDER (mainly grass and legumes), nowadays mainly in rotary drums, with a flow of hot air at about 150°C, as distinct from haymaking by natural drying. The dried material is cooled and may be baled, made into wafers of various sizes or milled to produce pellets. These products contain about 6–10% moisture and 10–20% CRUDE PROTEIN.

Grass Staggers, Grass Tetany. A condition, particularly of cattle, which usually occurs a few days after a herd has been turned out onto lush pasture in the spring after wintering indoors. It is due to inadequate magnesium in the bloodstream and is charac-terised by shivering and staggering. Ayrshire cattle are particu-larly susceptible. Also called hypomagnesaemia.

Perennial Ryegrass Italian Ryegrass Cocksfoot

Tall Fescue Meadow Fescue Timothy

The fully developed inflorescences of the main herbage grasses.

211

Grassland. Areas used for grazing stock, consisting mainly of grasses and clovers where cultivated, and also of mosses, lichens, heather, etc., where uncultivated or natural. Grassland in Britain is largely the result of human activity (e.g. forest clearing) and animal grazing, and is well suited to the climate. In the U.K. in 1979, about 1.9 million hectares (4.7 million acres) were under temporary grassland (i.e. under 5 years old); about 5.1 million hectares (12.6 million acres) were under more permanent grass (i.e. 5 years old and over), usually with a high proportion of PERENNIAL RYEGRASS; and about 6.8 million hectares (16.9 million acres) were classed as Rough Grazing (Source: Annual Review of Agriculture, H.M.S.O. 1980. ◗ LAND USE). A wide range of GRASSLAND SPECIES and varieties are used, over half the total supply of herbage seed being produced in Britain (◗ SEED CERTIFICATION).

Grassland production has been enhanced over recent years by the development and application of new techniques, notably the increased use of FERTILIZERS (especially nitrogenous), new methods of grazing control (◗ GRAZING SYSTEMS), improved herbage conservation for winter feed, and irrigation.

Grassland Species. The main species used in permanent or temporary grassland are as follows (most are dealt with alphabetically in the text):–

Grasses: Perennial Ryegrass, Italian Ryegrass, Cocksfoot, Meadow Fescue, Rough-Stalked Meadow-Grass, Timothy, and in some areas Crested Dogstail.
Clovers: Red Clover, White Clover, Alsike Clover, Trefoil, and Trifolium.
Other Legumes: Lucerne and Sainfoin.
Herbs: Burnet, Caraway, Chicory, Ribgrass, Sheep's Parsley, and Yarrow.

It is now common practice to sow a suitable mixture of grasses, clovers and herbs when establishing grassland (*see p. 211*). (◗ SEEDS MIXTURES)

Gratten ◗ STUBBLE.

Grave ◗ CLAMP.

Gravitational Water. That water which freely drains through the soil. (◗ CAPILLARY WATER, IMBIBITIONAL WATER)

Graze. To eat or feed on growing grass and other pasture plants.

Grazier. A person who keeps cattle on GRAZING LAND and fattens them for the market.

Grazing Land. Any area of PASTURE, MEADOW or other GRASSLAND available for stock to GRAZE.

Grazing Systems. There are various methods of pasture management all of which require an appropriate satisfactory STOCKING RATE, aimed at providing an adequate supply of herbage at an efficient rate of usage. (♦ FREE GRAZING, PADDOCK GRAZING, ROTATIONAL GRAZING, SET STOCKING, STRIP GRAZING, TWO-SWARD GRAZING, and ZERO GRAZING)

Grease. An inflammation of the skin of a horse's heels resulting in swelling, itching and the discharge of greasy pus.

Green Bacon. Bacon sold after curing with brine but before being smoked.

Green Belt. A term adapted from the ideas of the early idealist planner Raymond Unwin and given statutory force in the 1938 Green Belt (London and Home Counties) Act which made 'provision for the preservation from industrial or building development' of an area at the immediate fringe of the metropolitan built-up zone. The idea spread to other major urban areas and in 1955 the Government stated in a Ministry of Housing and Local Government circular that the purposes of green belt designation were to check urban growth, to prevent coalescence of neighbouring towns, and to preserve the special character of a town. Green belts are now often the focus of urban-rural conflict. There has never been official acceptance that green belts should serve to conserve farm land, although this is a common view of their current role.

Green Bottle Fly ♦ SHEEP MAGGOT FLY.

Green Crop ♦ FODDER CROP.

Green Feeding. The feeding of a crop to animals in the 'green' or unripe condition.

Green Forages. A group of FEEDINGSTUFFS comprising the grasses and young cereals, some legumes (e.g. CLOVER, SAINFOIN, LUCERNE, VETCHES, etc.) and certain BRASSICAS (e.g. KALE, CABBAGE, RAPE, etc.), green MAIZE, BEET tops, and SILAGE, etc. They characteristically have a higher protein content, particularly when very young, than succulent foods such as roots and tubers.

Green Furrows ♦ STANDING FURROWS.

Green Harvesting. The cutting of a crop in the 'green' or unripe state.

Green Manure. A crop (e.g. ITALIAN RYEGRASS, MUSTARD, etc.) specifically grown for subsequent ploughing-in, in order to provide HUMUS to the soil. Green manure crops are often planted in the autumn after a FALLOW period when no main ROTATION crop is to be planted, to prevent accumulated NITRATES from being leached out (♦ LEACH) during the winter.

Green Meat ♦ SOILING CROP.

Green Pound. The AGRICULTURAL REFERENCE RATE of the Pound sterling, i.e. the rate at which E.E.C. agricultural prices in UNITS OF ACCOUNT are converted to sterling. A notional currency used for E.E.C. agricultural trading purposes. (♦ MONETARY COMPENSATORY AMOUNT)

Green Rate ♦ AGRICULTURAL REFERENCE RATE, GREEN POUND.

Green Sickness ♦ CHLOROSIS.

Green Stage ♦ GANTRY.

Green Top Milk ♦ UNTREATED MILK.

Greenbag ♦ POKE.

Greenfly ♦ APHID.

Greenhouse ♦ GLASSHOUSE.

Greens ♦ GREENSTUFF.

Greensack ♦ POKE.

Greenstuff. Green vegetables for the table, especially CABBAGES. Also called greens.

Grey-back. An old breed of geese with a mostly white body and grey wings and upper legs.

Grey Leaf. A DEFICIENCY DISEASE of cereals, mainly of oats, but sometimes wheat and barley, due to inadequate manganese in the soil. Characterised by grey streaks on the leaves, causing many young plants to die and resulting in poor grain yield. Also called grey speck.

Grey Speck ♦ GREY LEAF.

Greyface. A crossbred sheep derived from a BORDER LEICESTER ram and a BLACKFACE ewe, produced mainly in Scotland. Greyface ewes are crossed predominantly with SUFFOLK rams, producing lambs with a high quality carcase.

Grice. A little pig.

Grid ♦ CATTLE GRID.

Grieve. An overseer on a Scottish farm.

Grinding Corn. Cereal grain used for grinding in a MILL to produce flour or MEAL.

Grip. 1. A drain, small ditch or furrow used to drain surface water off grassland. (♦ GRIPPING MACHINE)
2. A narrow opening made in the soil using a spade into which young plants or cuttings are placed.
3. A DUNGING CHANNEL in a COWHOUSE.

Gripping Machine. A machine, normally drawn by a CRAWLER TRACTOR, similar to a large plough, used to cut surface drains or grips in boggy conditions in hill areas or river valleys, and usually fitted with wide tracks to prevent sinking. The machine normally has an arched frame supporting the cutting mechanism, which removes a continuous strip of soil leaving a U-shaped furrow.

Grist. Cereal grain which is to be or has been ground in a MILL. Also MALT for brewing.

Grit. 1. Small fragments of various substances fed to poultry, viz, (*a*) soluble grit, such as limestone or oyster shell, provided to balance CALCIUM lost by laying birds in producing egg-shells, and (*b*) insoluble grit, such as granite or flint, provided to assist the GIZZARD to grind food.
2. Small woody parts of a pear.

Grits. Coarsely ground GRAIN, particularly oats.

Groats. Grain, particularly oats, after removal of the HUSKS.

Grooming. The cleaning, brushing, combing and general tending of the coat, tail, legs and feet of certain animals, particularly horses and cattle, especially when kept indoors. (◗ CURRY, DANDY BRUSH)

Gross Energy Value. The energy released as heat by a food when a given weight is completely oxidised, usually in a bomb calorimeter. Also known as energy of combustion. (◗ NET ENERGY VALUE, METABOLISABLE ENERGY)

Ground Chalk ◗ CHALK, GROUND LIMESTONE.

Ground Frost. A temperature at ground level of less than 0°C or 32°F. Most plant tissues are not destroyed until the temperature falls well below the freezing point of water.

Ground-keeper. A potato missed during harvesting, which subsequently grows up again.

Ground Limestone. Any of the natural forms of LIMESTONE ground to produce a powder for use in liming. The powder acts to rectify soil acidity more quickly than limestone applied in lumps. The Fertilizer and Feeding Stuffs Regulations 1973 require all ground limestones to carry an indication of the amount which passes through a B.S. 100-mesh sieve. To be officially designated as ground limestone at least 40% must pass through such a sieve, and contain not more than 3% magnesium. Ground magnesium limestone must satisfy the same conditions but contain in excess of 3% magnesium. Both types have a NEUTRALISING VALUE of 50–56. (◗ LIME, LIME REQUIREMENT)

Ground Nut Cake, Ground Nut Meal. CAKE or MEAL derived from the tropical ground nut (*Arachis hypogea*) and available either in the decorticated (♦ DECORTICATION) or undecorticated forms. Used to feed dairy cows and fattening cattle as well as pigs, poultry and sheep.

Ground Rock Phosphate ♦ ROCK PHOSPHATE.

Groundwater. Rain water which has percolated through to underground rock strata from the surface as distinct from draining to streams or rivers as run-off. Such water is the source of well and spring supplies.

Grower. 1. A loose term sometimes applied to producers of horticultural crops.
2. Young poultry between 8 and 20 weeks of age.

Growing-off. A condition of hops in which the BINES easily detach themselves from the rootstocks, due to Hop Canker.

Growing Point ♦ MERISTEM.

Growing Season. The period during the year roughly between spring and autumn when crops grow, usually from seed, to full maturity and ripen for harvesting. It corresponds to the period when mean daily temperatures are above 6°C (42.8°F) and is reduced by 1 day for every 8 m (25 ft) above sea level. A height of 225 m (675 ft) is generally regarded as the limit in Britain for general arable farming, the growing season being one month shorter than for sea level areas.

Growth Promoter. A substance included in an animal feed (♦ FEEDINGSTUFFS) which has the effect of increasing the efficiency of feed conversion (♦ FOOD CONVERSION RATIO) and/or leads to improved daily LIVEWEIGHT gains. Examples include ANTIBIOTICS such as flavomycin, virginiamycin and zinc bacitracin, and mineral copper sulphate.

Growth Regulator. A chemical applied to a plant, which regulates or alters its rate of growth, either by stimulation or retardation. Often applied to control weeds in crops (e.g. HORMONE WEEDKILLERS, TRANSLOCATED HERBICIDES) when growth is severely retarded and death usually results. These chemicals act in a similar manner to AUXINS, the natural plant growth regulating substances or HORMONES.

Grub. A term applied in the general sense to insect LARVAE such as CATERPILLARS and legless MAGGOTS, but sometimes used more specifically for the legged larvae of beetles.

Grub Out, Grub Up. To remove plants from the ground, particularly digging out the roots so that the land is cleaned. Especially applied to ORCHARDS.

Grubber. 1. A hand implement for grubbing out weeds, etc., or for clearing and stirring up the soil, consisting of a long stick with a forked tip behind which is fitted a curled metal pivot.
2. A heavy CULTIVATOR.

Guano. A FERTILIZER rich in NITROGEN and PHOSPHORUS, derived from the accumulated droppings of certain sea birds, found particularly on islands off the Peruvian coast. (◗ FISH GUANO)

Guaranteed Prices. Prices for specific agricultural commodities (currently potatoes, fat sheep and wool) guaranteed to British farmers, and determined each year at the ANNUAL REVIEW. Deficiency payments to producers make up the difference between the average market price and the guarantee price when the former falls below the latter. A similar mechanism applies to certain products (e.g. DURUM WHEAT, oilseeds and olive oil) in the E.E.C. as a whole in which the E.E.C. is not self-sufficient. (◗ INTERVENTION PRICE)

Guernsey. A breed of dairy cattle imported into England from Guernsey in the eighteenth century. Characteristically golden-red coloured (various shades exist) often with some white markings. Long-headed with short curved horns, and leanly built with a well developed udder. The average annual milk yield (◗ MILK YIELDS) in 1978/79 was 3 893 kg (3 792 litres; 833 gal) and the rich creamy milk contains 4.65% BUTTERFAT and is much used for butter-making. Similar in size to the AYRSHIRE, and slightly larger than the JERSEY, having a buff-coloured, as distinct from a black, nose (*see p. 90*). (◗ CHANNEL ISLAND BREEDS)

Guide Price. A TARGET PRICE applying in the E.E.C. to beef and veal. It also acts as a trigger or activator for import control measures and SUPPORT BUYING. (◗ COMMON AGRICULTURAL POLICY)

Gülle System. A practice originating in Switzerland involving the collection and spreading of diluted dung, urine and COWSHED washings onto GRASSLAND being grown for HAY or SILAGE, in order to maintain or increase yields.

Gunter's Chain. A chain, 22 yds long and divided into 100 equal lengths or links, used to measure land.

Gut. The ALIMENTARY canal.

Gypsum. ($CaSO_4 . 2H_2O$). Crystalline hydrated calcium sulphate. Often added to soil to correct saline conditions (e.g. following seawater flooding), the calcium displacing the sodium (♦ CATION EXCHANGE). SUPERPHOSPHATE contains a certain amount of gypsum.

H

Hack. A horse kept for hiring out, especially an old or poor one. Also an ordinary riding horse.

Hackles. Long, narrow shiny feathers on the neck of a COCK.

Hackney. A breed of light DRAUGHT HORSE, usually bay, brown or chestnut coloured.

Haemoglobin. The red pigment in blood cells with which oxygen combines during RESPIRATION. A protein compound containing iron similar in chemical construction to CHLOROPHYLL. Also found in the root NODULES of legumes.

Hag. A firm spot or higher spot in a marsh or bog. Also a place from which PEAT has been dug, often a hole filled with water.

Hagberg Test. A test used mainly by millers to determine the ALPHA AMYLASE content of MILLING WHEAT. A suspension of 7 gm of ground wheat in 25 ml water, in a special tube, is stood in a boiling water bath and stirred. The suspension thickens and subsequently thins if the enzyme is present. After 60 seconds the stirrer is allowed to fall under its own weight and its time of fall is measured. A good milling wheat has a falling time of over 200 secs. Wheats with falling times below 100 secs are not normally accepted for milling. Highly active wheats show falling times of less than 5 secs. Falling numbers are sometimes quoted. These include the total time in the water bath. Thus a falling time of 100 secs is equivalent to a falling number of 160 secs. (♦ FARRAND TEST)

Haggerel ♦ SHEEP NAMES.

Ha-ha. A fence, wall or hedge sunk between slopes or contained in a ditch, for retaining stock.

Hairst. A Scottish term for HARVEST. A hairst rig is a field ready for or being harvested, or a section of such a field. At one time the sections used to be cut in competition.

Hake Bar. The part of a trailed PLOUGH which is coupled to the draught linkage behind the tractor.

Half ◆ HALF-SIEVE.

Half-breed. Any animal of mixed breed. Specifically applied to a cross-bred (◆ CROSS-BREEDING) type of sheep resulting from a CHEVIOT ewe crossed with a BORDER LEICESTER ram.

Half Fallow ◆ BASTARD FALLOW.

Half-long. A cross-bred type of sheep derived from a CHEVIOT ram mated with a BLACKFACE ewe.

Half-sieve. A circular basket used to measure fruit with a capacity of half a BUSHEL.

Half-standard. A fruit tree, the lowest branches of which are about 1.2 m (4 ft) above the ground. (◆ STANDARD)

Halm ◆ HAULM.

Halter. A rope, leather or canvas device fastened to the heads of cattle and horses for holding and leading them.

Ham. The thigh of a pig, usually cured by salting and drying in smoke.

Hames ◆ COLLAR.

Hammel. 1. A ram lamb. (◆ SHEEP NAMES)
2. A small shed.

Hammer Mill. A machine usually housed in a barn, which grinds CEREALS and other FEEDINGSTUFFS. It consists of steel hammers which rotate at high speed in a casing and are surrounded by a perforated screen. When the grain is sufficiently reduced in size by the impact of the hammers it passes through the holes in the screen. Various mill sizes are available with differing screen mesh sizes, and throughput rates vary between 3 tonnes per hour for large powered mills and about 50 kg per hour for small automatic mills. (◆ CRUSHING MILL, PLATE MILL) (*see p.* 222)

Hampshire. A breed of black pig with a white belt or saddle and prick ears, developed in the U.S.A., and thought to have originated from exports of the BRITISH SADDLEBACK breed in the nineteenth century. It is widely used in CROSS-BREEDING programmes (*see p. 328*).

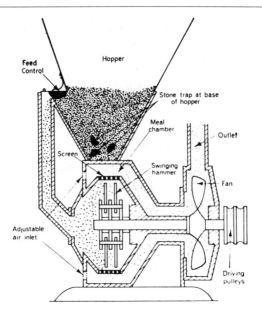

Section through a Hammer Mill.

Hampshire-Down. A large breed of polled sheep originating from crosses between SOUTHDOWN rams and the BERKSHIRE KNOT and WILTSHIRE HORN breeds. Characterised by a relatively large head, well covered with close wool over the forehead and cheeks and round the ears. The face and extremities of the woolly legs are dark brown, almost black. The rams mature quickly, are pre-potent, and used for crossing with other breeds for fat lamb production. The breed is known for its ability to gain weight relatively rapidly. It is numerically less important nowadays.

Hand. A measure of 4 inches (10.16 cm) used for describing the height of horses.

Hand-pin ◆ NIB.

Hanging. The practice of hanging animal and bird CARCASES, particularly game, on hooks at room temperature for a period so that the meat condition improves.

Hank. A length of wool equal to 512 metres (560 yds). (◆ BRADFORD WORSTED COUNT)

Haploid. A term applied to living cells having a single set of CHROMOSOMES in the nucleus. Characteristic of GAMETES. (♦ DIPLOID)

Hard Fruit ♦ SOFT FRUIT.

Hard Seeds. Clover seeds with very tough seed coats preventing GERMINATION and which require scratching in order to do so.

Harden Off. 1. To allow a nursery plant grown in a GLASSHOUSE to acclimatise to the more rigorous outside conditions by transferring first to a CLOCHE or frame, and subsequently to the open air.
2. To gradually reduce the temperature in a BROODER over a period of time until young chicks no longer need artificial heat to survive. Nowadays temperature is usually reduced by 5.5°C (10°F) per week over the first four weeks of life, from about 32°C to about 15.5°C (c. 90°F to 60°F).

Hardpan ♦ PAN.

Hardwoods ♦ BROADLEAF.

Hardy. A term applied to plants and animals capable of surviving the cold during winter.

Haricot Bean ♦ BEANS.

Harrows. Implements bearing TINES used for light, shallow, secondary cultivations, preparing SEED BEDS, covering seeds, and for destroying or collecting weeds. Harrows are similar to CULTIVATORS which have heavier tines and which penetrate the soil more deeply. Many types exist including CHAIN, DISC, DRAG, DUTCH, FRAME, POWER, SPRING-TINED and ZIG-ZAG harrows.

Harvest. The time when ripe or mature crops are cut (e.g. cereals), lifted (e.g. root crops) or picked (e.g. fruit, hops) and gathered in, the whole process being called harvesting (sometimes called ingathering). Sometimes applied loosely to the yield of a particular crop. Most arable crops nowadays are harvested by machine, e.g. COMBINE HARVESTER, POTATO LIFTER, SUGAR BEET HARVESTER. (♦ also FORAGE HARVESTER)

Harvest Year. The first, second, third, etc. year following the SEEDING YEAR of a LEY.

Harvester ♦ HARVEST.

Hassock. A tuft or tussock of tightly packed grasses, reeds or rushes, etc.

Hatchery. A term applied (*a*) to either a cabinet or a walk-in type of INCUBATOR, or (*b*) to premises specialising in the hatching of eggs and production of DAY-OLD-CHICKS, as distinct from the breeding or rearing poultry.

Hatching. The act of a young bird breaking out of an egg, aided by its EGG TOOTH. A HEN will normally sit on her eggs and incubate them to bring them to the point of hatching. Artificial incubation is used in commercial poultry rearing (♦ IN-CUBATOR). A hatch is also a BROOD of chicks.

Haugh A riverside MEADOW or flat alluvial (♦ ALLUVIUM) land in a river valley, particularly in Scotland.

Haulm. The stems and leaves of corn, peas, beans, potatoes, etc., especially after harvesting. Also called halm.

Haulm Killer. A chemical applied to a potato crop to kill the above ground foliage to make harvesting easier.

Hay. A term mainly applied to grasses, but also to LEGUMES, some HERBS, and occasionally CEREALS (♦ GRASSLAND SPECIES), which have been cut and dried (♦ HAYMAKING) and conserved for FODDER. Hay may be classified into two main types, viz, (*a*) seeds hay (sometimes called clover hay), usually derived from a 1- to 2-year-old LEY, at one time mainly from a RED CLOVER – RYEGRASS mixture, and (*b*) meadow hay, derived from permanent GRASSLAND or a long ley. Normally stored in BALE form in a barn nowadays.

Hay Bale Stack. BALES of HAY built into a stack for drying and storage, often under a DUTCH BARN, and usually on a base of rough timbers to raise the bales off the ground. Air spaces are left between the bales to allow moisture to escape and to allow lower bales to expand under the weight of higher ones. When chemical additives which prevent mould growth are introduced during baling, and hay is fully field-dried, bales may be more compactly stacked. Small temporary stacks are usually left in the field for a period to ensure full drying before barn stacking. Field stacks are occasionally thatched with hay or reeds, but more commonly a plastic sheet is used to keep out rain.

Hay Quality. Treatment and weather conditions during HAYMAKING can considerably affect HAY quality and feeding value. Rapid drying is best, reducing cell RESPIRATION and CARBOHYDRATE loss by oxidation. Considerable CAROTENE loss may occur due to excessive bleaching by the sun if the hay is left in the SWATH too long.

PROTEIN and ASH are more concentrated in leaves and are better conserved in hay harvested whilst still in leafy condition. Rain and dew can leach soluble nutrients from swaths, and MOULDING, rotting and heating can occur in hay stored in too moist condition (i.e. above 25%).

Hay Rack. A metal or wooden open frame in which HAY is placed and from which livestock may take what they require.

Haycock. A cone-shaped heap of loose HAY drying in a field. Now usually only found on hill farms and smallholdings where the use of BALERS is not practical. Also called pike.

Hayknife. A large broad-bladed knife with a cross-set handle at one end, used to cut HAY from a HAYSTACK, or SILAGE from a CLAMP.

Haylage. A FODDER CROP wilted in the field, after cutting, to between 50–60% moisture content and then chopped and blown into an airtight SILO. Carbon dioxide released by cell RESPIRATION inhibits bacterial activity and rotting. Drilage is almost identical except that wilting is only allowed to between 60% and 70% moisture content.

Hayloader. A trailed implement similar to a small ELEVATOR used to pick up loose HAY from a SWATH and deposit it onto a trailer for carting to a HAYSTACK. Now seldom used except on some hill or small farms where the use of a BALER is not practical.

Haymaking. The conserving of HAY for use as FODDER, involving drying and the maintenance of nutritive content (◗ HAY QUALITY). The moisture content of hay must be reduced from about 80%, when freshly cut (◗ MOWER), to below 20% to prevent MOULDING, rotting, heating, and possible combustion. There are many methods of haymaking, usually commencing by natural drying in SWATHS in the field, nowadays often turned frequently by a TEDDER to dry the under-surface. Drying may be accelerated by 'conditioning' the hay with a

CRIMPER or chopping with a flail mower. Swaths are sometimes windrowed (◗ WINDROW) at night if rain or a heavy dew is expected. Collection with a PICK-UP-BALER is now normal practice when moisture content is at or below 25% (judged by experience) with further drying in temporary stacks in the field or in a large stack under a DUTCH BARN (◗ HAY BALE STACK). The use of chemical additives introduced during baling, which prevent mould growth, is now becoming popular. Round bales are weatherpoof and are usually allowed to dry slowly in the field over several weeks. Alternatively, hay may be left in the swath for a shorter period and baled at 35–50% moisture content, and dried by various methods of barn drying (◗ BATCH, STORAGE AND TUNNEL DRYING), although power costs have made this an expensive practice. On some hill farms and small holdings haymaking by scythe-cutting, drying in HAYCOCKS and storage in HAYSTACKS is still practiced. On the Continent hay is still dried on small farms by suspending it on wooden frames or racks in the field (tripod hay).

Hay-rake. A hand-rake with a long handle and normally with wooden prongs, used to gather HAY.

Hayrick ◗ HAYSTACK.

Haysel. The haymaking season.

Haystack. A storage stack of loose HAY common before the introduction of the BALER and bale storage. Now usually only seen in hill areas and on small farms where topography precludes the use of a baler. Plastic sheets, as protection against rain, have now replaced thatch roofs on such stacks. Some stacks of baled hay (◗ HAY BALE STACK) are referred to as haystacks. Also called hayrick. (◗ HAYMAKING)

Hay-sweep. An implement, either horse-propelled or fitted to the front of a tractor (and sometimes to an old car), once used commonly to pick up hay from SWATHS or HAYCOCKS and to transport it to a stack. It consisted of a series of wooden prongs about 3 m (10 ft) long and spaced at 30 cm (1 ft) intervals in a frame. Often used in conjunction with an ELEVATOR or stationary BALER. Nowadays most hay is collected using a PICK-UP-BALER.

Hayward. A person in the Middle Ages who had responsibility for fence and hedge repairs and preventing stock breaking through.

Haywire. Twine, normally polypropylene nowadays, used in a BALER to tie bales. Now usually called baler twine.

Hazel. A common shrub (*Corylus avellana*) coppiced in oakwoods and scrub, and in certain areas used for making HURDLES.

He Teg ♦ SHEEP NAMES.

Head Corn. The largest grains in a cereal sample as distinct from the smallest grains, or tail corn.

Headland, Headrig. The strip along the border of a field where a plough is turned during ploughing, and which is itself ploughed when the rest of the field is completed (*see p. 339*). Also called forracre.

Heart. A term applied to fertile soil capable of producing good crops (in good heart) or infertile soil in poor condition (in poor heart).

Heart Rot. 1. The decay of the heartwood of a tree as a result of fungal disease.
2. A disease of BEET due to BORON deficiency, resulting in the death and browning of the root centre and death of young leaves. Similar to BROWN HEART.

Heat Treated Milk. Milk subjected to heating, to destroy pathogenic bacteria and those organisms which cause souring. The method of heat treatment differs for the various designations of milk. (♦ PASTEURISED MILK, STERILISED MILK, ULTRA HEAT-TREATED MILK and FARM BOTTLED MILK)

Heath. An area of open, uncultivated, barren country, with poor ACID SOIL, often sandy or gravelly, and with a characteristic vegetation cover of low shrubs, dominated usually by HEATHER and other ericaceous plants.

Heather. An ericaceous low shrub (*Calluna vulgaris*), characteristically found on HEATHS and MOORS and colouring them purple in August. Favoured as food by the red grouse. Also called ling.

Heavy Cutter ♦ PIG.

Heavy Land, Heavy Soil. Land or soil with a high CLAY content, a high DRAWBAR pull, and harder to cultivate than LIGHT SOIL, and thus sometimes called man's land.

Hectare. A metric unit of land measure. 100 ares or 10 000 sq. metres. Equivalent to 2.4711 ACRES.

Hedge. 1. A close row of shrubs, bushes or small trees forming a fence or field boundary, and often planted or maintained so as to be stock-proof. Hawthorn (*Cretaegus sp.*) is commonly used. Hedges last much longer than wire fences but are much more expensive.
2. To plant, maintain, trim (◗ HEDGE CUTTER) or layer (◗ LAYERING) a hedge.

Hedge Cutter, Hedge Trimmer. A tractor-mounted machine used to trim hedges. There are three main types, viz. (*a*) flail type (◗ FLAIL HEDGER), (*b*) circular-saw type, with a heavy saw blade mounted on an articulated arm driven hydraulically by p.t.o., and (*c*) reciprocating cutter-bar type, also usually operated by p.t.o. There are also various hand-held machines for light or awkward work.

Heeder ◗ SHEEP NAMES.

Heel-in. To temporarily store plants until required for permanent planting by placing them in a trench and packing soil around the roots.

Hefted Sheep. Hill or mountain sheep which are allowed to graze pasture in the same area in which they were born without fencing and which do not stray. A heft is such a group of sheep. (◗ ACCLIMATISED SHEEP)

Heifer. A term usually applied to a female cow over 1 year old which has not borne two calves. A maiden heifer is one that is still virgin. An in-calf heifer is a pregnant one. (◗ FIRST-CALF HEIFER, HEIFER CALF)

Heifer Calf. A female cow less than 1 year old. Also called quey calf or cow calf.

Hemiptera. An order of insects with piercing and sucking mouth parts, including APHIDS.

Hen. A female FOWL over 18 months of age, having completed the first LAYING PERIOD. Usually applied to domestic poultry. Also any female bird.

Heptachlor. An INSECTICIDE of the CHLORINATED HYDRO-CARBON type, similar to and sometimes incorporated in CHLORDANE, used for seed dressing. (◖ DRESS)

Herb. A vascular plant distinguished from a tree or shrub by having a non-woody stem, often used in medicine or for providing scent or flavouring.

Herbage. Herbaceous vegetation. Generally applied to GRASS-LAND SPECIES.

Herbage Seed ◖ SEED CERTIFICATION.

Herbage Seed Subsidy. A subsidy paid under the COMMON AGRICULTURAL POLICY to producers of Basic Seed or Certified Seed (◖ SEED CERTIFICATION) of certain herbage species, field beans and field peas, grown under contract with a merchant or under a special declaration, and harvested between 1st July and 31st December. It is administered by the INTERVENTION BOARD FOR AGRICULTURAL PRODUCE.

Herbicide. A PESTICIDE which kills weeds. The action of a herbicide may be either selective, killing the weeds only and leaving the crop unharmed, or non-selective (or total), in which all vegetation is killed. The latter type are usually applied to fields before planting. Herbicides are normally applied either as foliar sprays (◖ CONTACT HERBICIDE, TRANSLOCATED HERBICIDE) or direct to the soil where they have a residual effect. Applications may be before crop sowing (pre-sowing), after sowing but before crop emergence (pre-emergence), or after crop emergence (post-emergence). (◖ AGRICULTURAL CHEMICALS APPROVAL SCHEME)

Herbivore. An animal which eats grass and other HERBAGE (e.g. cattle, sheep, rabbits, etc.) as distinct from a flesh-eating carnivore or an OMNIVORE.

Herd. 1. A group of animals kept together for management, particularly cattle.
2. A person who looks after a herd or flock, e.g. cowherd, SHEPHERD, etc. Also an abbreviated form of shepherd.

Herd Book. A register of animals conforming to a breed type, compiled by a BREED SOCIETY. The herd book is said to be 'open' when the Society is newly formed and animals are being registered. When the Society becomes firmly established the herd book is 'closed', after which only the progeny of registered animals may be entered.

Herding ♦ HIRSEL.

Herdsman. A person in charge of a herd of cattle. Also a COWMAN'S assistant responsible for rearing calves.

Herdwick. 1. A very hardy mountain breed of sheep, restricted mainly to the Lake District. The face, legs and much of the fleece of lambs are black, but with ageing they become almost white. A stocky, short-legged breed with a long coarse fleece used mainly for carpeting. Only rams are horned, and have a characteristic mane on the neck.
2. A sheep pasture in the North of England.

Heredity. The transfer of characteristics from generation to generation of plants or animals. (♦ GENE)

Hereford. A heavy and hardy breed of cattle, characteristically deep red with white markings on the head, back, chest, belly, legs and tail tassel. A relatively early maturing breed, with short, strong legs, and a short, broad head. Sometimes polled nowadays. A dominant breed throughout the world, known for its grazing ability (*see p. 88*).

Herringbone Parlour ♦ MILKING PARLOUR.

Heterozygous. A genetical term for a plant or animal having a DOMINANT gene for a particular characteristic (e.g. hair colour) derived from one parent and a recessive GENE from the other, as distinct from two identical genes (when termed HOMOZYGOUS).

Hexham Leicester ♦ BLUEFACED LEICESTER.

High Farming. Farming to a high standard with good management and producing good yields, as distinct from low farming in which yields are low, due to poor management or infertile land.

High Temperature, Short Time Method ♦ PASTEURISED MILK.

Higher Voluntary Standard ♦ SEED CERTIFICATION.

Highland. 1. A hardy breed of beef cattle native to and common in the Highlands (◗ 2) and Western Islands of Scotland. Variable in colour, with red, fawn, brown, black and BRINDLED being common. Characterised by thick, long shaggy hair shed in summer to reveal a dense soft undercoat. The breed has long sweeping horns borne on a short broad head, short legs, and a broad deep body. Slow maturing and normally kept on poor mountain grassland although often finished on good pasture. Continuous crossing with the BEEF SHORTHORN has resulted in a new breed, the LUING. (◗ KYLOES)
2. Mountainous or hilly land, particularly in North-West Scotland.

Hile. 1. ◗ STOCK.
2. A local term for an AWN, particularly on barley.

Hill. An individual hop plant, particularly when cultivated on a small mound of soil. Also called stock.

Hill Grazing. Poor quality hill or mountain grassland used to graze sheep and cattle at very low stocking rates. Characterised by various grasses including BENT, FESCUES, *Molinia sp.*, and *Nardus sp.*, and by various other plants including cotton grass (*Eriophorum sp.*), gorse (*Ulex sp.*) and HEATHER.

Hill Land. In general terms any land on hills, mountains, moorlands, etc.. Under the Hill Livestock Compensatory Allowance Order it is officially classified as 'land not less than 3 hectares, being land – situated in an area which is included in the list of LESS-FAVOURED FARMING AREAS .... and consists predominantly of mountains, hills or heath, and – which is, or by improvement could be made, suitable for use for the breeding, rearing and maintenance of sheep or cattle but not, in the opinion of the appropriate Minister, for the carrying on to any material extent, of dairy farming, the production, to any material extent of fat sheep or fat cattle or the production of crops in quantity materially greater than that necessary to feed the number of sheep or cattle capable of being maintained on the land'.

Hilum. The point of attachment of a SEED in a POD, marked after removal or shedding by a small scar.

Hilt ◗ GILT.

Hind. 1. An old term, still sometimes used, for a farm worker or BAILIFF occupying a farm cottage, and in the past bound to supply a female to do field-work for the farmer (♦ BONDAGER). 2. A female deer.

Hinge. A term applied to the uncut soil left by the SHARE or wing of a plough which does not cut a full FURROW width.

Hinny. The progeny of a stallion and female ASS; hardy, but sterile. (♦ MULE)

Hiped Cattle. Those injured by the horns of other cattle.

Hirsel. A group or HEFT of sheep. Also the area over which such a heft is grazing, often under the charge of a SHEPHERD. Sometimes called a herding.

Hitch. The mechanism by which trailers and trailed implements are connected to a tractor for towing. The two main types are the DRAWBAR and the PICK-UP HITCH.

Hive ♦ BEEHIVE.

Hock. The projecting middle joint, of elbow-like appearance, on the hind legs of animals.

Hoe. A trailed implement used to cultivate the soil between row crops to control weeds, consisting of horizontal blades drawn just below the soil surface. Various types exist including TRACTOR HOES, STEERAGE HOES, and front- or mid-mounted hoes which are rigidly connected to the tractor and are raised or lowered hydraulically, and can be operated by the tractor driver. Hand hoes comprising a thin blade on a long handle are used by gardeners to loosen soil and to chop and remove weeds.

Hog. 1. A castrated male pig. Also called a bar or barrow pig. 2. ♦ CLAMP.

Hog Cholera. A term for SWINE FEVER.

Hogg, Hoggerel, Hogget. A male or female sheep between being weaned and being shorn for the first time. Sometimes spelt hog. Also called teg. (♦ SHEEP NAMES)

Hogg Lamb ♦ SHEEP NAMES.

Holder Method ♦ PASTEURISED MILK.

Holderness ♦ DURHAM.

Holding ♦ AGRICULTURAL HOLDING.

Holdover. The practice of a tenant farmer, on quitting an AGRICULTURAL HOLDING, being allowed to temporarily retain possession of some land (e.g. a BOOSEY PASTURE) and parts of the buildings, in order to harvest and market growing crops (♦ AWAY-GOING CROPS), to feed the remaining hay, straw and other FEEDINGSTUFFS to his stock on the holding, and sometimes for lambing. In many cases such customary rights are commuted for cash by agreement with the incoming tenant.

Hollow Land. Recently ploughed land which has not settled and compacted again.

Holm. Flat, fertile land adjacent to a river.

Holstein. FRIESIAN cattle first imported into Canada in 1881 from Holland and bred there to produce a variety with dairy-like characters (i.e. less flesh) and giving high milk and BUTTERFAT yields (♦ MILK YIELDS). Characteristically black and white in colour. Also called the Holstein-Friesian or Canadian Holstein.

Holt. 1. An ORCHARD.
2. A wooded hill.

Holy Grass ♦ SAINFOIN.

Home Farm. A term usually applied to the principal farm on a large estate, often farmed by the estate owner himself, and usually attached to or situated near to the owner's residence. It is distinguished from the other farms on the estate which are usually let to tenants. Also called mains in Scotland.

Home Grown Cereals Authority. A statutory body established under the Cereals Marketing Act 1965, with the aim of improving the marketing of home grown cereals (wheat, barley, oats, rye, and maize since 1970 for certain non-trading functions). The Authority's main non-trading functions include improving market intelligence and statistics and promoting research and development. Trading functions include INTERVENTION BUY-ING, storage and disposal of cereals and oilseed rape, and DENATURING wheat, as an executive agent for the INTERVENTION BOARD FOR AGRICULTURAL PRODUCE. The Authority is financed both by Government and a levy on cereal producers, and consists of representatives of growers, merchants, and users, and independent members.

Home Grown Grain. Cereal crops produced in the U.K. as distinct from those imported. Also cereal crops grown and used on the farm (♦ HOME MIXING) as distinct from those bought-in.

Home Mixing. The mixing of CONCENTRATES and HOME GROWN GRAIN on the farm for feeding to livestock, as distinct from feeding bought-in, ready-mixed COMPOUND FEEDS.

Homogenised Milk. Milk made more digestible by breaking up the fat globules so that they are evenly distributed instead of becoming concentrated as cream.

Homozygous. A genetical term for a plant or animals having two identical GENES for a particular characteristic (e.g. hair colour), one gene derived from each parent. The genes may be both DOMINANT or both recessive. (♦ HETEROZYGOUS)

Honey. A thick, sweet fluid prepared by worker bees in a BEEHIVE from nectar collected from flowers and stored in the honeycomb (♦ COMB) for feeding to the LARVAE. Different honeys with characteristic flavours are produced from nectar derived from different plant species.

Honey Chamber ♦ BEEHIVE.

Honeycomb ♦ COMB.

Honeycomb Stomach ♦ RETICULUM.

Honeydew ♦ APHIDS.

Honeydew Honey. HONEY produced wholly or mainly from secretions of, or found on, living parts of plants other than the flowers, and which is light brown, greenish brown, black or of any intermediate colour.

Hoof and Horn Meal. A FERTILIZER, rich in PROTEIN, consisting of dried and ground animal hooves or horns (or a combination of both). It has a NITROGEN content of 12–14% which is rapidly released when applied to warm soils, normally before planting or sowing. Also called Hoof Meal, and sometimes Horn Meal.

Hoose ♦ HUSK.

Hoover. A type of potato digger with a chain elevator.

Hop. A climbing plant (*Humulus lupulus*) of the mulberry family, characterised by a long, rough twining stem and rough vine-like leaves (hop bine), bearing clusters of bitter, catkin-like fruits or cones (the hops) used for flavouring beer. Hop plants are usually grown in special fields (known as 'gardens' in the South-East and as 'yards' in the West Midlands) where the bines are trained to grow up wires supported by poles (*see p. 237*).

Hops for the brewing industry are grown in Kent, East Sussex, Hampshire and Hereford and Worcester. They occupy only about 6 000 hectares (*c.* 15 000 acres) but have a very high yield (about 1.8 tonnes per hectare in 1979) and value by weight. The area grown is declining slowly as more productive varieties are introduced and as the improved utilisation of hops in brewing has led to a reduction in demand from the brewing industry in Britain.

(Sources: Annual Review of Agriculture, H.M.S.O., 1980; and 'Agriculture in Britain', Central Office of Information, R5961/1977).

Hop-acre. A unit of hop production consisting of 1 200 individual hop plants (HILLS), and sometimes 1 000 plants, depending on the planting distance.

Hop-dog. A hand tool with a wooden handle and serrated iron jaws used to remove poles (♦ HOP) from the ground.

Hop Factor. A dual agent of the HOPS MARKETING BOARD and individual hop growers. They carry out much work on the Board's behalf including providing specially illuminated showrooms where hops are laid out for grading and valuation, and they assist the Board in allocating hops to sales contracts and invoicing these hops to merchant buyers. Advice is also provided to growers on husbandry matters, quota transactions and other issues.

Hop Garden ◗ HOP.

Hop Mildew. A fungal disease (*Sphaerotheca humuli*) of hops, causing white spots on leaves, and attacking the female flowers and preventing CONE formation. Sometimes late attacks occur causing cones to become reddish coloured (red mould).

Hop Pocket ◗ POCKET.

Hoppus Foot. A unit of measurement of round timber equal to 1.273 cubic ft. The extra 0.273 ft compensates for surplus wood assumed lost during sawing to produce squared planks.

Hops Marketing Board. The first Marketing Board to be set up under the Agricultural Marketing Act 1931. It commenced operations in 1932 as a producer-controlled, independent organisation, functioning under the Hops Marketing Scheme. Hops are grown by registered growers under a quota system which matches supply to demand, and which encourages the production of specific varieties, mainly on forward contracts with buyers, which attract priority payments. Hops are weighed, sampled, examined, and certified for quality at Board warehouses. The Board represents the English hop industry in the E.E.C. and in the affairs of the International Hop Growers Convention. It also encourages research, variety improvement and advancement in cultivation and harvesting techniques.

Hop Yard. A hop garden. (◗ HOP)

Horizon ◗ SOIL HORIZONS.

Hormones. Organic substances produced by plants and animals in minute quantities. Plant hormones (e.g. AUXINS) are involved in growth regulation. Animal hormones (e.g. adrenalin) are usually secreted by various endocrine glands into the blood stream, and affect behaviour and a variety of body functions. (◗ GROWTH REGULATOR)

Hormone Weedkiller. A synthetic HORMONE applied to growing crops which acts selectively by causing distorted growth and eventual death to weeds whilst leaving the crops unaffected, e.g. 2, 4-D, M.C.P.A. Usually called a GROWTH REGULATOR.

Horn Meal ◗ HOOF AND HORN MEAL.

Horse Bean ◗ BEANS.

A hop garden, showing hop plants growing up strings supported by wirework and poles.

Horse Fly. A general name for the blood-sucking, two-winged tabanid flies, e.g. BREEZE FLY, CLEG, GADFLY.

Horticultural Capital Grant Scheme. A scheme begun in 1974 and varied subsequently, administered by the MINISTRY OF AGRICULTURE, FISHERIES AND FOOD, providing grant aid to horticultural production businesses towards capital expenditure for improvements to land (e.g. orchard grubbing, land levelling, provision of roads, etc.) and for buildings (except dwelling-houses). Usually abbreviated to H.C.G.S.

Horticultural Marketing Inspectorate. A part of the AGRICULTURE SERVICE of A.D.A.S. responsible for the implementation of the E.E.C. Quality Standards for fresh fruit, vegetables, bulbs, and flowers. The Inspectorate also acts in an advisory role concerning grading, packaging, presentation and labelling. It is responsible for collecting and disseminating market prices and intelligence in respect of the major wholesale markets. Inspectors operate at all points in the distribution chain from the grower to retail outlets, including packhouses, supermarket depots, etc., and also at ports, airports and inland clearance depots.

Horticulture. In strict terms the scientific cultivation of fruit, vegetables, flowers and shrubs. Also used to describe the commercial production of such crops, some on general farms, but mostly on specialised holdings where soil, climate, skilled labour and irrigation can produce maximum yields of high-quality crops and where access to markets or good roads enable their sale at economical prices. Most horticultural enterprises are increasing output per unit area with the help of cultivation and environmental control, and the widespread use of machinery. On some farms the shortage and high cost of labour have led to the introduction of 'pick your own' harvesting for direct sales to consumers.

Hosier System. A system of dairy cattle management pioneered by Mr Hosier, a Wiltshire farmer, in which the cows are grazed on outlying fields and milked *in situ* in a portable milking shed or BAIL. During milking, CONCENTRATES are provided. The bail is relocated every few days and the cows remain in the fields day and night. The system is efficient for outlying fields and obviates the need and expense of transporting FARMYARD MANURE back to the land, and the need for expensive buildings. It cannot be used in unfavourable climatic conditions. (♦ BAIL MILKING)

Host. A plant or animal infected by a PARASITE. Some parasites require two hosts: the primary or definitive host, in which sexual maturity is reached, and a secondary, alternate or intermediate host, in which the rest of the life cycle is completed, e.g. various mud-snails in the case of the LIVER FLUKE.

Houdan. A black and white, five-toed breed of fowl, originating from Houdan in France, with slightly pinkish legs, feathery tufts on the crown of the head and beneath the beak, and producing white eggs.

Housey Bine. Hop BINES which are excessively developed so that the CONES are enclosed in foliage, making spraying and picking more difficult.

Hoven ♦ BLOAT.

Hover. A canopy in a poultry house under which chicks are kept warm by artificial heating.

Hull. The outer covering or HUSK of seeds. Also the pod containing peas and beans.

Huller. A mobile machine which operates like a THRESHING MACHINE and is used to separate the seed from the head of clovers, etc.

Hummel, Humlie. 1. An old term for a hornless cow. Also to remove a cow's horns.
2. To remove the AWNS from barley.

Humus. Decomposed and partly decomposed ORGANIC MATTER in the soil, derived from plant and animal remains as a result mainly of bacterial action, giving a dark colour to the upper SOIL HORIZONS. Humus has colloidal (♦ COLLOID) properties and plays an important role in CATION EXCHANGE, and assists in giving CRUMB structure to soil.

Humus-nucleus. A stable complex of LIGNIN and PROTEIN derived from decaying plant remains, present in the soil.

Hungry Gap Kale. A frost-resistant hybrid variety of KALE with a long thin stem and many side shoots in the leaf axils which, like RAPE KALE, produces young growth in the late spring and is useful as FORAGE before spring grass is fully available. It is susceptible to CLUB ROOT and MILDEW. It yields better than rape kale although it is somewhat less hardy.

Hungry Soil. One requiring plentiful supplies of FERTILIZER to produce good crops, due to lack of ORGANIC MATTER and relative infertility. (♦ SANDY SOILS)

Hurdle. A mobile wooden or metal frame used for temporary fencing, particularly for sheep.

Husbandry. The economic management of a farm.

Husk. 1. The dry, thin, outer covering of certain fruits and seeds, e.g. the GLUMES of cereal seeds.
2. A disease of cattle and sheep due to parasitic nematode worms, and typified by a husky cough and, when severe, difficult breathing.

Hybrid. The offspring of parents of different species, varieties or breeds of plants or animals. They may be fertile or sterile. The likelihood of sterility is increased the greater the difference between the GENOTYPES of the parents, due to the increased chance of imperfect CHROMOSOME pairing.

Hybrid Vigour. Qualities in a HYBRID not present in either parent, e.g. increased hardiness, improved growth rate.

Hydrated Lime ♦ CALCIUM HYDROXIDE.

Hydrogen. (H). An odourless, colourless, tasteless gas, occurring in water combined with OXYGEN, and in all ORGANIC compounds (e.g. CARBOHYDRATES).

Hydrophobia ♦ RABIES.

Hydroponics. The growing of plants (e.g. tomatoes) in nutrient solution without soil.

Hygienic Quality Scheme ♦ MILK QUALITY SCHEMES.

Hypomagnesaemia ♦ GRASS STAGGERS.

I

Icker. An EAR of corn.

Ile de France. A French breed of sheep derived from the MERINO crossed with the DISHLEY LEICESTER breed. Characterised by a white face and pinkish nose, with wool covering the top of the head. Rams are used to provide cross-bred lambs for meat production.

Imago. A sexually mature adult insect; the final stage of METAMORPHOSIS.

Imbibitional Water. Moisture internally absorbed by the clay-humus complex in the soil. (♦ CAPILLARY WATER, GRAVITATIONAL WATER)

Immunisation. The inducing of resistance to an infectious disease or poison in an animal, normally by injecting a dose of VACCINE or ANTITOXIN.

Immunity. The ability of a plant or animal to resist an infectious disease or the effects of a poison.

Improved Dartmoor. A longwool breed of sheep developed by crossing old moorland DARTMOOR sheep with the DEVON LONGWOOL breed. Found in localised areas only, mainly in semi-lowland situations.

Improvement Grants ♦ GRANT AID.

In-. A prefix used descriptively for a pregnant female of the species suggested by the added name of the appropriate young animal, which is being carried in the womb, e.g. in-calf, in-lamb, in-pig, etc.

In Burr. A stage in the growth of hop BINES when the female flowers are fully developed and the STIGMAS are protruding.

In Lay. A hen during an egg LAYING PERIOD is said to be in lay.

In Milk. A cow during a LACTATION period is said to be in milk.

In Season. A female animal in the state ready for mating. Also called 'on heat'.

Inarching. The grafting of a growing branch to a STOCK, without separating it from its parent stem. (♦ GRAFT)

In-breeding. The mating of closely related animals (or plants), e.g. parent with offspring, brother with sister, etc., to increase the number of individuals bearing specific highly desired characteristics. The practice tends to increase the number of HOMOZYGOUS gene pairs, but genetically linked undesirable characteristics are also accumulated and progeny with such characteristics are usually culled. (♦ BREED IN, CROSS-BREEDING, LINE BREEDING, OUT-CROSSING)

In-bye Land. Enclosed fields, particularly those near the farm buildings. Especially applied to hill farms. (♦ OUT-RUN LAND)

Inclosure ♦ ENCLOSURE.

Incubation. The hatching of eggs, either naturally by a hen sitting on them, or artificially by keeping them in an artificially heated INCUBATOR. (♦ INCUBATION PERIOD)

Incubation Period. 1. The period of time required for a newly laid egg to develop to the stage when the young bird is ready to HATCH out. For chicken eggs the period is 21 days.
2. The period between an animal becoming infected by disease-causing germs (e.g. BACTERIA) and the symptoms developing, during which germ numbers increase dramatically.

Incubator. An apparatus used to hatch eggs by providing artificial heat, usually maintained at 37.2°C (99°F) and reduced for the last 2 days of the INCUBATION PERIOD to about 36.1°C (97°F). Humidity is maintained at about 60%. Types vary from small flat ones holding 50–100 eggs, used for a small breeding flock, to large cabinet types holding up to 40 000 eggs in trays or specially designed walk-in rooms holding up to 80 000 eggs, normally used at a HATCHERY. Trays are frequently tilted to turn the eggs before they are transferred for the final 2 days of incubation to a hatching section.

Indian Game. A heavy table breed of poultry, variously coloured, including dark, white, white laced, or mottled (red, white and black), with yellow legs, skin and flesh, and producing tinted eggs. Known for its meat quality and used in crossing. Also called Cornish.

Indigenous ▶ ENDEMIC.

Infected Place, Infected Area. A place (e.g. a surgery, building, garage, etc.) defined in a notice issued by a Local Authority DISEASES OF ANIMALS INSPECTOR or an inspector of the Ministry of Agriculture, Fisheries and Food, which requires the animal(s) therein to be kept isolated. The notice is served when an animal is suspected or known to have disease, and its purpose is to stop the disease from spreading by contact. The period of isolation enables observations and tests to be made to establish the existence of disease or to make arrangements to deal with the diseased animal.

An Infected Area is similar to an Infected Place but the area covered by the notice is much larger (perhaps a whole farm). Similar restrictions apply and its purpose is the same. Some animal movements may be allowed out of the Area at the discretion of a Ministry Veterinary Inspector and under a MOVEMENT LICENCE. (▶ CONTROL AREA)

Infectious Bulbar Paralysis ▶ AUJESZKY'S DISEASE.

Infectious Diseases. Those diseases capable of transmission from animals suffering from the disease to disease-free animals.

Infertile Soil. A soil which is deficient in plant nutrients and requires plentiful applications of FERTILIZER to produce good crops.

Infield. An old Scottish term for a field near the farm buildings under continuous arable cultivation and constantly manured. (▶ OUTFIELD)

Ingathering ▶ HARVEST.

Injurious Weeds. Certain harmful weeds (i.e. spear thistle, creeping thistle, curled dock, broad-leaved dock, and ragwort) which the Ministry of Agriculture, Fisheries and Food may require an occupier to prevent from spreading, by written notice issued under the Weeds Act 1959. Also called Scheduled Weeds.

Inoculation ◗ SEED INOCULATION.

Inorganic. A term applied to a substance of mineral origin, as distinct from an ORGANIC one which contains CARBON.

Ins. The point where a plough is dropped into work at a HEADLAND mark, as distinct from the outs, or point where it is lifted out of work.

Insect(s). A class of ARTHROPOD with bodies divided into a head, thorax and abdomen, the head bearing a pair of feelers or antennae, the thorax three pairs of legs and wings. Adulthood is preceded by three stages: egg, LARVA and PUPA. Insects include FLIES, fleas, BEETLES, APHIDS, BEES, etc.

Insecticide. A pesticide which kills INSECTS. Most of those in farm use today are synthetic organic compounds and include CHLORINATED HYDROCARBONS, ORGANO-PHOSPHOROUS COMPOUNDS and CARBAMATES. They are available in granular, liquid and powder forms and may be applied in various ways. (◗ DUSTING, SEED DRESSING, SPRAYS, also AGRICULTURAL CHEMICALS APPROVAL SCHEME)

Inseminate. 1. To sow seed.
2. To artificially transfer POLLEN to the STIGMA of a plant.
3. ◗ ARTIFICIAL INSEMINATION.

Inside Bats. Poles between the STRAINING POLES in a HOP garden which support the TOP WIRE.

Institutional Prices. The various support prices for agricultural products operating under the COMMON AGRICULTURAL POLICY, e.g. BASIC PRICE, GUIDE PRICE, INTERVENTION PRICE, TARGET PRICE, etc.

Intake. Hill pasture taken-in (fenced) and improved.

Intensive Farming. A method of farming in which the aim is to produce the maximum number of crops per year, of high yield, from the amount of land available and to maintain a high STOCKING RATE for livestock. (◗ EXTENSIVE FARMING)

Intensive Livestock Production. The keeping of certain livestock (e.g. beef, pigs, poultry, etc.) mainly indoors, often in relatively large numbers, with the aim of maximising efficiency by reducing per capita costs (e.g. labour, equipment, feed, etc.) and the area required. A wide variety of management systems exist. (♦ BATTERY HENS, PIGGERIES)

Inter Body Clearance. The diagonal distance between the bodies (see BODY) of a PLOUGH.

Intermediate Host ♦ HOST.

Intermediate Stem Piece ♦ DOUBLE WORK.

Internal Drainage Board or District. Normally, an independent Board established under the Land Drainage Act 1930, with powers to manage the drainage (including maintenance and improvements) of a low-lying district or level (an Internal Drainage District), financed by a rate levied on landowners and tenants, who elect the members of the Board. Usually abbreviated to I.D.B.

Certain I.D.B.'s are incorporated in Regional Water Authorities, in which cases the LOCAL LAND DRAINAGE COMMITTEE looks after drainage management of the Internal Drainage District concerned.

International Federation of Agricultural Producers. The international representative body for agricultural producers, formed in London in 1946, and recognised by the U.N. for consultation on all matters affecting agricultural production and trade in farm products. Its present membership consists of 56 Farmers' Unions and Co-operative Unions from 48 countries. Policy is determined by biennial general conferences, and various standing committees cover the main agricultural commodities.

Internode. The stem of a plant between leaf junctions or NODES.

Intervention Board for Agricultural Produce. A statutory executive body established in 1972, responsible for implementing the regulations of the COMMON AGRICULTURAL POLICY in the U.K. Its functions include the licencing of trade in a range of agricultural produce with THIRD COUNTRIES, INTERVENTION BUYING (sometimes through the agency of other organisations, e.g. HOME GROWN CEREALS AUTHORITY, MEAT

AND LIVESTOCK COMMISSION), and the operation of other market support arrangements. The Board is financed both by Government and the E.E.C. Similar bodies function in other E.E.C. member states.

Intervention Buying. An E.E.C. price support mechanism for agricultural produce (also called support buying) designed to ensure that farmers' returns are not unduly depressed by the existence of occasional surplus production within the E.E.C. It involves 'intervening' in the market, when the free market price of a commodity falls some way below the TARGET PRICE level (or BASIC PRICE for fruit and vegetables), by the purchase and storage of the particular agricultural commodity by the E.E.C. (in Britain through the INTERVENTION BOARD FOR AGRICULTURAL PRODUCE). The commodity is subsequently resold either within the E.E.C. or to a THIRD COUNTRY, is disposed of as food aid or in some cases is destroyed, in such a manner as to avoid depressing the free market price. (♦ COMMON AGRICULTURAL POLICY, COMMON PRICES, INTERVENTION STOCKS)

Intervention Centre. A place where INTERVENTION STOCKS are kept in storage, e.g. a cold store for meat, a general store for cereals, etc.

Intervention Price. The price at which INTERVENTION BUYING in the E.E.C. takes place. The price is set at a given level at the ANNUAL REVIEW for individual commodities, and in effect puts a 'floor' in the market.

Intervention Stocks. Various commodities (e.g. beef, butter, skim milk, barley, etc.) which, after satisfactory sampling and testing, have been accepted and purchased, in Britain by the INTERVENTION BOARD FOR AGRICULTURAL PRODUCE through various organisations acting as agents, e.g. HOME-GROWN CEREALS AUTHORITY, MEAT AND LIVESTOCK COMMISSION. Such INTERVENTION BUYING is a mechanism of the COMMON AGRICULTURAL POLICY. (♦ MOUNTAINS, WINE LAKE)

Intestinal Juice. A DIGESTIVE JUICE secreted by the walls of the small intestine containing a variety of ENZYMES, e.g. erepsin, sucrase, lactase, peptidases, LIPASE, etc. Also called *succus entericus*.

246

Intrascope. An optical probe used to measure backfat thickness in the classification of pig carcases. (◊ CARCASE CLASSIFICATION)

Iodophor. An Iodine containing liquid used for TEAT DIPPING to prevent the spread of MASTITIS in dairy cows.

Ion. An electrically charged atom, group of atoms (e.g. the sulphate radical, SO_4) or molecule. Positively charged ions (cations) have fewer electrons, and negatively charged ions (anions) have more electrons, than are needed for electrical neutrality. (◊ CATION EXCHANGE, CATION EXCHANGE CAPACITY)

Ion Exchange ◊ CATION EXCHANGE.

Iron. (Fe). A TRACE ELEMENT which, if deficient in the soil (particularly calcareous soils), can cause DEFICIENCY DISEASE (◊ CHLOROSIS) especially in fruit crops. Also essential for HAEMOGLOBIN formation (deficiency causes anaemia in animals) and for a variety of metabolic (◊ METABOLISM) needs in both plants and animals.

Iron Pan ◊ PAN.

Irrigation. The application of water to soil to provide an adequate supply for crop needs, to increase crop yields or to aid their establishment. Mainly used for grass, vegetables, potatoes and sugar-beet in Britain. Water may be supplied by the mains, by abstraction under licence (◊ ABSTRACTION LICENCE), from a watercourse, or from on-farm reservoirs. Application is measured in inches per acre (◊ ACRE-INCH) or millimetres per hectare, the amount and frequency being related to various considerations including SOIL MOISTURE DEFICIT, soil type (LIGHT SOILS contain less READILY AVAILABLE MOISTURE than HEAVY SOILS), rainfall pattern and EVAPOTRANSPIRATION. The greatest need for irrigation in Britain is in the South-East, where potential evaporation is relatively high and rainfall relatively low. Irrigation methods include the use of rotary sprinklers, rain guns, spraylines, surface channels and underground pipes. (◊ SPRAY IRRIGATION)

Irrigation Efficiency. The ratio of the amount of water used by irrigated crops to the amount actually supplied.

Italian Ryegrass. A very productive biennial grass species (*Lolium multiflorum* var. *italicum*) similar to PERENNIAL RYEGRASS, which grows luxuriantly, particularly in the spring, and is often used to provide an EARLY BITE, after which it quickly recovers. Commonly used for short duration LEYS. Hybrids with perennial ryegrass are also popular (*see pp. 211–12*). (♦ WESTERWOLDS RYEGRASS)

Itch ♦ SCABIES.

J

Jacob. A breed of multi-horned sheep of uncertain origin, usually with two or four horns, although six-horned and polled types are known. Characterised also by a white fleece with large black or brown patches, and normally a white blazed face. The wool is much used for spinning. Once rare, considerable numbers are now found on the Scottish islands (*see p. 394*).

Jersey. A relatively small, early maturing breed of dairy cattle imported into England from Jersey towards the end of the eighteenth century. Various coat colours exist from light fawn to darkish red, almost black, sometimes with white markings. Characterised by a lightish coloured ring encircling the black muzzle, with a small head and dished face. Leanly built with a large well-developed udder. The average annual milk yield (◗ MILK YIELDS) in 1978/79 was 3727 kg (3630 litres; 798 gal.) and the rich creamy milk contains 5.14% BUTTERFAT and is much used for butter making (*see p. 92*). (◗ CHANNEL ISLAND BREEDS)

Jersian. A hybrid type of beef cattle derived from a JERSEY bull crossed with a FRIESIAN cow. Also called an F–J hybrid.

Johne's Disease. An infectious bacterial disease (*Mycobacterium paratuberculosis*), mainly of cattle (other animals including sheep are sometimes affected), characterised by severe inflammation of the intestines, continuous diarrhoea, weakness and rapid loss of condition leading to emaciation.

Jointer. A term for a skim COULTER.

Joint-ill, Joint-felon. A disease of young livestock caused by various BACTERIA entering the body via the unhardened navel, and characterised by abscess formation at the navel and swellings in some limb joints. Also called navel-ill.

June Return. Answers to a questionnaire which farmers must provide, by law, to the MINISTRY OF AGRICULTURE, FISHERIES AND FOOD, giving details about areas and type of crops grown, numbers and kinds of livestock kept, etc.

K

Kade, Kaid ◗ KED.

Kainit. A low-grade FERTILIZER containing 12–30% potash and about 50% common salt, used mainly on sugar beet. Sometimes also containing about 10% magnesium.

Kale. A crop of the CABBAGE group characterised by a strong tall stem and open curled leaves, grown as a FODDER CROP. Fed either in the field or after cutting, or made into SILAGE. There are several types, viz, HUNGRY GAP KALE, MARROWSTEM KALE, RAPE KALE, and THOUSAND-HEADED KALE.

Keb. A ewe that has given birth to a lamb prematurely.

Kebbing ◗ ENZOOTIC ABORTION.

Ked. A brown, flat, wingless, hairy, blood-sucking fly (*Melophagus ovinus*), parasitic on sheep, and causing constant irritation. Also called cade, kade, kaid, sheep louse and sheep tick.

Kedassia Milk. KOSHER MILK supervised by the Rabbinate of the Union of Orthodox Hebrew Congregations. 'Kedassia' is a registered trade mark given by this organisation to products and establishments under its supervision.

Keel. 1. The central vein on the underside of a leaf.
2. The lower petals of pea and bean plants.
3. The ridge of the breast bone of a bird to which the wing muscles attach.
4. To plough with a prominent ridge.
5. ◗ RUDDLE

Keep. Grass or other FODDER CROPS on which livestock are grazed. Under a grass keep arrangement the owner of an area of land grants a licence to another person to occupy the land for the purpose of grazing animals. The licensee is responsible for the animals, unlike AGISTMENT when the owner of the land is responsible. Unless a grass keep licence is for a specified period of less than a year, the arrangement may be converted into a tenancy from year to year under the AGRICULTURAL HOLD-INGS ACT, 1948.

Kemp. Coarse, brittle, white fibres, shed during the growth of a FLEECE, which are difficult to dye. Red kemp is often found in Welsh wool.

Kent ▶ ROMNEY.

Keratin. A tough, sulphur-containing, fibrous PROTEIN. The major constituent of the outer layer of skin and horn.

Kerry. 1. A breed of leanly built dairy cattle once dominant in Ireland, now found mainly in South-West Ireland. Characteristically black, sometimes with a little white on the udder, and with slightly upward curving horns. Tolerant of cold, tough conditions and often kept in areas of poor fertility. Milk yield per lactation averages 3 100 l (675 gallons) with a BUTTERFAT content of 4.0%.
2. ▶ KERRY HILL

Kerry Hill. A breed of sheep developed in the Kerry Hills on the Welsh border, with a black and white speckled face and legs (the markings being sharply defined), a black nose, and wool covering the poll and cheeks. Ewes are polled but rams may sometimes be horned. A highly productive breed with ewes often used for crossing, particularly with rams of DOWN BREEDS, for fat lamb production. Also well known for its soft wool. Sometimes called Kerry.

Key Money. A premium or additional sum of money, over and above the average current market rent, paid to secure a farm tenancy. Sometimes called gate money.

Kibbled. Broken into coarse fragments, e.g. kibbled maize, kibbled lime.

Kid. A young goat less than 1 year old.

Kidney Bean ▶ BEANS.

Kieserite. ($MgSO_4 . H_2O$). Magnesium sulphate, a greyish white crystalline powder containing about 16% magnesium, used as a FERTILIZER. Regarded as a concentrated form of EPSOM SALTS, having less water for crystallisation.

Killing Out Percentage. The weight of a dressed CARCASE as a percentage of the LIVEWEIGHT of the animal (♦ DEAD-WEIGHT). An indication of the amount of meat on a carcase after the removal of the head (except in pigs), limbs, hide, blood and offal. For beef cattle the average is 54–57%.

Kine (also Scottish **Kye**). Old term for cattle.

Kip. The hide of a young animal.

Kirn. 1. A CHURN. Kirnmilk is a Northern term for BUTTERMILK. 2. A Scottish term for the final corn cut in the harvest.

Knacker. An animal slaughterer.

Knives. Reciprocating sections which slide to and fro across the flat faces of the FINGERS on the CUTTER BAR of a MOWER, BINDER or COMBINE HARVESTER.

Knock Up ♦ YELM.

Knotter. A mechanism on a BALER or BINDER which knots the twine binding the SHEAVES. Also called knotter bill.

Kohl Rabi. A fodder crop of the CABBAGE group with a much swollen, turnip-shaped, scarred stem, from which grow long-stemmed leaves. Mainly grown in the East and South-East of England, often as a substitute for the SWEDE.

Kosher Milk. Milk permitted for Jewish consumption. It is super-vised from the commencement of milking by an accredited representative appointed by the Beth Din or acknowledged Rabbinical Authorities, and also during separating, processing, handling and filling of lorry containers and churns, and subsequent sealing in separate bottles. This supervision is to ensure purity and cleanliness according to Jewish custom. The London Beth Din and provincial Rabbinical Authorities arrange for such supervision especially for the Festival of Passover, and during the year where possible. (♦ KEDASSIA MILK)

Kye ♦ KINE.

Kyloes. A mainly black, shaggy haired, small breed of cattle once found on the Hebridean islands and which contributed, with mainland cattle, to the development of Highland cattle. Sometimes used to mean HIGHLAND cattle.

L

Lace. A lighter or darker coloured border round the edges of some feathers.

Lactation. The period during which a female animal is secreting milk from the MAMMARY GLANDS. In cows and goats kept for milk production the period is greatly increased by regular milking, without which secretion ceases. The optimum lacation for dairy cows is 10 months (305 days) from calving to when twice daily milking ceases, thus allowing a 2 month rest (dry period) prior to calving again. Milk yield during lactation varies with a peak at about 10–12 weeks after calving. Various factors affect the yield and quality of milk including breed, diet management and length of dry period. Average yields per lactation for the main dairy breeds are given under MILK YIELDS.

Lactic Acid. An organic acid formed by bacterial action on LACTOSE in milk, causing souring.

Lactory. A MILKING PARLOUR.

Lactose. A disaccharide sugar (♦ CARBOHYDRATES) found in milk. A compound of GLUCOSE and galactose molecules, into which it is split by the enzyme lactase contained in INTESTINAL JUICE.

Laevulose ♦ FRUCTOSE.

Laid Crop. One, particularly a cereal crop, flattened in the field by the weather conditions, e.g. wind or rain, or because the stem is not strong enough to support the grain vertically. Also called a lodged crop.

Laid Hedge. A HEDGE that has been layered.

Laid Up Field. One reserved for a particular purpose, e.g. kept ungrazed for HAY production.

Lairage. A place where livestock are housed, particularly at markets and docks while awaiting slaughter or export.

Lamb. A young sheep under 6 months old, or the meat derived from it. Also to give birth to lambs.

Lambing Percentage. The number of lambs born to 100 EWES.

Lamina. 1. The blade of a leaf.
2. A sensitive layer lying immediately inside the horny wall of a hoof. (◗ LAMINITIS)

Laminitis. Inflammation of the LAMINA in an animal's hoof, particularly cattle and horses, thought to be an allergic reaction to overfeeding with barley, or to toxic products from afterbirth breakdown or bacteria.

Land Agent. A professional person employed to let farms, collect rents, etc., particularly in respect of large estates.

Land Army. An organisation set up in Britain in 1917 comprising enrolled women who assisted farmers during the First World War. At one time it numbered 20 000 women, each of whom was given a uniform, received training and a wage. It was revived for the Second World War.

Land Classification. The classification of land into categories, variously called grades, quality classes, or CAPABILITY CLASSES, based on the properties of the land and/or its potential agricultural use. In Britain the MINISTRY OF AGRICULTURE, FISHERIES AND FOOD'S classification map recognises five main grades nationally, from grade 1 (land with very minor or no physical limitations to agricultural use) to grade 5 (land with very severe limitations due to adverse soil, relief or climate, or a combination of these). Grade 3 embraces a wide spectrum

Agricultural Land Grades in England and Wales

Grade	%	Comment
1	3	Very versatile land, capable of growing wide range of crops. Mainly located in E. Midlands and
2	14	S.E. England
3	49	Wide spread of agricultural uses.
4	20	Mainly suitable for cattle and sheep production.
5	14	Mainly rough grazing

of land quality into which the majority of land in Britain falls, and the Ministry is now preparing maps showing three sub-divisions (sub-grades 3a, 3b, and 3c). The limitations of the grades may affect the range of crops that can be grown, yield level or consistency, and growing costs (see table *Agricultural Land Grades*). The maps also indicate urban areas, woodland and public open spaces. (◗ LAND USE)

Land Commission ◗ AGRICULTURAL LAND COMMISSION.

Land Drainage. 1. The construction of DRAINS in or under a field to remove surplus water from the land to a DITCH. Such drainage stabilises soil structure, improves aeration and root development, allows the soil to warm up more rapidly, promotes early crop growth, and lengthens the growing season for grass and arable crops. It also reduces the incidence of certain plant and animal diseases, and weed growth.
2. The removal of surplus water from land via ditches, streams and rivers, which may be under the responsibility of an INTERNAL DRAINAGE BOARD or a LAND DRAINAGE COMMITTEE.

Land Drainage Committee. A statutory committee of a Water Authority responsible for LAND DRAINAGE and sea defences. Regional committees have jurisdiction for whole Water Authority regions and local committees over specific river catchment areas. Membership is representative of County Councils and agricultural interests, with the majority of the former. (◗ INTERNAL DRAINAGE BOARD)

Land Drainage Service. A division of the AGRICULTURAL DEVE-LOPMENT AND ADVISORY SERVICE concerned with LAND DRAINAGE (including flood protection) from field to sea, with agricultural water supply and with fisheries harbours. Advice, mostly in connection with work receiving Government grant aid, is given to landowners, farmers and horticulturalists, and the Service works closely – also mostly on grant work – with public authorities. In May 1980 the Minister of Agriculture, Fisheries and Food announced proposals to merge this Service with the LAND SERVICE.

Land Quality ◗ LAND CLASSIFICATION.

Land Service. A division of the AGRICULTURAL DEVELOPMENT AND ADVISORY SERVICE providing advice to landowners, farmers and horticulturalists on aspects of rural estate management and resource planning. Advice covers the design and layout of farm buildings and other fixed equipment, grant aid under various capital grant schemes, land use, land economics and conservation. Research and development work is undertaken in these areas. The Service is also responsible for the estate management of land held by the Ministry of Agriculture, Fisheries and Food. In May 1980 the Minister announced proposals to merge this Service with the LAND DRAINAGE SERVICE.

Land Settlement Association Ltd. An organisation which manages, on behalf of the Minister of Agriculture, Fisheries and Food, ten estates of statutory SMALLHOLDINGS which are situated in various parts of England. They are organised under a scheme providing centralised services for the tenants, such as purchasing and marketing, on a commercial basis. Production is by the intensive cultivation of horticultural crops, mainly under glass. The Association also manages ORCHARDS in East Anglia on behalf of the Minister.

Land Use. The main uses of land in the U.K. are shown in the following table:

U.K. Land Use in 1971

Use	England & Wales %	Scotland %	Britain %	U.K. %
Agriculture	77	81	78	78
Cropland	38	16	31	30
Permanent Grass	26	6	19⎫	
Rough Grazing	13	59	28⎭	48
Woodland	7	9	8	8
Urban Land	11	3	8	8
Other Land	5	7	6	6
Total Land Area (m.ha)	15	8	23	24
Population (millions)	49	5	54	56

(Source: Dr. R. Best, Wye College, University of London. Figures are for 1971, the year of most recent comprehensive data.)

M.A.F.F. estimates in 1975, based on JUNE RETURN data over the previous 6 years, indicated that the annual average net transfer of land out of agriculture in England and Wales was 30 800 ha (76 100 ac.), of which 15 300 ha (37 800 ac.) was to urban use.

Types of Farming: The increasing use of intensive methods of production both in crops and in animal husbandry has led to greater specialisation. Three-fifths of the full-time farms in Britain are devoted mainly to dairying or beef cattle and sheep, one in six is a cropping farm and the remainder specialise in pigs, poultry or horticulture or are mixed farms. In England the farms devoted primarily to ARABLE CROPS are mainly in the eastern part of the country – in East Anglia, Kent, Lincolnshire, Humberside and the eastern parts of the northern counties. In Scotland the rich lowlands of the east coast are also primarily arable. Potato and vegetable growing on a substantial scale marks the farming of the FENS (in South Lincolnshire and Cambridgeshire), the alluvial areas around the rivers Thames and Humber and the peaty lands in South Lancashire. EARLY POTATOES are an important crop in Dyfed and South-West England. Elsewhere, horticultural crops are widely ‘dispersed amongst agricultural crops. The areas under these various crops are shown in *The Use of Agricultural Land in the U.K. (see p. 258).*

Dairying occurs widely, but there are concentrations in South-West Scotland, the western parts of England and South-West Wales, where the wetter climate encourages the growth of good grass. Sheep and cattle are reared in the hill and moorland areas of Scotland, Wales and Northern and South-Western England. Beef fattening takes place partly on better grassland areas and partly in yards on arable farms.

In Northern Ireland dairying is the main occupation on 40% of the full-time farms, while a further 35% concentrate on beef and sheep production. The remainder are either specialised cropping and horticultural holdings, intensive pig and poultry units or mixed farms with no predominant single enterprise. Oats and barley are widely grown, mainly for livestock feeding, and the only important cash crop is potatoes.

Grassland: The British climate suits grassland farming. GRASSLAND forms an important section of the economy of most farms, whether as permanent grass, mostly occupying

257

land less suitable for cultivation, or as sown grassland (LEYS), often part of the arable ROTATION. Rough grazings remain as semi-natural grassland used for extensively-grazed flocks and herds, producing young animals for fattening elsewhere. The areas under grass are shown in the table (◗ LAND CLASSIFICATION)

The Use of Agricultural Land in the U.K.

At June each year	Average of 1968–70	1975	1976	1977	1978	1979 (provisional)
A. Crop areas ('000 hectares)						
Total area	19 374	18 978	18 987	18 840	18 846	18 804
of which: Wheat	940	1 034	1 231	1 076	1 257	1 371
Barley	2 352	2 345	2 182	2 400	2 348	2 347
Oats	380	232	235	195	180	133
Mixed corn	63	35	28	24	17	16
Rye	4	6	8	10	9	6
Total cereals (b)	3 740	3 652	3 684	3 705	3 811	3 873
Potatoes	266	204	222	232	214	203
Sugar beet	187	198	206	202	209	214
Oilseed rape	..	39	48	55	64	74
Hops	7	7	6	6	6	6
Vegetables grown in the open	190	198	206	221	211	197
Orchard fruit	68	53	52	50	47	47
Soft fruit (c)	18	17	17	16	17	18
Ornamentals (d)	15	15	14	13	12	13
Total horticulture (e)	291	285	289	302	289	277
Total tillage (f)	4 946	4 816	4 821	4 863	4 932	4 965
All grasses under five years old (g) (h)	2 335	2 138	2 154	2 124	2 069	1 918
Total arable	7 281	6 954	6 975	6 986	7 001	6 883
All grasses five years old and over (i)	4 959	5 074	5 081	5 003	5 002	5 099
Rough grazing: Sole right	5 840	5 429	5 386	5 191	5 169	5 125
Common (estimated)	1 126	1 126	1 126	1 209	1 206	1 212
Other land (j)	..	395	419	451	467	485

(Source: Annual Review of Agriculture. H.M.S.O. 1980.)

(a) The coverage for 1973 and onwards includes all known holdings in the United Kingdom with 40 standard man-days or more (a standard man-day (smd) represents 8 hours' productive work by an adult male worker under average conditions). All holdings with less than 40 smd in Scotland are excluded but in England and Wales and Northern Ireland holdings with less than 40 smd are excluded only if they have less than 4 hectares of crops and grass and no regular whole-time worker. The same criteria applied in Great Britain in the years 1970 to 1972, and in England and Wales in the years 1968 and 1969, except that the threshold for standard labour requirements in those years was 26 smd.

The 1968–69 figures for Scotland related to all known agricultural holdings exceeding one acre (0.4 hectares) in extent. The figures for Northern Ireland for these years related to holdings of one acre (0.4 hectares) or more, except for numbers of livestock which were collected from all owners, irrespective of the size of the holding, as well as from land-less stockholders.

The introduction of the changes of definition in Northern Ireland in 1973, following similar changes in Great Britain which excluded some 14 000 statistically insignificant holdings in 1970 and about 8 000 in 1973, had the net result of eliminating about 6 000 or so holdings from the Northern Ireland census.

(b) Cereals for threshing, excluding maize.

(c) Includes small area of soft fruit grown under orchard trees in England and Wales.

(d) Hardy nursery stock, bulbs and flowers.

(e) Most of the difference between total horticultural area and the sum of individual sectors is made up by the glasshouse area.

(f) Includes area of other crops and bare fallow not shown in the table.

(g) Includes lucerne.

(h) Before 1975 collected as:
    In England and Wales—'clover, sainfoin and temporary grasses';
    In Scotland   —'grass under 7 years old';
    In Northern Ireland —'1st, 2nd and 3rd year'.

(i) Before 1975 collected as:
    In England and Wales—'permanent grass';
    In Scotland   —'grass 7 years old and over';
    In Northern Ireland —'4th year or older'.

(j) Returns of 'other land' were collected for the first time in England and Wales in June 1969. From June 1969 to June 1973 'other land' in Great Britain was collected as woodland and areas under roads, yards, buildings, etc., the use of which was ancillary to farming of the land; in Northern Ireland it included land within agricultural holdings which was under bog, water, roads, buildings, etc., and waste land not used for agriculture. In June 1974 the definition was changed in England and Wales to include all other land forming part of the holding and in Scotland it was extended to include ponds and derelict land. The Northern Ireland definition is unchanged.

Landrace. A breed of pig native to Denmark and imported into Britain from Sweden in 1949 for breeding. It is white, with lop ears, a small head and a long body. Second only to the LARGE WHITE numerically and kept mainly for bacon production. One of the main breeds used in CROSS-BREEDING (*see p. 328*).

Lands. Strips or divisions marked out in a field due to be ploughed by dividing ridges or RIGS, each subsequently ploughed systematically (▶ CASTING, GATHERING). The width of each land is determined according to tractor and plough size, e.g. 40–50 m wide for a 4-furrow plough.

Landside. A long piece of metal fitted to the FROG of a plough which bears against the furrow wall on the unploughed side and resists the force of the MOULDBOARD as it turns the furrow, thus providing lateral accuracy and stability. The landside sometimes includes a secondary spring-loaded, wheel-like part which rolls and reduces wear (*see p. 338*).

Lanolin. A grease derived from sheeps' fleeces used in preparing ointments, cosmetics, etc.

Large Black. A dual-purpose, hardy breed of large pig, with fine silky black hair, and long lopped ears which obscure vision. Sows are prolific, producing blue or blue and white offspring when crossed with a white boar. A docile breed which feeds well on grass, but is seldom seen nowadays.

Large White. The most popular breed of pig in Britain, and widely exported. Large with adult sows weighing 204–272 kg (450–600 lbs). White in colour, with a long, deep, smooth, flat-sided body, forward-pointing upright ears, a long nose and slightly dished face. Sows are prolific with good milk production. Mainly used for bacon production and as one of the basic breeds in all CROSS-BREEDING programmes. Also called Yorkshire or Large Yorkshire (*see p. 329*).

Large White Butterfly ♦ CABBAGE WHITE BUTTERFLY.

Large Yorkshire ♦ LARGE WHITE.

Larva. A pre-adult form of certain animals. An active stage in METAMORPHOSIS. In insects the egg hatches into a larva (♦ MAGGOT) e.g. CATERPILLAR, and subsequently develops into a PUPA.

Late-flowering Red Clover ♦ RED CLOVER.

Laterals. Side shoots which develop from lateral buds on the branches or stems of trees or shrubs, as distinct from terminal shoots which develop from TERMINAL BUDS.

Lattermath ♦ AFTERMATH.

Layer. 1. A bird that is IN LAY.
2. ♦ LEY.

Layering. 1. Part of the process of hedging in which partly cut stems are bent over and woven with others to produce a new hedge. Also called plashing or pleaching.
2. The pegging in the ground of a branch or half-cut branch from the stem of a plant so that it produces new roots and gives rise to a new plant. A common method of PROPAGATION.

Laying Period. The period or season during which poultry are IN LAY, normally of about 50 weeks duration, and usually commencing at 22–24 weeks of age for egg-producing strains, and 26–28 weeks for BROILER strains. Most egg-producing FOWL are slaughtered after the first laying period as fewer eggs are laid in the second period.

Lea. Open country such as meadow, pasture, or arable land left FALLOW or under grass. Another term for LEY.

Lea Plough. A type of plough with a long pointed SHARE, keeping the cutting edge well in front of the slightly twisted convex MOULDBOARD, thus producing a continuous smooth furrow slice, as distinct from the rough broken slice of the DIGGER PLOUGH. This type only ploughs to about 15 cm (6 in.) depth and is rarely used now.

Leaching. The removal of nutrients in solution from the soil, particularly under heavy rain.

Lead Feeding ◗ CHALLENGE FEEDING.

A layered hedge.

Leaf Roll. A virus disease of potatoes transmitted by APHIDS causing the leaves to curl upwards and inwards, to thicken and turn yellow, and the plants to remain stunted. Yields are reduced by up to 90%.

Leam. An open DRAIN, particularly used in LAND DRAINAGE systems in the FENS.

Leather-jacket ♦ CRANE FLY.

Ledgers ♦ FINGERS.

Legbar. An AUTO-SEX-LINKED breed of poultry produced by crossing the brown LEGHORN and the barred ROCK. Barred in colour with yellow legs, a single comb, and producing white eggs.

Leghorn. A small, nervous, laying breed of poultry, variously coloured including black, blue, brown, black and white, gold and silver, and white, the last being the most popular. Yellow-legged, single-combed, and producing large numbers of white eggs.

Legume. A plant of the pea family (Leguminosae) which produces seed in pods and is characterised by five-petalled flowers and by root NODULES capable of NITROGEN FIXATION. Some are cultivated for their protein-rich seeds (e.g. PEAS and BEANS) for human consumption and for feeding to stock. Others are sown with grasses to produce mixed LEYS (e.g. CLOVER, LUCERNE, etc.). Legumes are often grown as a break between cereal crops to enrich the nitrogen in the soil.

Lehmann System. A system of pig feeding designed to utilise relatively cheap roots or other bulk foods in quantity. The pigs are fed a basic fixed ration of MEAL (a mixture of barley and fish meals) and, as they grow, increasing amounts of roots, mainly cooked potatoes, according to appetite.

Leicester, Leicester Longwool ♦ ENGLISH LEICESTER.

Lemma. The lower of the two BRACTS which enclose the flower of a grass. (♦ PALEA)

Less Favoured Areas. Land, in relation to applications for GRANT-AID, (*a*) situated within the current E.E.C. schedule of such areas, mainly mountains, hills and HEATH, or (*b*) considered by the Government to be, or by improvement, capable of use for breeding, rearing and maintaining sheep or cattle, but not for use for dairy farming, fat sheep or fat cattle production, or crop production other than for livestock maintenance. (◆ HILL LAND)

Let Down. The release of milk from a cow's UDDER during milking activated by the HORMONE oxytocin, the secretion of which can be stimulated by various events such as washing the udder before milking, introducing feed to the mangers, etc. It is a conditioned reflex. Let down normally lasts about 7–10 minutes after which milk extraction from the udder becomes increasingly difficult.

Level. In general terms a flat, usually low-lying area of land. Specifically applied to a DRAINAGE DISTRICT or lowland area in a river valley or FEN with more or less uniform drainage characteristics and usually managed as one unit, e.g. Pevensey Levels, Romney Marsh.

Lewing. A screen of material, e.g. coir mesh, plastic-coated netting or plastic strips, erected on the exposed boundaries of hop gardens (◆ HOP) and ORCHARDS where natural windbreaks do not exist.

Ley. Land temporarily sown to grass (◆ GRASSLAND SPECIES) and ploughed after 1–3 years (short duration ley) or after a longer period, up to 10 years in some cases (long duration ley). SEED MIXTURES are available to suit particular conditions and the type of ley required. Leys may be established by (*a*) UNDERSOWING, either by broadcasting (◆ BROADCAST) or preferably by drilling (◆ DRILL), (*b*) DIRECT SEEDING, (*c*) DIRECT RESEEDING, sometimes with a COMPANION CROP, to encourage early grazing, and (*d*) direct drilling into old grassland previously treated with a herbicide. Short duration leys are an important element in crop ROTATION.

Ley Corn. A cereal crop which follows a LEY in a ROTATION.

Ley Farming. A system of farming in which several fields (and sometimes an entire farm) are cropped as LEYS with regular reseeding, or, alternatively, any ROTATION or cropping system which includes leys. Also called Alternate Husbandry.

Lichen. A lowly type of plant comprising an alga (♦ ALGAE) and a FUNGUS living together symbiotically (♦ SYMBIOSIS). Commonly found on tree trunks, old walls, bare rock, etc. Lichens are primary colonisers of rocks and important in soil formation.

Lift. To harvest root crops and potatoes etc., by digging them from the ground, either mechanically (♦ POTATO HARVESTER, SUGAR BEET HARVESTER) or by hand, as distinct from harvesting by cutting.

Lifting Shares. A device commonly used on a SUGAR BEET HARVESTER, consisting of a pair of triangular-shaped steel SHARES, which lift the beet from the ground onto the elevator section to be deposited in a trailer.

Light. Livestock are said 'to go light' in various ways, e.g. (*a*) a dairy cow when a teat dries up, (*b*) an animal when it becomes lame in one foot, (*c*) a fowl when it develops tuberculosis.

Light Soil. Soil with a high proportion of SAND, a low DRAWBAR pull, and easier to cultivate than a HEAVY SOIL, and thus sometimes called boy's land. (♦ SANDY SOIL)

Light Sussex. A variety of SUSSEX poultry characterised by white feathers and black marks on the neck, wing tips and tail. A dual-purpose breed producing good quality white meat and brown eggs.

Lignin. A complex ORGANIC substance deposited in plant tissue, particularly in woody plants, and often combined with CELLULOSE, giving rigidity and strength to stems and tree trunks. It comprises about 25–30% of the wood of trees.

Lime. Various CALCIUM containing materials applied to the soil to raise pH and correct acidity (♦ ACID SOIL) and mainly derived from natural deposits of CHALK and LIMESTONE. The main types include GROUND LIMESTONE, ground chalk, BURNT LIME, hydrated lime (♦ CALCIUM HYDROXIDE) and various by-products and waste materials mainly containing CALCIUM CARBONATE. Applying lime to soil improves soil structure, provides calcium (and sometimes magnesium) for plant nutrition, and makes other nutrients (e.g. NITRATES, PHOSPHATES and POTASH) freely available to plants, although if over-supplied some minor nutrients (e.g. BORON, MANGANESE) can be made unavailable. (♦ LIME REQUIREMENT, NEUTRA-LISING VALUE)

Lime Requirement. The amount of LIME needed to raise the pH in the top 15 cm (6 in.) of soil to about 6.5. It is expressed in tonnes per hectare (or cwt per acre) of CALCIUM OXIDE or CALCIUM CARBONATE. The amount varies according to the acidity and type of soil, e.g. HEAVY SOILS require more lime than LIGHT SOILS. (♦ NEUTRALISING VALUE)

Limestone. 1. Various types of sedimentary rock with a high CALCIUM CARBONATE content, as much as 99% in the case of CHALK. Most types have an organic origin, particularly chalk, containing the hard calcareous remains of small marine organisms. Some types are relatively hard, e.g. the carboniferous limestone of the Derbyshire Pennines; others are softer, e.g. the magnesian limestone of Yorkshire. All are soluble in water containing carbon dioxide. Limestone is an important source of various types of LIME.
2. A moorland breed of sheep, indigenous in the Southern Pennines, which contributed to the development of the DERBYSHIRE GRITSTONE. Now numerically unimportant.

Limousin. A French breed of beef cattle recently imported into Britain. Red in colour with somewhat lighter tanned legs. Large bodied, rapid growing, and noted for producing a carcase with a high meat to bone ratio (*see p. 89*).

Linch ♦ LYNCHET.

Linch Pin. One of the pins used to lock an implement onto the three-point-linkage of a tractor.

Lincoln Curly Coat. A now extinct breed of pig with a curly white coat, lopped ears and a heavy face which contributed to the development of the CHESTER WHITE.

Lincoln Longwool. The largest and heaviest breed of sheep in Britain. Noted for its long shiny wool which grows over the forehead and falls over the eyes and is much used for lustre yarns and sheepskin rugs. A polled, white-faced breed, which has been widely exported.

Lincoln Red. A large breed of uniformly deep-red coloured cattle developed in Lincolnshire, and closely related to the SHORTHORN, with short, forward and downward curling horns and a short, broad head and with a long muscular body and short stocky legs. Formerly regarded as a dual-purpose breed, it is now popularly crossed with dairy breeds to produce beef calves.

Line Breeding. The mating of related animals, but not those closely related such as cousins or parents, with off-spring. (♦ IN-BREEDING, OUT-CROSSING, CROSS-BREEDING)

Liner. A synthetic rubber tube which lines the TEAT CUP of a milking machine.

Ling ♦ HEATHER.

Link. A measure of length equal to 7.92 in. A 1/100th part of a CHAIN.

Linkage. The association of two or more GENES, usually near each other on the same CHROMOSOME, so that they tend to be inherited together. The characteristics passed on by such genes are said to be linked.

Linnaen System. A binomial system of naming plants and animals developed by the Swedish naturalist, Carl Linnaeus, in which the first name represents the GENUS and the second the SPECIES, e.g. *Hordeum sativum* (barley).

Linseed. Varieties of the plant *Linum usitatissimum*, selected and grown for their oil-rich seeds, which are sometimes fed after crushing, or after scalding or cooking, particularly to young calves and sick animals, but more usually as Linseed CAKE which puts a 'bloom' on fattening cattle. (♦ FLAX)

Lipase. An ENZYME which splits FATTY ACIDS into alcohol and acid. It is present in INTESTINAL JUICE and PANCREATIC JUICE.

Liquid Manure. The urine of farm animals, DUNG LIQUOR and drainings from farm buildings containing valuable NITROGEN and POTASH, at one time mostly collected in tanks and distributed on fields or over solid dung heaps. Nowadays urine and faeces are commonly mixed with litter to produce FARMYARD MANURE. SLURRY is another type of liquid manure.

Liquid Nitrogen Fertilizers. Various nitrogen FERTILIZERS applied in liquid form, usually less concentrated than the granulated, powder and crystal forms, e.g. a solution of UREA and AMMONIUM NITRATE containing 25–30% nitrogen, and often applied as a TOP DRESSING, particularly to grassland and winter cereals. Other nitrogenous fertilizers used in liquid form include gas liquor and pressurised anhydrous ammonia. (◗ AMMONIA).

Litter. 1. Bedding for livestock, e.g. straw, shavings, sawdust, etc. (◗ DEEP LITTER)
2. All the young animals born to a female at one time, particularly applied to pigs. The average litter size is 11 piglets, but can be as many as 20 or more, and over 10% die in their first few days due to cold or crushing. Well-managed breeding sows produce between 2.0 and about 2.6 litters per year depending on the age of weaning of piglets.

Liver Fluke, Liver Rot. A parasitic flatworm (*Fasciola hepatica*), the common fluke, which infests the liver of various animals, particularly sheep and cattle, causing bile duct inflammation, diarrhoea, wasting, and reduced milk yield in dairy cows. Eggs pass out with the faeces, hatching to produce larvae which infect various species of mud-snails (the alternative HOST), and after multiplication they leave the snails to form cysts on grass, to infect grazing animals. A very widespread disease of cattle and sheep. Also called bane and cord.

Livestock. Domesticated animals such as cattle, horses, pigs, poultry and sheep, etc. Defined in the AGRICULTURE ACT 1947 as including 'any creature kept for the production of food, wool, skins or fur, or for the purpose of its use in the farming of the land'.

Livestock Units. The different classes of grazing animals at varying ages and levels of production can be related to a standard Livestock Unit, which is defined by the Ministry of Agriculture, Fisheries and Food as 'the food energy necessary for the maintenance of a mature 625 kg FRIESIAN cow plus the production of a 40–45 kg calf and 4500 litres of milk at 3.6% BUTTERFAT and 8.6% SOLIDS-NOT-FAT per year'. The equivalent for all other classes of grazers is calculated from the appropriate feed energy requirements, and is expressed as the

Livestock Unit value. Stocking Rate is then expressed in terms of Livestock Units per acre or hectare. The grazing Livestock Unit factors currently used by A.D.A.S. are as follows:

Livestock Class	Livestock Unit Factor
Dairy cows	1.0
Bulls for dairy herd	0.65
Beef cows	0.75
Bulls for beef herd	0.65
Other cattle:	
0–12 months	0.34
12–24 months	0.65
Over 24 months	0.8
Ewes and ewe replacements:	
Under 50 kg	0.06
50–70 kg	0.08
Over 70 kg	0.11
Rams	0.08
Lambs:	
Birth to store	0.04
Birth to fat	0.04
Birth to hoggets	0.08
Store lambs (purchased and agisted)	0.04
Horses	0.08

Liveweight. The weight of a live animal as distinct from the weight of the DRESSED CARCASE of the animal. (♦ DEADWEIGHT)

Liveweight Certification Centre. A livestock market or place, approved by the MEAT AND LIVESTOCK COMMISSION, where fat sheep or CLEAN CATTLE are sold live, and presented for certification by an M.L.C. fatstock officer as eligible, respectively, for a fatstock guarantee payment (♦ FATSTOCK GUARANTEE SCHEME) or a beef premium payment (♦ BEEF PREMIUM SCHEME). The animals are then sold by auction or privately, or are subsequently slaughtered for sale in the producer's own butchery business. (♦ DEADWEIGHT CERTIFICATION CENTRE)

Llanwenog. A breed of sheep very similar in appearance to the CLUN FOREST, native to South-West Wales, derived from a cross between the SHROPSHIRE DOWN, and the Llanllwni (a black-faced, horned breed).

Lleyn, Llŷn. A breed of small, polled, white-faced sheep, native to the Lleyn peninsula of North Wales, noted for its prolificacy and high milk yield.

Loam. A soil with a balanced soil particle mixture (approximately 25% CLAY, 40% SAND and 35% SILT), having most of the advantages of CLAY SOILS and SANDY SOILS and few of their disadvantages. Loams are easily cultivated, but can be sticky when wet, and are usually drained. (♦ LAND DRAINAGE)

Local Land Drainage Committee ♦ LAND DRAINAGE COMMITTEE.

Lockjaw ♦ TETANUS.

Locks. A type of wool oddment purchased by the BRITISH WOOL MARKETING BOARD comprising small pieces of wool detached from a fleece during shearing.

Lode. An open DITCH or DRAIN in the FENS.

Lodged ♦ LAID CROP.

Long Dung ♦ FARMYARD MANURE.

Longhorn. A hardy breed of beef cattle, characterised by its long, forward and downward curving horns. Coat colour is variable including roan, red and dark plum, and many intermediate shades. There is usually a white ridge along the spine and tail, and white is also normally found on the face and thighs. A long-bodied, short-legged breed, producing a lean carcase, and milk high in BUTTERFAT, at one time much used for cheese making.

Long Manure. Another term for long dung. (♦ FARMYARD MANURE)

Longwool Breeds ♦ SHEEP.

Lonk. A mountain breed of sheep found in the Pennines in Lancashire and Yorkshire, closely related to the SWALEDALE, and similar to the SCOTTISH BLACKFACE, although generally larger, with a shorter, denser fleece, longer legs and tail, and white markings on the face.

Loose-box. A stable or part of a stable in which animals are kept untied with no other fittings other than a hay-rack, manger, water-bowl, and tying ring(s). Often several are grouped together in a row with two-halved doors, and insulated concrete floors with a slight drainage slope. Conveniently used for isolating diseased animals.

Lop Ears. Ears which hang loosely and often cover the eyes.

Lop-grass ♦ BROME GRASS.

Louping Ill. A paralytic virus disease of sheep, carried by a TICK (*Ixodes ricinus*), common on mountain pastures, particularly in northern England and Scotland. The sheep are caused to twitch, quiver, and stagger, and eventually to leap erratically before paralysis occurs. Also called staggers.

Louse. 1. A wingless, flat bodied, parasitic insect, with short legs, found on the skins of livestock, mostly sucking blood, e.g. KED (sheep louse).
2. ♦ APHID

Low Farming ♦ HIGH FARMING.

Lowland. Low-lying land in a region or district as distinct from hill areas and mountains. (♦ HIGHLAND)

Low Loader. A trailer with its bed near to the ground facilitating easy loading.

Lucerne. A perennial leguminous plant (*Medicago sativa*) grown for both FODDER and FORAGE. Characterised by deep roots (with NODULES) giving it drought resistance, trifoliate leaves, distinct purple flowers, and twisted pods containing kidney-shaped seeds. It is intolerant of acid or poorly drained soils. Also called alfalfa.

Luing. An early-maturing, hardy breed of beef cattle, recently developed from crosses between the BEEF SHORTHORN and HIGHLAND breeds. Red coloured with some roan, gold or white.

Lump Lime ♦ CALCIUM OXIDE.

Lumpy Jaw ♦ ACTINOMYCOSIS.

Lumpy Wool ♦ WOOL ROT.

Lunky ♦ SMOUT.

Lupins. A leguminous plant (*Lupin sp.*) sometimes grown as a GREEN MANURE crop for ploughing in, particularly on poor light land in the eastern counties deficient in LIME and HUMUS. White, yellow and blue flowered varieties exist. The sweet lupin has been used for FORAGE, being the only non-toxic variety.

Lymph. A watery colourless liquid derived from blood filtering through tissues from capillaries and collected in lymph vessels to be returned to the bloodstream. It carries nutrients to the body's tissues.

Lynchet. A boundary ridge or unploughed strip. Also a terrace or bank formed on a slope due to continuous CONTOUR PLOUGHING. Also called linch.

M

Macronutrient. A chemical element needed by crops in relatively large amounts (normally greater than 1 ppm), normally applied as a FERTILIZER or liming material. (♦ LIME, TRACE ELEMENT)

Mad Itch ♦ AUJESZKY'S DISEASE.

Maggot. A legless LARVA of certain two-winged FLIES of the order Diptera, with a very tiny head. The maggots of many flies attack livestock and crops, e.g. BLOW FLY, CRANE FLY, FRIT FLY, WARBLE FLY, etc.

Magnesium. (Mg). A metallic element found in many rock types (e.g. dolomite, magnesite) and essential in life for META-BOLISM. A constituent of CHLOROPHYLL. Deficiency in the soil can cause CHLOROSIS in plants and can lead to GRASS STAGGERS in livestock, and is corrected by applying magnesian limestone (♦ GROUND LIMESTONE), EPSOM SALTS or KIESERITE, etc. Other FERTILIZERS containing magnesium include FARMYARD MANURE, KAINIT and BASIC SLAG.

Maid. A ewe teg (♦ SHEEP NAMES) shown to be barren after being served by a ram.

Maiden Heifer ♦ HEIFER.

Maiden Seeds. A LEY which has not been grazed or mown.

Maiden Tree. A fruit tree, about 1 year old (♦ FEATHERED MAIDEN). Also a fruit tree developed from a seed as distinct from one developed from a GRAFT.

Mailing. A Scottish term for a rented farm.

Maincrop Potatoes ♦ POTATO.

Maine-Anjou. A large breed of beef cattle imported into Britain from France. Coat colour includes roan, and red and white. Also produces plentiful milk.

Mains. A Scottish term for a HOME FARM.

Maintenance Ration. The amount of food which when supplied to an animal will keep it healthy and maintain a constant live-weight, with no loss or gain. (♦ PRODUCTION RATION)

Maize. A cereal plant (*Zea mays*) now commonly grown in Britain for grain, SILAGE, FORAGE and FODDER. It requires good drainage, grows best on lightish soils and is frost-sensitive. Characterised by a thick, stout, tall stem, and numerous long leaves arising alternately on each side of the stem. The male flower (tassel) forms at the top of the shoot, and several female flowers form in the axils of some of the middle-stem leaves and develop into long cylinders (cobs) of tightly packed grain. Maize is extensively grown in North America and is imported into Britain for the manufacturing of breakfast cereals, e.g. corn flakes, and for incorporation into FEEDINGSTUFFS. Maize is rich in CARBOHYDRATE but low in utilisable PROTEIN and FIBRE. It is much used as a MEAL for fattening livestock but is also fed flaked (♦ FLAKED MAIZE) (*see p. 95*). Also called corn on the cob. (♦ SWEET CORN)

Malathion. An organophosphous insecticide and ACARICIDE available either alone or in mixtures (i.e. with D.D.T.). Used in liquid or dust form, particularly to control skin parasites of cattle, and to control various crop pests.

Malt ♦ MALTING.

Malt Culms, Coombs. A by-product of the MALTING process, used as a FEEDINGSTUFF for dairy cows, sheep and horses, after thorough soaking.

Malt Sugar ♦ MALTOSE.

Maltase An ENZYME which breaks down MALTOSE into GLUCOSE. It occurs in YEAST and is an important element of the brewing process.

Malting. The process whereby barley or other grain is soaked in water and allowed to partly geminate (during which stored STARCH is broken down to MALTOSE by the enzyme Diastase), and is then kiln dried, after which the roots and shoots are removed (malt culms) and the remaining grain (the malt) is used in brewing. The malt is steeped in water and the maltose extracted into solution (the wort) which is run-off, leaving the residual grain which is used as an animal FEEDINGSTUFF. (◗ BREWERS' GRAINS)

Malting Barley. The best quality BARLEY sold for MALTING, representing about one fifth of the British crop, for which a premium price is paid. It has a high STARCH content (required for obtaining MALTOSE) and a low PROTEIN content (determined during testing by measuring Nitrogen content). Good quality malting barley, after being harvested well, should be dry, and should smell 'sweet' (with no suggestion of mustiness). Also the grains should be evenly sized with few broken or damaged ones, evenly coloured with wrinkly surfaces, and free from weed seeds and disease. These qualities are important for malting, and tests with strict criteria are carried out by millers before grain is accepted from farmers.

Maltose. A disaccharide sugar (◗ CARBOHYDRATES) comprising two GLUCOSE molecules, formed by STARCH breakdown by the enzyme DIASTASE, particularly during digestion and seed gemination (◗ MALTING). Also called malt sugar.

Maltsters. Those who make malt (◗ MALTING) for sale to breweries.

Mammals. A class of animals characterised mainly by hair, and the secretion of milk to feed the young (◗ MAMMARY GLANDS), which includes all farm livestock, except poultry.

Mammary Glands. Milk-producing glands, characteristic of female MAMMALS. Located in the UDDER of cows. Milk production and secretion is controlled by various HORMONES. (◗ LACTATION)

Mammitis ◗ MASTITIS.

Management Unit ◗ FIELD.

Mane. 1. Long hair on the back of a horse's neck.
2. A narrow strip of grass or cereal not cut by a MOWER, or by a BINDER or COMBINE HARVESTER.

Manganese. (Mn). A metallic TRACE ELEMENT required for metabolic processes (♦ METABOLISM) as an ENZYME activator. It can be deficient in NEUTRAL or ALKALINE SOILS rich in ORGANIC MATTER, causing DEFICIENCY DISEASES in crops (e.g. SPECKLED YELLOWS in beet, GREY LEAF in cereals), which is usually rectified by spraying with manganese sulphate. Cattle infertility may result if it is deficient in pasture.

Mange. A skin disease of animals caused by MITES which results in the hair dropping out in patches or, due to persistent itching, being rubbed away by the animal. (♦ SCABIES, SHEEP SCAB)

Mangel, Mangel-Wurzle. A plant of the beet family (*Beta maritima*), a cultivated form of wild sea beet. Grown, mainly in the East and South-East of England, as a FODDER crop. Characterised by its swollen root and long, wide, glossy leaves borne on stalks. It has a larger root than FODDER BEET, less buried in the soil, and with a lower DRY MATTER and sugar content. Mangels may be classified into low dry matter (9–12%) and medium dry matter (13–15%) varieties. The former may be subdivided into globe and intermediate (including tankard-shaped) varieties, the latter having more root below ground. Also called mangold and mangold-wurzel. (♦ also SUGAR BEET)

Manger. A trough in which food for cattle and horses is placed, often located in a COWHOUSE or STABLE.

Mangold ♦ MANGEL.

Manioc. A tropical plant (*Manihot esculenta*) producing a TUBER with a high STARCH but low PROTEIN content, used as a cereal substitute in animal FEEDINGSTUFFS. Also called cassava or tapioca.

'Man's Land' ♦ HEAVY LAND.

Manure. A general term for anything added to the soil to increase its content of plant nutrients (♦ FERTILIZERS). Normally applied to organic, usually bulky, materials, mostly derived from farm and animal waste products, such as COMPOST, DUNG, FARMYARD MANURE, SLURRY, SEWERAGE SLUDGE, etc. They are slower acting than inorganic fertilizers but have the advantage of adding ORGANIC MATTER to the soil, increasing its capacity to retain water, and improving its structure. (♦ GREEN MANURE)

Manure Spreaders. Various types of machine used to distribute solid or liquid MANURES on the land. Also called muck spreaders. The main types include:

(*a*) Self-emptying trailers containing a slatted conveyor in the base, delivering the (mainly solid) manure to a revolving mechanism at the rear which shreds and flings it on the field. Also called trailer-spreaders.

(*b*) Cylindrical, open-topped tanks with a p.t.o driven, central, axial shaft which operates a set of attached flail chains, which throw the manure (usually containing SLURRY) out to one side of the tank.

(*c*) Gun mechanisms which distribute, under pressure, slurry conveyed via a pipe from a nearby fixed tank or from a mobile vacuum tanker. The latter are sometimes used to spread slurry directly onto the land.

A self-emptying trailer manure spreader.

Manurial Residues. That proportion of a MANURE added to the soil but not used by the crop, or residues left in the soil by a crop, which are available to future crops.

Manurial Value. Those nutrients in a FEEDINGSTUFF which an animal does not take in and pass out with the faeces so as to enrich the dung. Also called the residual value or fertilizer value of a feedingstuff.

Manyplies ♦ OMASUM.

Maran. A heavy breed of poultry with BARRED feathers, white legs, a single comb, producing dark brown eggs. Also used as table birds.

Marchigiana. An all-white breed of beef cattle developed in the Marche region of Italy from ancient stock imported by the Barbarians. A thrifty, docile breed adapted to a wide range of environmental conditions. Recently imported into Britain for crossing (♦ CROSS-BREEDING), mainly with FRIESIAN cows, to improve calf quality.

Marden Layer. A term for a young LEY.

Mare. A female horse of 5 years of age or more. (♦ FILLY)

Marek's Disease. A virus disease of poultry causing lameness, progressing to leg paralysis, drooped wings, neck twisting and greying of the iris in both eyes.

Marginal Land. Poor quality land suffering from various permanent disadvantages such as steep slopes, bad drainage, poor thin soil, harsh environmental conditions, etc. About 1 million ha (2.5 million acres) of enclosed land in the U.K. (0.6 million ha, 1.5 million acres, in England and Wales) can be classified as marginal land, mostly located near hills, moors and mountains, It is mostly under grass and used for livestock production, being unable to grow worthwhile crops economically. Even as grass it requires tending, more use of FERTILIZERS, and produces lower yields than better land.

Marked Coat. An animal's coat having two or more specific colours as distinct from a single colour (self-coloured).

Market Garden. Land used to grow produce, mainly vegetables and fruit, for sale (usually to a market, e.g. COVENT GARDEN MARKET) for direct human consumption, as distinct from for processing purposes. The term is often restricted to the production of vegetables on a small scale.

Marketing. Agricultural products are marketed mainly through private trade channels such as corn merchants, livestock auctions and markets, and bacon factories, or through producers' co-operative organisations. For certain commodities, however, marketing is conducted by MARKETING BOARDS. For certain other commodities there are broadly-based organisations representing producer, distributor and independent interests. (◗ HOME-GROWN CEREALS AUTHORITY, MEAT AND LIVESTOCK COMMISSION, and EGGS AUTHORITY)

Marketing Boards. Boards responsible primarily for the marketing of specific agricultural commodities, and operating under the Agricultural Marketing Act 1958 (which consolidated earlier legislation) or the Agricultural Marketing Act (Northern Ireland) 1964. The schemes under which they are constituted and operate have been approved by Parliament and, except in Northern Ireland, each confirmed by a poll of producers. Most of the members of the boards are elected by producers, but a minority are appointed by the Minister of Agriculture, Fisheries and Food, or Ministers concerned. The interests of consumers and others are protected by further safeguards.

Marketing boards fall into two broad categories: first, those which are sole buyers of the regulated product from all producers not specially exempted or those which exercise a comparable influence by controlling all contracts between producers and first buyers; and, secondly, boards which maintain only a broad control over marketing conditions, leaving producers free otherwise to deal individually with buyers. The boards for hops, milk and wool and the Northern Ireland pigs and seed potato boards fall into the first category and the Potato Marketing Board into the second.

The role and function of each board is described alphabetically elsewhere in the text. (Source: Agriculture in Britain. Central Office of Information, R5961/1977).

Marl. A soft, unconsolidated clay soil containing considerable quantities of LIME. Marl is sometimes spread on LIGHT SOIL to improve its texture (marling).

Marrowstem Kale. A variety of KALE with a nutritious, thick, pithy stem, bearing large leaves on long stalks. It is not frost-tolerant and is grown mainly as a FODDER CROP for dairy cows.

Marsh. An area of usually low-lying land, often flooded by fresh or saline water, with poorly drained, impermeable mineral soil, usually silt or clay (as distinct from the organic soils of FENS) which is continuously waterlogged. Characterised by sedges and rushes. Many marsh areas have been artificially drained (♦ LAND DRAINAGE) and protective embankments established to prevent flooding, with subsequent improvement of their grazing value, and in some cases the introduction of arable cropping.

Mash. A mixture of animal foods, crushed and stirred to a soft pulpy state with water (wet mash) or without it (dry mash).

Masham. A hardy crossbred type of sheep resulting mainly from a TEESWATER ram mated with a SWALEDALE ewe. Sometimes rams of other breeds are used. Face colour includes shades from black to almost white. They are able to graze relatively poor land. Wether lambs (♦ SHEEP NAMES) of this cross are fattened to produce mutton, and ewe lambs are used for breeding with rams of DOWN BREEDS.

Mashlum, Maslin. A mixture of oats and barley and sometimes wheat, sown to provide grain for feeding to livestock. Occasionally beans or peas may be added to the mixture but ripening is uneven and they are not suited to being harvested by a COMBINE HARVESTER. Also called dredge, maslin, or meslen.

Mass Selection ♦ PHENOTYPIC SELECTION.

Mast. The fruit of Beech, Chestnut, Oak and other trees eaten by pigs as PANNAGE.

Mastitis. Inflammation of a cow's UDDER, due to bacterial infection, resulting in the secretion of clots of milk, and sometimes pus when the infection is acute. Also called mammitis, garget and udder-clap.

Mat. A closely interwoven or tangled mass of dead grass which develops when a pasture is undergrazed, particularly on ACID SOILS.

Maw. The stomach, particularly the fourth stomach or abomasum of RUMINANTS, Also the CRAW of a bird.

Maybug ♦ COCKCHAFER.

Mazzard ♦ GEAN.

M.C.P.A. An abbreviation for 2-methyl-4-chloro-phenoxy-acetic acid, a TRANSLOCATED HERBICIDE or HORMONE WEEDKILLER, used to control many broad-leaved annual and perennial weeds. Often also included in mixtures with other translocated herbicides.

M.C.P.P. An abbreviation for 2,methyl-4-chlorophenoxypropionic acid. A hormone HERBICIDE used mainly to control WEEDS in cereals. Also called Mecoprop.

Mead. 1. A shorter version of MEADOW.
2. Wine produced from honey.

Meadow. Strictly applied to an area of GRASSLAND used to produce HAY or SILAGE rather than for grazing (♦ PASTURE). Also used loosely for rich waterside fields which are frequently flooded by river water, either naturally or via sluices and which are grazed (water meadows).

Meadow Fescue. A loosely tufted perennial grass (*Festuca pratensis*) forming large tussocks, with narrow, bright green, hairless leaves, shiny on the underside, and loose, green or purple PANICLES. Valuable for HAY and grazing, and particularly found on rich, well-drained, fertile CLAY SOILS, and often abundant in water meadows (♦ MEADOW). A very palatable grass which grows well mixed with white CLOVER or TIMOTHY (*see p. 211*).

Meadow Hay ♦ HAY.

Meadow Soft Grass ♦ YORKSHIRE FOG.

Meal. 1. A type of FEEDINGSTUFF consisting of a single or a mixture of finely ground ingredients, e.g. CEREALS, OIL SEEDS, meat, bone, fish, etc. Mainly used for pigs and poultry. (♦ BLOOD, BONE, FISH and MEAT MEALS)
2. Another term for oil seed CAKE. (♦ CONCENTRATES)

Measles. A condition of the carcases of cattle and pigs characterised by the presence of TAPEWORM cysts in the meat.

Meat and Livestock Commission. An independent organisation established by Act of Parliament in 1967, with 10 Commissioners appointed by Ministers and statutory arrangements for consultation with industry organisations. It is financed by a levy on all cattle, sheep and pigs slaughtered in Great Britain and by various agency fees and charges for services.

M.L.C. services include the evaluation of breeding stock potential by central testing and on-farm testing and recording, pig A.I., carcase evaluation and classification, and the provision of market information, trends and forecasts. It initiates and supports a wide range of research projects, and operates a computer-based livestock and meat data bank. M.L.C. acts as an agent for the Agricultural Departments and the INTERVENTION BOARD FOR AGRICULTURAL PRODUCE in operating national and E.E.C. support schemes for cattle, sheep and pigs. (◗ COMMON AGRICULTURAL POLICY)

M.L.C. fatstock staff operate in abattoirs and auction markets certifying animals and carcases, and providing independent weighing and CARCASE CLASSIFICATION.

Through the Meat Promotion Executive the M.L.C. promotes British meat by direct advertising, advice and information.

Meat Meal, Meat-and-bone Meal, Meat Guano. MEAL derived from waste meat and bone trimmed off animal carcases and steam cooked until moisture content is below 10%. Surplus fat is drained off and most of the rest expelled by pressing the residue. By law meat meal must contain at least 55% PROTEIN, meat-and-bone meal 40–55% protein. In both types salt content must not exceed 4%. Also called tankage.

Mechanical Grazing ◗ ZERO GRAZING.

Meiosis. A process of cell division in which the number of CHROMOSOMES are halved so that DIPLOID cells are reduced to HAPLOID cells. In animals, GAMETES are formed in this way. (◗ MITOSIS)

Merchants' List ◗ FARMERS' LIST.

Mere. 1. A small lake or pond (particularly in Cheshire and Shropshire) or a marshy area (particularly in the FENS).
2. A landmark or boundary such as a path.

Merino. A fine-woolled breed of sheep with a white face and pink muzzle, with horned rams but polled ewes. Developed in Spain by the Moors and subsequently spread throughout the world. Particularly important in Australia. The breed contributed to the development of DORSET HORN and WHITEFACED WOODLAND sheep.

Meristem. An area of active cell division in a plant (e.g. stem and root tips, CAMBIUM, etc.) which gives rise to permanent tissue. Also called a growing point.

Meslen ▶ MASHLUM.

Metabolisable Energy. That proportion of the energy in a food which is utilisable by an animal in its METABOLISM. Energy losses (ignoring thermal losses) occur as undigested food (faeces), as alimentary gases (particularly methane in RUMINANTS), and as incompletely oxidised nitrogenous compounds expelled in urine. Metabolisable energy thus equals the energy of food (GROSS ENERGY VALUE) less the combustion value of the excretion products. It is the metric replacement for STARCH EQUIVALENT for indicating the feeding value of a food when calculating livestock ration composition, and is measured in Joules. The daily M.E. needed in a dairy cow's MAINTENANCE RATION is about 60 million Joules, i.e. 60 Megajoules (M.J.) Also called physiological heat value. (▶ NET ENERGY VALUE)

Metabolism. The chemical processes occurring in living organisms, including those which break down complex organic compounds to more simple ones (catabolism), e.g. oxidation (▶ RESPIRATION), and those which build up relatively simple compounds into more complex ones (anabolism), e.g. PHOTOSYNTHESIS. Most of these processes are controlled by ENZYMES.

Metamorphosis. The change in form of an organism through various growth stages from LARVA to adult, e.g. a CATERPILLAR into a PUPA and subsequently into a fully developed insect.

Metered Chop Harvesters. Two kinds of FORAGE HARVESTER, producing a uniform short chop, having either a flywheel or cylinder type of chopping mechanism. With both types the chop length and rate of delivery by a feeder to the chopping mechanism can be controlled. These harvesters are usually trailed, but self-propelled machines are available. Also called precision chop harvesters (*see p. 192*).

Methane. (CH$_4$). A colourless hydrocarbon gas which can be formed in large quantities in a cow's RUMEN (♦ BLOAT). Now being developed as a source of fuel on farms, produced by fermenting SLURRY.

Meuse-Rhine-Ijssel (M.R.I.). A dual-purpose breed of cattle, taking its name from three rivers in Holland where it was developed. Imported into Britain in 1970, particularly for increasing SHORTHORN productivity. The breed has a similar milk yield to the FRIESIAN but better meat production. A large-bodied breed, characterised by a patched red and white coat, with short white legs, white belly and tail, and sometimes a white blaze on the broad head, and by short, forward and inward curving horns.

Microbes. Microscopic organisms such as BACTERIA.

Micronutrient ♦ TRACE ELEMENT.

Micropyle. A tiny pore in a seed coat made by the pollen tube entering the OVULE, and through which water enters prior to GERMINATION.

Midden. A DUNG heap.

Middle White. A rare, early maturing breed of pig developed by crossing the LARGE WHITE and the now extinct Small White. A smallish, white-skinned, compact breed, with a short, dish-faced head, upright ears, turned-up nose, heavy jowls and short legs. Usually kept for pork production.

Middlings ♦ WEATINGS.

Mids. Potatoes sold for human consumption, which are of 'WARE' standard, and which will not pass through a 20 mm horizontal mesh when manipulated, but will pass through a 45 mm mesh. (♦ BAKERS)

Milch Cow. A cow giving milk or kept for milking.

Mildew. Various fungi, which are parasitic on plants, causing disease. The threads or hyphae of some PHYCOMYCETE fungi penetrate deeply (Downy Mildews), whilst other types of ASCOMYCETE fungi grow more on the leaf surface with a powdery white appearance (Powdery Mildews). Also called moulds.

Milk. A white liquid secreted by female mammals (♦ MAMMARY GLANDS) for the nourishment of the young, containing valuable essential food constituents, e.g. PROTEINS, CARBOHYDRATES, FATS, etc. (♦ MILK COMPOSITION). Milk produced by dairy cattle for liquid consumption and for the manufacturing of various products (e.g. butter, cheese, yogurt, etc.) is marketed under the control of the MILK MARKETING BOARD. Whole, fresh milk sold for public consumption must contain at least 3.0% BUTTERFAT and is tested for quality on purchase from the farm (♦ MILK COMPOSITIONAL QUALITY SCHEME). Several types of liquid milk are available including UNTREATED MILK, PASTEURISED MILK (including HOMOGENISED MILK), STERILISED MILK, and ULTRA HEAT-TREATED MILK. (♦ also SKIM MILK, SEPARATED MILK)

In England, Scotland and Wales, about half the milk produced goes for liquid consumption and the remainder for manufacture, but in Northern Ireland the greater part of the milk is used for manufactured products (♦ MILK PRODUCTS). Average consumption of liquid milk per head is about 2.56 litres (about 4.5 pints) a week. (♦ MILK YIELDS)

Milk Clipping. The clipping (♦ SHEARING) of ewes which are still suckling lambs.

Milk Composition. Cow's milk is a balanced food containing all the nutrients (including VITAMINS) necessary to humans and is particularly fed to children and young animals. The composition can vary according to breed (♦ CATTLE), age, level and type of feeding, stage of LACTATION, etc. The following are average figures:

	Per cent	
Protein	3.40	mainly CASEIN
Sugar	4.75	mainly LACTOSE
Fat	3.75	♦ BUTTERFAT
Mineral Matter	.75	♦ ASH
Water	87.35	
	100.00	

(♦ MILK COMPOSITIONAL QUALITY SCHEME).

Milk Compositional Quality Scheme. A MILK MARKETING BOARD scheme, operated in co-operation with milk distributors, for paying dairy farmers on the quality of the milk produced. A monthly sample is analysed for total solids, BUTTERFAT and SOLIDS-NOT-FAT (as a fair indication of the total food content of the milk), and the data is collated by the Board. Each producer's milk is categorised according to its 12 monthly average composition (reviewed every 6 months) and appropriate payment per gallon made. Payment for certain milk, e.g. that of producer retailers, new producers, farmhouse cheesemakers, etc., is made on an 'unclassified' basis. A premium for butterfat-rich milk from CHANNEL ISLAND BREEDS or the SOUTH DEVON breed is paid. (◗ GOLD TOP MILK)

Milk Cooler. Since daily bulk collection (◗ BULK MILK) has become universal in Britain, milk is usually temporarily stored on-farm in stainless steel bulk tanks in which it is cooled (to prevent bacterial multiplication) by a jacket of chilled water.

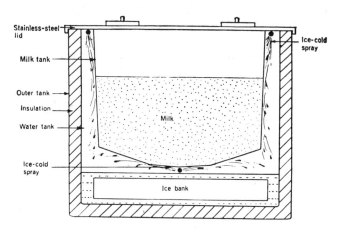

Section of a bulk milk tanker.

Other types of milk cooler include surface coolers in which the milk runs as a thin film over parallel steel plates between which chilled water is carried, and in-can or in-churn coolers which fit on the top of churns and allow the cooling water to flow down the outside. The latter type are now obsolete in Britain where CHURN COLLECTION has now ceased.

Milk Fever. A disease of various animals, particularly dairy cows, usually occurring a few days after calving. It is associated with a fall in the blood CALCIUM content the reasons for which are not fully understood. It results in initial excitement, then loss of balance and later of consciousness, with the neck in a typical twisted position. Not an actual fever, body temperature often falling a few degrees below normal. Treated by injecting calcium borogluconate.

Milk Marketing Board. An autonomous producer-controlled co-operative organisation established in 1933 under the Milk Marketing Scheme to control sales of milk in England and Wales. All producers must register with the M.M.B. which undertakes the marketing of all the milk they wish to sell wholesale, and arranges for its collection and transport to buyers who are allocated supplies (ensuring the supply of milk to all markets, when needed, at the lowest possible cost). It negotiates the wholesale price of manufacturing milk centrally with the DAIRY TRADE FEDERATION. The maximum retail and wholesale price of milk sold for processing as liquid is fixed by the Government. Producers are paid an average price based on the total returns to the M.M.B. from all milk sold for manufacturing or as liquid, but on scales under the MILK COMPOSITIONAL QUALITY SCHEME. A number of producer-retailers operate under M.M.B. licences (◊ UNTREATED MILK). Milk for wholesale is collected daily from farms or collection points by bulk tankers owned by the M.M.B. or dairy companies, or by contracted independent hauliers, and is delivered to dairies or creameries. The M.M.B. itself owns and operates a number of dairies and creameries, and in 1979 obtained a greater degree of control over the milk market by purchasing 16 cheese, butter and skim milk powder factories from Unigate Foods Ltd. The M.M.B. offers certain fee-paying services including MILK RECORDING, milking machine testing, costings and consultancy services, and comprehensive A.I. and other veterinary services. It also carries out sales promotion work, and both undertakes and supports research. Registered producers elect 15 of the 18 Board members and also the members of the 11 Regional Committees.

Similar M.M.B.s were established in the three milk-producing areas in Scotland in 1933 and 1934, and in N. Ireland in 1955. (◊ MILK PRODUCTS)

Milk Non-marketing and Dairy Herd Conversion Scheme. An E.E.C. scheme introduced in 1977, replacing the 1973 Dairy Herd Conversion Scheme, with the aim of reducing E.E.C. milk product surpluses by providing grants to milk producers who undertake to cease production for 5 years (non-marketing premia). Higher rates (conversion premia) are paid to those producers willing to convert to beef or sheep production and to produce no milk for 4 years.

Milk Products. In 1978/79, in the U.K., 49% of the milk produced was sold for liquid consumption under the designations shown in the table *Designated Milk Sales*, whilst 51% was sold for manufacturing as shown in the table *Sales of Manufacturing Milk*.

Designated Milk Sales in the U.K. in 1978/79

Type of milk	% of total milk sold
Untreated.	3
Pasteurised (including homogenised pasteurised).	89
Sterilised	7
Ultra Heat-Treated.	1
	100

(Source: Milk Marketing Board).

Sales of Manufacturing Milk in the U.K. in 1978/79

Type of product	% of manufacturing milk sold
Butter	46.9
Cheese	29.0
Cream	13.2
Condensed Milk	3.9
Chocolate Crumb	3.3
Milk Powder	2.6
Other Products	1.1
	100.0

(Source: Milk Marketing Board).

Milk Quality Schemes. Various schemes are operated in Britain jointly by the MILK MARKETING BOARD and the DAIRY TRADE FEDERATION for paying farmers on the quality of the milk produced. Payment is made on the basis of compositional quality (◗ MILK COMPOSITIONAL QUALITY SCHEME) with price penalties for failing to reach required standards. Milk may be rejected if BUTTERFAT is below 3%. Under the Hygienic Quality Scheme milk is rejected as unmarketable if it is sour, tainted or contains blood, and price deductions are made if samples fail hygienic tests (the resazurin test). Price penalties are applied if milk contains antibiotics (measured by the B.C.P. TEST) or has a low SOLIDS-NOT-FAT content. In certain cases the Board may arrange separate collection of milk failing sediment tests.

Milk Recording. The recording of the milk yield of each cow in a dairy herd by measuring the volume or weight of milk given during milking sessions. Usually the quantity can be read off a scale on the collecting jar into which the MILKING MACHINE initially extracts the milk. Under the MILK MARKETING BOARD'S voluntary National Milk Records Scheme, milk producers may have samples analysed monthly for fat and protein content, and receive a computerised monthly statement. This information and milk yield records enable the economical rationing of CONCENTRATES and is very useful in the selection and breeding of cows producing high yields of good quality. (◗ MILK YIELDS)

Milk Sinus. The cavity in each of the teats of a cow's UDDER into which milk is secreted and from which is is extracted during milking.

Milk Vein. One of two large veins visible on a cow's belly running from the UDDER.

Milk Vetch. Various leguminous plants (*Astragalus spp.*) grown on lime-rich soils for FODDER and supposed to increase the milk yield of dairy cattle.

Milk Yields. The average yield of dairy cows has been steadily increasing since the war and the 1978/79 average milk yield per dairy cow per LACTATION (in England and Wales) was 5 335 kg (5 196 litres; 1 142 gallons). The total production of milk for human consumption in the U.K. in 1979 was 15 364 million litres (3 376.7 m. gal). (Source: Annual Review of Agriculture, H.M.S.O. 1980).

The yield and composition of milk given by dairy breeds varies considerably as indicated in the table *Average Milk Yields*.

Average Milk Yields, Butterfat and Protein Percentages for Dairy Breeds in England & Wales

Breed	% of total dairy cows	AVERAGE YIELD Lactation			AVERAGE YIELD Annual			Butter-fat Lactation %	Protein Lactation %
		Kg	Litres	Gal	Kg	Litres	Gal		
Ayrshire	3.0	4863	4737	1041	4813	4688	1030	3.94	3.36
Holstein	1.3	6218	6056	1331	6151	5991	1316	3.76	3.19
Dairy Shorthorn	0.7	4730	4607	1012	4743	4620	1015	3.62	3.30
Dexter	–	2198	2141	470	2008	1956	430	4.06	3.48
Freisian	85.4	5441	5299	1164	5511	5368	1179	3.79	3.25
Guernsey	2.4	3941	3838	843	3893	3792	833	4.65	3.60
Jersey	2.5	3776	3678	808	3727	3630	798	5.14	3.83
Northern Dairy Shorthorn	–	4084	3978	874	4069	3963	871	3.64	3.35
Red/Poll British Dane	–	3904	3802	835	3880	3779	830	3.80	3.31
South Devon	–	3538	3446	757	2955	2878	616	3.84	3.43
Welsh Black	–	3165	3083	677	2919	2843	625	3.82	3.27
Mixed	4.7	5135	5001	1070	5186	5051	1100	3.83	3.29
All Breeds	100.0	5335	5196	1142	5391	5251	1154	3.83	3.27

(Figures according to the Milk Marketing Board's National Milk Records for the Milk Recording Year ending September 1979).

Milking. The extraction of milk from a cow's UDDER by hand or MILKING MACHINE. Milk secretion is actuated by the LET DOWN mechanism, and the squeezing action of the hand or the squeezing-suction action of the TEAT CUPS withdraws the milk from each MILK SINUS. (◗ MILKING PARLOURS)

Milking Machines. There are three main types of milking machines all of which, through a pulsator mechanism, apply an intermittent partial vacuum to the teats of the cow's UDDER, simulating the sucking action of a calf. They vary in the design of collecting the extracted milk and include (*a*) bucket plants, used mainly in COWSHED milking, in which milk is extracted into covered buckets and carried to the MILK COOLER, (*b*) in-churn plants, in which the milk is conveyed directly into CHURNS, in which it is cooled and transported to the dairy (◗ CHURN COLLECTION), and (*c*) releaser plants, in which the milk is extracted into a collecting jar (in which each cow s yield is quantified) and from which it is sucked via vacuum piping to be 'released' into the cooler, but without disturbing the vacuum system. This system is now almost universal. The abbreviation m/m is sometimes used for milking machine.

Milking Parlour. A building containing specialised MILKING MACHINES. Many designs exist, the choice depending on herd size, average yield, available labour, feeding method, etc. There are two main categories of parlour, having either one or two stalls per milking unit, the latter having a greater through-put at about 10 cows per unit per hour. CONCENTRATES are often automatically fed to the cows whilst in the stalls. Layout may be
(*a*) Abreast, with stalls side by side, sometimes slightly raised.

Abreast-type milking parlour.

(b) Tandem, with stalls end to end around a pit, so that cows stand head to tail. In the simplified chute type there are two stalls to each side of the pit and cows enter via steps or ramps and leave together after milking.

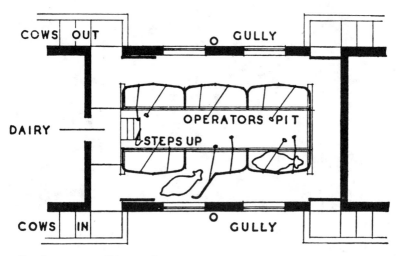

Tandem-type milking parlour.

(c) Herringbone, in which there are no stalls and the cows stand side by side at an angle and to each side of a central pit (*and see p. 292*)

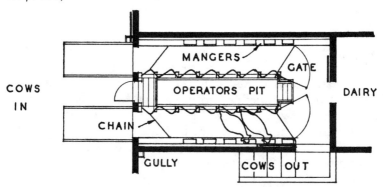

Herringbone milking parlour.

(d) Rotary, in which any of the preceding formations move round the milker on a raised, slowly moving platform.

291

A herringbone milking parlour showing central operator's pit and collecting jars.

Mill. A machine or a building equipped with machinery for grinding (by crushing) cereal grain or other dry FEEDINGSTUFFS to produce flour or MEAL (◆ CRUSHING MILL, HAMMER MILL, PLATE MILL). Also to remove seed HUSKS.

Milling Wheat. The best quality wheat purchased from producers by the flour-milling industry for bread making, etc. Such wheat, normally grown under contract, must reach specified standards, and samples are subject to visual and qualitative tests before acceptance by the miller, e.g. GRAIN MOISTURE CONTENT, cleanliness and purity, damaged and diseased grain content, odour, bushel weight, protein content, GLUTEN quality, ALPHA-AMYLASE content, etc. (◆ HAGBERG and FARRAND TEST)

Mineral. A term mainly applied to inorganic substances, usually occurring naturally in the soil or rocks as distinct from those substances which are ORGANIC, being derived from living matter. (◆ MINERAL MATTER)

Mineral Deficiencies ▶ DEFICIENCY DISEASES.

Mineral Lick ▶ MINERAL MIXTURES.

Mineral Matter. Mineral elements and compounds, found as a constituent of the soil and living organisms, and essential to METABOLISM. Mineral matter in the soil consists of fragments of rock in various stages of degradation and their decomposition products, e.g. quartz, clays, iron oxides, etc., intimately mixed with ORGANIC MATTER. Some of the mineral elements may be present in plants and animals as mineral salts or combined in organic compounds such as PROTEINS. Certain parts of plants have a higher mineral content (e.g. roots, vegetative parts), and as GREEN CROPS mature the mineral content of the DRY MATTER diminishes. The useful mineral matter content of FEEDINGSTUFFS (expressed as ASH) of animal origin is normally higher than in those of vegetable origin.

Mineral Mixtures. Minerals, including TRACE ELEMENTS, supplied to animals to ensure an adequate supply for METABOLISM and to prevent DEFICIENCY DISEASES, infertility, lack of thrift, or poor production, etc. Often added to CONCENTRATE mixtures or CAKES. Also given loose as mineral mixtures in field troughs, or as hard slabs (mineral licks or salt licks) for livestock to lick at will.

Mineral Phosphate ▶ ROCK PHOSPHATE.

Minorca. A light breed of poultry of Mediterranean origin, black or white in colour, with usually dark slate-coloured legs, a large single comb and producing white eggs.

Mites. A sub-class of the ARACHNIDA allied to TICKS, found in many habitats (e.g. the soil, decaying ORGANIC MATTER) and parasitic on both plants and animals.

Mitosis. A process of cell division in which the CHROMOSOME number remains constant. Most cells in living organisms are DIPLOID (i.e. have a double set of chromosomes) and growth in plants and animals is achieved as a result of such cells reproducing by mitosis. Certain cells (e.g. GAMETES) are HAPLOID (i.e. have a single set of chromosomes) and are produced by MEIOSIS.

Mixed Farm. A term usually applied to a farm on which a range of activities, including crop production (arable, grassland, fruit, etc) and livestock keeping, are practiced, as distinct from one specialising in a particular activity (e.g. poultry production).

Mixed Fertilizer ♦ COMPOUND FERTILIZER.

Mixen. A DUNG heap. Also called a midden.

Moisture Content ♦ GRAIN MOISTURE CONTENT, HAYMAKING, SOIL MOISTURE DEFICIT.

Molasses. A thick treacle produced as a by-product of sugar refining, containing 20–30% water, and rich in soluble sugars (particularly SUCROSE). Used to improve the palatability of FEEDINGSTUFFS, and as a binding agent in the production of CAKE. Also used as a laxative and as a wilting agent in SILAGE making.

Mole Drain. A subsurface LAND DRAINAGE channel formed by drawing a MOLE PLOUGH across a field, commonly 7.5–10 cm (3–4 in.) in diameter at a depth of 50–60 cm (20–24 in.). Mole draining is usually carried out on heavy clay soils. On lighter soils the drains tend to collapse.

Mole Plough. An implement used to form MOLE DRAINS, consisting of a torpedo-shaped cylinder (a mole or cartridge) attached to a vertical, knife-edged COULTER, which is itself fixed to a metal beam suspended between two wheels. Attached directly behind the mole is an expander (a tapered cylinder of greater diameter) which slightly enlarges the drain being cut, smoothing and compressing the drain walls as the mole is drawn through the soil.

Molinia ♦ FLYING BENT.

Molluscicide. A chemical used to kill slugs and snails.

Molybdenum. (Mo). A metallic TRACE ELEMENT required by LEGUMES for NODULE formation. When present in excess (e.g. the TEART soils of the West Country) it can cause SCOURS in RUMINANTS, particularly cattle.

Monetary Compensatory Amount. An instrument of the COMMON AGRICULTURAL POLICY whereby a subsidy is paid or a levy charged on imports and exports of those agricultural products for which there are COMMON PRICES. M.C.A.s are designed to prevent distortion of E.E.C. agricultural trade due to differences in actual farm prices in member states. They are based on the percentage difference in value between the market exchange rate of a currency and its fixed AGRICULTURAL REFERENCE RATE or 'green rate' (e.g. GREEN POUND), less an arbitrary 'franchise' of 1.5%. Positive or negative M.C.A.s can apply depending, respectively, on whether the market rate is above or below, respectively, the 'green rate'. Thus, an E.E.C. country to which a net positive M.C.A. applies in relation to another E.E.C. country, receives a subsidy on exports and charges a levy on imports. Conversely, in a country in which a net negative M.C.A. applies, a levy is charged on exports and imports receive a subsidy.

Monocotyledon ♦ COTYLEDON.

Monoculture. The growing of the same crop on a field year after year.

Monoecious. A term applied to plants having both separate male and female flowers on the same individual plant, e.g. hazel. (♦ DIOECIOUS)

Mono-Pitich Fattening Piggery ♦ PIGGERY.

Monosaccharide ♦ CARBOHYDRATES.

Moor, Moorland. Open country, usually on high ground, having acid soil (mostly PEAT) and covered with HEATHER, coarse grasses, sedges, bracken, etc.

Mor. An infertile acid soil containing undecayed plant remains, found particularly on sandy HEATHS. (♦ MULL)

Morbidity. The percentage of a herd, flock, etc., that becomes infected by a particular disease.

Mosaic Diseases. Various virus diseases of crops causing light and dark mottling on affected leaves, e.g. VIRUS YELLOW.

Moss. 1. A northern term for a BOG.

2. A class of small lower plants (Bryophyta) with simply constructed leaves, and lacking woody tissue.

Mould. 1. Loose soil with a good tilth. Also soil rich in HUMUS.

2. A general term for a variety of minute Fungi which produce a powdery or downy growth. Also called MILDEW. (♦ PHYCO-MYCETES)

Mouldboard. A curved steel plate on the BODY of a plough which turns over the FURROW slice. Also called breast or shell-board. Design varies according to the type of plough. The GENERAL PURPOSE MOULDBOARD is almost flat and is twisted along its length, producing a continuous almost unbroken furrow slice. (♦ DIGGER, SEMI-DIGGER and LEA PLOUGH). When worn, the leading edge or shin of the mouldboard may be replaced (*see p. 338*).

Mountains. A loose descriptive term for E.E.C. structural surpluses in certain commodities, e.g. butter mountain, purchased and stored as INTERVENTION STOCKS under the COMMON AGRICULTURAL POLICY. (♦ also WINE LAKE)

Movement Licence. A licence to move animals in accordance with certain conditions. Licences are issued by Local Authority DISEASES OF ANIMALS INSPECTORS and by Ministry of Agriculture, Fisheries & Food inspectors (depending on the nature of the movement) and are aimed at controlling the movement of animals to aid disease control.

Not all animals are subject to licensing requirements. All movements out of an INFECTED AREA, INFECTED PLACE or CONTROL AREA are subject to licence as are many movements within such areas. Requests for licences can be refused if the movement is considered to constitute a disease risk or is contrary to the law.

Movement Record. A record required to be kept under the Movement of Animals (Records) Order 1960 (as amended) showing details of each animal movement onto and off the premises to which the record relates. Certain cattle movements are exempt and Movement Records need not be kept by certain persons (e.g. market auctioneers, slaughterhouse managers and exhibitors).

The movement record, which shows such details as the date

of the movement, the number, breed and sex of the animals, and their destination, is the main means by which animal movements are traced, and hence animal contacts identified, so that disease can be isolated and eradicated. The proper completion of records is essential to disease control.

Mow. 1. To cut grass or a forage crop. (♦ MOWER)
2. A pile or stack of hay or corn, particularly in a barn. Also the place in a barn where such a stack is put (sometimes called mowhay).

Mower. An implement used to cut a grass crop. The main types are:
(*a*) CUTTER BAR mowers, incorporating a high speed reciprocating knife mechanism (♦ FINGERS). They may be fully mounted or partially supported on wheels, and driven by p.t.o. or gearing from the wheels. The crop receives a single cut, without further laceration and is deposited in a neat SWATH.
(*b*) Rotary mowers. The flail type consist of flail knives mounted on a horizontal axle, which lacerate the grass, assisting drying (♦HAYMAKING), and are useful for cutting laid crops. The vertical spindle type consist of one to four vertical drums containing hinge-mounted knives which rotate horizontally at high speed.
(*c*) Cylinder mowers, usually comprising a gang of three independent horizontal cylinders, each bearing a number of cutting blades. Normally tractor-drawn and used for cutting short grass. Rarely used on farms nowadays.

Mowhay ♦ MOW.

Muck. A general term for FARMYARD MANURE.

Muckspreader ♦ MANURE SPREADERS.

Mug. A woolly-faced sheep.

Mulch. Material such as straw, leaves, sawdust, grass cuttings, loose soil, etc., applied to the soil surface to protect the soil and plant roots from drying out and from the effects of heavy rain, freezing, etc.

Mule. 1. A cross-bred sheep derived from a BORDER LEICESTER ram and a hardy hill ewe. Mule ewes are used to produce fat lambs.
2. The hybrid offspring of a male ass and a MARE. (♦ HINNY)

Mull. A soil containing a good mixture of partly decayed ORGANIC MATTER, HUMUS and MINERAL particles, and containing plentiful earthworms. (◗ MOR)

Multiple Suckling. A system of managing cattle in which dairy breed NURSE COWS are used to suckle several beef calves at a time. Between suckling sessions the cows are put out to graze. (◗ DOUBLE-SUCKLING, SINGLE-SUCKLING)

Muriate of Potash. A FERTILIZER containing at least 60% potassium chloride, and as much as 3% sodium chloride or common salt. Available as a straight granular fertilizer but mostly incorporated in COMPOUND FERTILIZERS.

Murray Grey. A breed of polled beef cattle developed in Australia in the Murray Grey Valley. A very gentle breed, silver to grey in colour. Cows have good milking qualities and calve easily, producing quick-growing calves.

Mustard ◗ BLACK MUSTARD and WHITE MUSTARD.

Mycelium. The web of threads or hyphae which constitute the vegetative feeding part of fungi.

N

National Agricultural Centre. Founded by the Royal Agricultural Society of England in 1963 and situated at Stoneleigh, Warwickshire, the site of the Royal Show. Its objects are:

(*a*) to demonstrate, both at the Royal Show and throughout the year, modern developments in agricultural practice and science, embracing crops, livestock, machinery and equipment, through the results of research;

(*b*) to provide the facilities and environment where progressive farmers, traders, manufacturers, etc., can meet regularly throughout the year to keep abreast of new developments and discuss their application;

(*c*) to stimulate a continuing interchange of ideas;

(*d*) to provide a shop window for British Agriculture and;

(*e*) to encourage continuing co-operation between the agricultural industry and the manufacturing and distribution industries associated with it.

Many organisations connected with agriculture and rural life are represented at the Centre, and it acts as a focus for expression of developments in agriculture of benefit to industry. Facilities and activities include livestock and cereal demonstration units, exhibition centres, demonstrations, conferences, seminars and education.

National Association of British and Irish Millers. A trade association founded in 1878, now collectively representing the flour milling industry of the U.K. on all matters affecting it, e.g. legal, technical, commercial, scientific, transport, industrial relations, health and safety, policy affecting the use and purchase of home-grown and imported wheat, and the production of milling by-products.

National Farmers' Union. A voluntary organisation established in 1908 in Lincolnshire with the aim of promoting the interests of farmers and providing a forum for the exchange of ideas and information concerning farming. The N.F.U. now represents farmers and growers in England and Wales in negotiations with

Government and, through C.O.P.A., with the E.E.C., on all matters relating to agriculture (♦ ANNUAL REVIEW); and in liaison with a wide variety of public and statutory authorities, official bodies and interest groups, etc. Based in London, there are branches in all counties (some on an amalgamated basis), each comprising a number of more local branches. Commodity and other committees at national and branch level, attended by member delegates, discuss and determine policy. N.F.U. services include information and advice through national, county and local branch staff. The latter also provide a specialised farm insurance service through the closely associated N.F.U. Mutal Insurance Society Ltd. Separate but similar Farmers' Unions exist in Scotland, N. Ireland, Isle of Man and the Channel Islands.

National Institute of Agricultural Botany. An independent organisation established in 1919 and financed by grant aid from the Ministry of Agriculture, Fisheries and Food (M.A.F.F.), providing impartial advice on all aspects of seed quality. The N.I.A.B. tests all new varieties of most agricultural and vegetable crops, assessing field characters, quality, yield, distinctness, uniformity and stability. The results are used by M.A.F.F. in compiling the U.K. NATIONAL LISTS, and with those of further tests by the N.I.A.B. to produce Recommended Lists of varieties, widely used by farmers. These lists contain the better varieties from those available on the National Lists and are regularly revised. The N.I.A.B., on behalf of M.A.F.F., also supervises the technical operations required for the U.K. Seed Certification Schemes in England and Wales and trains inspectors to examine crops entered for certification (♦ SEED CERTIFICATION). The Official Seed Testing Station carries out the statutory tests on seed samples required for certification, tests samples of farmers' own seed, and issues reports on germination, purity and presence of disease.

National Lists (of Seed Varieties). Under the Plant Varieties and Seeds Act, 1964, the seeds of most agricultural and horticultural crops (i.e. cereals, fodder plants, beet, oil and fibre plants, and vegetables) marketed or sold must comply with prescribed standards for varietal and analytical purity, germination and weed seed content. Since joining the E.E.C. it has become an offence to sell or offer for sale seed of plant varieties not included in either (*a*) the U.K. National Lists of seeds meeting the prescribed standards, which are compiled and maintained by the Ministry of Agriculture, Fisheries and Food, or (*b*) the E.E.C. Common Catalogue of varieties. (◗ NATIONAL INSTITUTE OF AGRICULTURAL BOTANY)

National Milk Records Scheme ◗ MILK RECORDING.

National Pig Breeders' Association. Established in 1884 to look after the interests of all the major breeds of pedigree pigs found in the U.K. It is now also responsible for the work of the Breeding Companies which are of increasing importance.

All breeds within the Association maintain accurate breeding records. The N.P.B.A. also undertakes political and promotional work on behalf of all pig breeders, and organises sales and carcase competitions in which breeders can exhibit their animals.

National Proficiency Tests Council ◗ PROFICIENCY TESTS.

National Seed Development Organisation. An organisation, incorporated as a limited company in 1967, the primary function of which is to develop in the U.K. and abroad, plant varieties resulting from research carried out with the aid of public funds. The N.S.D.O. is the commercial agent for and distributor of BASIC SEED of ABERYSTWYTH VARIETIES and is responsible for such matters as growers' prices, contracts, cleaning charges, etc., with the advice and assistance of the British Herbage Seeds Committee.

National Union of Agricultural and Allied Workers. A trade union organisation founded originally as the National Agricultural Labourers Union in 1906. It now has a membership of some 75 000 members engaged in agriculture, horticulture, forestry and allied rural industries. The National Union of Agricultural and Allied Workers, as it became known in 1968, functions through its 2 500 branches based on villages throughout England and Wales, but also has a few industrial branches based on food processing factories in Scotland.

The Union's Rule Book states that it aims to advance the social condition of its members, and exists for their mutual aid and protection. Through its staff the N.U.A.A.W. represents members in connection with all kinds of employment and housing matters, and in addition through the legal aid scheme instructs solicitors in a wide variety of common law claims arising from industrial and road accidents as well as breaches of contracts of employment.

Navel-ill ◊ JOINT-ILL.

Navy Bean ◊ BEANS.

Neat. A cow, heifer, bull or steer.

Neck Collar. A broad strap which fits round the neck of a cow or horse, securing it in a stall.

Necrobacillosis ◊ ORF.

Neep. A Scottish term for a TURNIP.

Negative M.C.A. ◊ MONETARY COMPENSATORY AMOUNT.

Nematodes ◊ ROUNDWORMS.

Nematodirus Disease. A disease of lambs caused by parasitic worms (*Nematodirus spp.*). They are transmitted by eggs in the faeces which later hatch in pasture to infect the following year's lambs, causing diarrhoea and lack of thrift, and sometimes anaemia.

Nest Boxes. Boxes in which poultry kept under a system of DEEP LITTER are able to lay eggs. They may be of the single type designed to accommodate 5 birds, or the communal type accommodating up to 100 birds. On large poultry farms the boxes are often arranged so that newly laid eggs roll on to a conveyor belt for delivery to a packing room.

Net Energy Value (N.E.). The METABOLISABLE ENERGY value of an animal food less that fraction lost in the form of heat. This value is theoretically a more accurate indication of the utilisable energy in a food than its M.E. value. N.E. values vary according to the physiological state of an animal, e.g. whether it is being fattened or is on a MAINTENANCE RATION, etc. The determination of N.E. values is expensive and time consuming and they are available for only a few foods.

Nettlehead. A virus disease of HOPS resulting in weak deformed BINES with little ability to climb or fruit.

Neutral Soil. A soil in which the surface layer is neither acid nor alkaline in reaction, i.e. its pH is about 7.0 (strictly, neutrality is pH 7.0). (♦ ACID SOIL, ALKALINE SOIL)

Neutralising Value. An indication of the ability of a liming material (♦ LIME) to neutralise soil acidity compared to the effect of CALCIUM OXIDE as a common factor. The n.v. of a liming material is the number of kgs of calcium oxide which would have the same neutralising effect as 100 kg of the material. (♦ LIME REQUIREMENT)

New Hampshire. A heavy, dual-purpose breed of poultry originating in America, red in colour, with a single comb, yellowish legs and feet, and producing brown eggs.

New York Dressed Poultry ♦ TRADITIONAL FARM FRESH POULTRY.

Newcastle Disease. A highly infectious NOTIFIABLE DISEASE of poultry caused by a virus. Symptoms include reduced yield of often soft and misshaped eggs, lack of appetite, troubled breathing, nasal discharges and foul smelling, yellowy, watery diarrhoea. High mortality amongst young birds is common. Control is by voluntary vaccination. The disease resembles fowl plague (rare in Britain) and both are also known as fowl pest.

Nib. 1. One of the hand grips on the shaft of a SCYTHE. Also called dole, hand-pin, thole.
2. A wedge of HOPS removed from a POCKET for examination, as distinct from a cube-shaped sample taken for valuation and sale purposes.

Nick. A term used of two particular animals found to produce consistently good offspring when mated together. They are said to nick well.

Nidget. A horse-drawn hoe once used for cultivating the ALLEYWAYS in hop gardens.

Nitram. A granular form of AMMONIUM NITRATE containing about 34.5% nitrogen.

Nitrate. A general term for any nitrogenous FERTILIZER. The term is specifically applied to salts containing the nitrate ION (NO_3). The nitrogen in many COMPOUND FERTILIZERS is in the form of the ammonium ion (NH_4), which is quickly changed to the nitrate form by soil bacteria (◗ NITROGEN CYCLE). Nitrates ions are more readily absorbed by some plant roots (e.g. cereals) whilst others (e.g. potatoes, grass, etc.) absorb nitrate and ammonium ions equally well.

Nitrate of Potash. (KNO_3) Potassium nitrate. A FERTILIZER containing about 13% NITROGEN and 44% POTASH used particularly by horticulturists. Also called saltpetre. (◗ POTASH NITRATE)

Nitrate of Soda. ($NaNO_3$) Sodium nitrate. A FERTILIZER containing about 16% NITROGEN derived from natural deposits of 'Caliche' in Chile (also called Chile saltpetre). A fast acting fertilizer, sometimes used as a TOP DRESSING, particularly by horticulturists.

Nitrification. The conversion by various soil BACTERIA of organic nitrogenous compounds which are not available to plants, to NITRATES which they can readily absorb. Organic matter from decaying animal and plant remains in the soil is reduced by enzyme action to amino compounds. These are broken down by *Pseudomonas* to ammonia and ammonium compounds, then to nitrates by *Nitrosomonas*, and eventually to nitrates by *Nitrobacter*. (◗ NITROGEN CYCLE)

Nitrobacter ◗ NITRIFICATION.

Nitrochalk. A granular FERTILIZER containing AMMONIUM NITRATE blended with ground CHALK, containing nowadays about 25% NITROGEN. The chalk content reduces the acidifying effect on the soil of the ammonium nitrate.

Nitrogen. (N). An odourless, colourless, gaseous element, forming about 78% of the atmosphere. It is present in the soil in inorganic forms such as ammonium compounds (◗ AMMONIA) and NITRATES, and in various organic forms such as AMINO ACIDS and protein complexes as constituents of HUMUS, which are eventually broken down by bacterial action to nitrates (◗ NITROGEN CYCLE). As a major plant nutrient nitrogen is applied to the soil in the form of FERTILIZERS or MANURES. Nitrogen is essential to life, mainly as a constituent of PROTEIN and nucleic acids (e.g. DNA), and in plants is vital for vegetable growth. Nitrogen deficiency results in stunted plants with yellowish leaves and poor root development. Excess produces dark green succulent foliage, but if prolonged may retard the maturing of crops, and can cause cereals to lodge (♦ LAID CROP) due to the growth of long weak straws.

Nitrogen Cycle. The circulation of nitrogen atoms and compounds in nature, mainly caused by living organisms. Inorganic forms of nitrogen (mainly NITRATES) are absorbed by plants and assimilated into complex organic compounds (e.g. PROTEINS). Plants either die and decay, or are eaten by animals, which produce excreta and also eventually die and decay. The nitrogenous products of excretion and decay are progressively degraded in the soil, mainly by bacterial NITRIFICATION eventually to nitrates. In addition there is secondary circulation of atmospheric nitrogen which is 'fixed' and assimilated into proteins by certain bacteria living symbiotically in root NODULES of various leguminous plants and by various free-living bacteria in the soil (♦ NITROGEN FIXATION). A certain amount of nitrogen is released from nitrogen compounds by denitrifying bacteria.

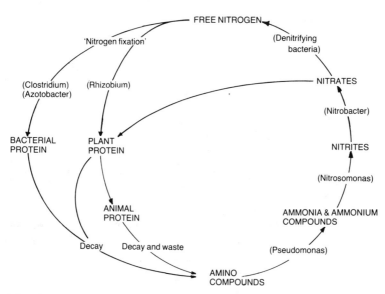

The Nitrogen Cycle.

Nitrogen Fixation. The combination of free atmospheric nitrogen into nitrogenous compounds by various BACTERIA in the soil. Certain aerobic bacteria (*Rhizobium spp*) living symbiotically in the root NODULES of leguminous plants are able to 'fix' nitrogen using energy supplied by the plants. A number of free-living bacteria (e.g. aerobic *Azotobacter* and anerobic *Clostridium*) can also 'fix' nitrogen using energy derived from soluble carbohydrates in HUMUS. A certain amount of nitrogen in the atmosphere is combined into organic form by lightning. (♦ NITROGEN CYCLE)

Nitrogen-free Extract. One of the fractions measured by the PROXIMATE ANALYSIS of FEEDINGSTUFFS, mainly containing soluble CARBOHYDRATES such as STARCH and SUGARS.

Nitrogenous Fertilizers ♦ FERTILIZERS.

Nitrosomonas ♦ NITRIFICATION.

N/K Ratio. The ratio of the NITROGEN content of a FERTILIZER to its POTASH content. (♦ PLANT FOOD RATIO)

Noctuid Moth ♦ CUTWORM.

Node. The point on the stem of a plant from which a leaf or BRACT arises.

Nodules. Swellings on the roots of leguminous plants (♦ LEGUME) containing bacteria (*Rhizobium spp*) living symbiotically with the root tissue and capable of converting atmospheric NITROGEN into nitrogenous compounds (♦ NITROGEN FIXATION) using energy derived from CARBOHYDRATES supplied by the plant. These compounds are used to assimilate PROTEIN. Some are released to the soil and converted to NITRATES which may be used by other plants such as the grass in a mixed clover-grass LEY. The stored protein in legume crops is released to the soil on death or ploughing-in, and is degraded to nitrates (♦ NITROGEN CYCLE). Legumes such as beans or clover are thus often grown prior to cereal crops to enrich the soil with nitrogen. The various strains of bacteria which form nodules are usually naturally present in the soil, but to achieve nodulation with LUCERNE the seeds are usually coated or 'inoculated' with the appropriate bacterial culture before sowing. The bacteria enter the plant via the root hairs.

Norfolk Rotation. A system of crop ROTATION which dominated English farming from the mid-eighteenth century to the start of the twentieth century. It was based on four courses in the following order, (*a*) roots (e.g. turnips, swedes, mangels or kale), (*b*) barley (sometimes oats), (*c*) LEY, mainly red clover (often mixed with ryegrass), although beans, peas, sainfoin and trefoil were sometimes used, and (*d*) wheat. Sheep were folded (◗ FOLD) on the root crops and grazed on the leys. This system maintained soil fertility (ARTIFICIAL FERTILIZERS were not available), controlled pests and weeds and provided continuous employment.

North Country Cheviot ◗ CHEVIOT.

North Devon ◗ DEVON.

North Holland Blue. A heavy breed of poultry, dark barred in colour, with white legs, a single comb, and laying tinted eggs. A rapid growing breed once mainly produced for the table.

Northern Dairy Shorthorn. A dairy breed of cattle developed in the dales of the northern Pennines, from DURHAM and AYRSHIRE cattle and suited to the harsh upland conditions. Mainly light roan in colour, but red, white and red and white are also known. The characteristic horns curve forwards and upwards. The average annual MILK YIELD in 1978/79 was 4 069 kg (3 963 litres; 871 gal) with a 3.64% BUTTERFAT content.

Notifiable Diseases. (*a*) Those diseases of animals and poultry which, under the DISEASES OF ANIMALS ACT 1950, must be reported to the police when the owner or his veterinary surgeon suspect that an outbreak has occurred on a farm. The Divisional Veterinary Officer of the MINISTRY OF AGRICULTURE, FISHERIES AND FOOD is notified by the police and arrangements made for the animals to be examined. The list of diseases currently includes the following:

African horse sickness
Anthrax
Aujeszky's disease
Cattle plague in ruminating animals and swine
Dourine in horses, asses, mules or zebras
Enzootic bovine leukosis (*continued overleaf*)

Epizootic lymphangitis in horses, asses and mules
Equine infectious anaemia
Equine encephalomyelitis
Foot-and-mouth disease in ruminating animals and swine
Fowl pest (fowl plague and Newcastle disease) in all kinds of
    poultry
Glanders and farcy in horses, asses and mules
Parasitic mange in horses, asses and mules
Pleuro-pneumonia in cattle
Rabies
Sheep pox
Sheep scab
Sheep fever
Swine fever
Swine vesicular disease
Teschen disease of pigs
Tuberculosis (certain forms only) in cattle

(*b*) Those diseases and pests of plants, the existence or suspected existence of which must, by law, be notified to the Ministry of Agriculture, Fisheries and Food. Currently included are:

Colorado beetle
Wart disease of potatoes
Progressive wilt disease of hops
Red core disease of strawberries
Fireblight disease (in certain fruit and ornamental trees)
Plum pox (Sharka disease) (in certain fruit and ornamental
    trees)

Nott ◗ BERKSHIRE KNOT.

N.P.K. An abbreviation for NITROGEN, PHOSPOROUS, and POTASSIUM in describing the composition of a FERTILIZER. (◗ UNIT OF FERTILIZER, PLANT FOOD RATIO)

Nubian. A short-haired breed of goat which originated in North-East Africa and was crossed with short-haired British goats to produce the ANGLO-NUBIAN breed. Characterised by long drooping ears, a Roman nose, and an arched face with prominent forehead. Various colours exist, the most common being red, black and tan, often combined with white. Noted for its milk rich in BUTTERFAT.

Nucleus. A body in most living CELLS, surrounded by PROTOPLASM, which contains the CHROMOSOMES and exercises control over the activity of the cell, particularly during cell division. (♦ MEIOSIS, MITOSIS)

Nucleus Herd ♦ PIG IMPROVEMENT SCHEME.

Nucleus Multiplier Herd ♦ PIG IMPROVEMENT SCHEME.

Nurse Cow. A cow used to suckle a calf produced by another cow.

Nurse Crop ♦ COVER CROP.

Nursery. A place or piece of ground where young plants are cultivated for transplanting or sale.

Nut. 1. A small cube or cylindrical piece of compressed MEAL. (♦ CAKE)
2. A type of FRUIT, dry, one-seeded, and usually with a hard woody shell, e.g. hazel nut.

Nutritive Ratio (N.R.). The relationship in a food ration between the amount of DIGESTIBLE CRUDE PROTEIN and the fat-producing constituents (digestible CARBOHYDRATES, oil and FIBRE present). Also called Albuminoid Ratio. This ratio is important in ensuring a balance in animal feed between digestible protein and other digestible non-proteins. Too much starch may cause digestion troubles, too much protein stresses the kidneys. Well balanced foods have ratios of 6:1. (♦ also BALANCED RATION, PROTEIN EQUIVALENT, STARCH EQUIVALENT)

O

Oast. A kiln in which hops are dried. An oast-house is a building in which hops are dried and pressed into POCKETS, consisting of one or more kilns and a cooling room containing a press.

Oatmeal. An animal FEEDINGSTUFF consisting of ground oats with the HUSKS removed to varying extents. The husks provide a roughage element to the diet. Fed either dry or together with other constituents in a wet mash. Also fed as a porridge (sometimes called brose) to animals which are ill or recovering from sickness. (◗ SUSSEX GROUND OATS)

Oats. A cereal crop with a high FIBRE content, the grains of which are used for stock-feeding and for making bread, porridge and OATMEAL. Grown mainly in the cooler and wetter areas of North and West England, Wales and Scotland, although to a decreasing extent as it yields less well than BARLEY and is more difficult to harvest. Most cultivated oats are of the species *Avena sativa* although some *A. strigosa* is grown in the Welsh Hills and West Scotland. Oats are distinguished from other CEREALS by having a flowering head in the form of a tall PANICLE with clusters of slender spreading branches bearing SPIKELETS (*see p. 95*). (◗ WILD OATS)

Obligatory Parasite ◗ PARASITE.

Occupation Licence. A licence issued by a DISEASE OF ANIMALS INSPECTOR or an inspector of the Ministry of Agriculture, Fisheries and Food, in circumstances where a road severs a farm. It permits the movement of cattle across the road and is required only where Infected Area (◗ INFECTED PLACE) restrictions are in force in respect of FOOT AND MOUTH DISEASE. Licences are issued only in exceptional circumstances due to the risk of contamination of the road by the cattle.

Oestrogen. A sex hormone produced by female animals which controls the development of female characteristics including the MAMMARY GLANDS and controls the oestrus or 'heat', the period during which females are receptive to mating with males. (♦ PROGESTERONE)

Oestrus 1. ♦ OESTROGEN.
2. ♦ SHEEP NOSTRIL FLY.

Off-going Crop ♦ AWAY-GOING CROP.

Official Seed Testing Station ♦ NATIONAL INSTITUTE OF AGRICULTURAL BOTANY.

Oil Cake ♦ CAKE.

Oil of Vitriol ♦ SULPHURIC ACID.

Oilseed Crops. Various crops grown for their seeds, rich in oil and extracted after crushing for the manufacture of margarine, cooking oil and salad oil, etc. The residues are usually processed to produce CAKE, rich in PROTEIN and valued for stock-feeding. Only LINSEED, OILSEED RAPE and SUNFLOWER are grown in Britain. Various cakes derived from tropical oilseed crops are also commonly used (e.g. GROUNDNUT CAKE, SOYA BEAN CAKE).

Oilseed Rape. Certain varieties of RAPE grown under contract for their seeds, rich in oil and used for the manufacturing of margarine, cooking and salad oils. The crop is often used as a useful BREAK between cereals and is conveniently harvested with the same machinery. A valuable CAKE is a by-product after seed crushing and oil extraction.

Old English Game. A heavy table breed of poultry of extremely variable body and leg colour. The breed produces considerable quantities of high quality breast meat and is used in crossing. Hens lay relatively few white eggs.

Old Gloucester ♦ GLOUCESTER.

Olland. The AFTERMATH following the harvesting of HAY from a clover LEY.

Omasum. The third stomach of a RUMINANT in which food is ground following partial digestion in the second stomach or reticulum. Also called manyplies or psalterium.

311

Omnivore. An animal which eats both vegetable and animal foods (e.g. the pig), as distinct from a flesh-eating carnivore or a HERBIVORE.

Once-grown Seed. 'Seed' from one further multiplication of plants grown from a certified grade of seed (♦ SEED CERTIFICATION, SEED POTATOES). For most species the U.K. crop seed regulations only permit the sale of certified seed. Farmers may use, for their own needs, once-grown (or twice-grown, etc.) seed from certified seed.

One-way Plough ♦ REVERSIBLE PLOUGH and PLOUGHING.

On-off Principle ♦ ROTATIONAL GRAZING.

Oof ♦ ORF.

Open Furrow. A type of FURROW in which the furrow slices are turned in opposite directions so that they lie facing away from each other. Open furrows are used when completing the ploughing of a LAND or a field (the finish).

Opening ♦ RIDGE.

Open-textured Soil ♦ SOIL TEXTURE.

Opposite Leaves. Leaves attached in pairs to the stem of a plant but on opposite sides of the same NODE. (♦ ALTERNATE LEAVES)

Orchard. An area of land used for growing fruit trees (e.g. apples, pears, plums, cherries, etc.) either on a small scale or on a more extensive commercial scale, the latter, in the U.K., being mainly confined to areas within the southern half of England. Nowadays commercial fruit varieties are normally grown on the more dwarfing rootstocks selected for their ability to control the size, growth and cropping characteristics, since it is now economically desirable to harvest early in the life of the trees and to do so from the ground or short steps. In addition tree shape is greatly affected by the pruning system used. At the end of their useful life orchards are grubbed, the tendency being to grub trees on dwarf rootstocks at a younger age than those on vigorous rootstocks. Orchards producing desert and culinary fruit are mainly operated by specialist growers, whilst a certain number of orchards producing cider fruit are still combined with livestock grazing, mainly in the westerly parts of England. Similarly in the South-East, mainly Kent, some cherry orchards are associated with sheep grazing.

Orchard Pig ♦ GLOUCESTER OLD SPOT.

Orf. A disease mainly of sheep, but also affecting cattle and goats, due to a primary attack by a virus which causes pustules. These are secondarily infected by a fungus (*Fusiformis necrophorus*) and develop into ulcers and scabs on the lips and mouth, and on other parts including the face, legs, tail, genitals, teats, etc. Also called contagious pustular dermatitis, contagious ecthyma, necrobacillosis, oof and ulcerative stomatitis.

Organic. A term applied to a substance containing CARBON combined with hydrogen, and very often also with nitrogen, oxygen and other chemical elements, forming complex molecules. (♦ INORGANIC, ORGANIC MATTER)

Organic Fertilizers. Those FERTILIZERS derived from animal products (e.g. DRIED BLOOD, HOOF AND HORN MEAL, SHODDY, etc.) and from plant residues (e.g. malt culms, castor seed meal) all of which have a considerable NITROGEN content. Also applied to more bulky organic materials such as MANURES. (♦ ARTIFICIAL FERTILIZERS)

Organic Matter. 1. The ORGANIC substances in the soil, including animal and plant residues in various stages of decomposition (♦ HUMUS), cells and tissues of soil organisms (e.g. bacteria, fungi, insects, earthworms, etc.), those organic compounds synthesised by such organisms, and organic materials added to the soil as fertilizers, e.g. MANURES. The organic fraction of the soil is intimately mixed with the MINERAL MATTER. The organic matter content of soil varies according to soil type. On a dry weight basis it is usually in the range 2–15%, but certain organic soils consist almost entirely of organic matter (e.g. peats, black FEN soils).
2. One of the two main components of the DRY MATTER of a FEEDINGSTUFF (the other being Mineral Matter) and consisting mainly of CARBOHYDRATES, FATS and PROTEINS.

Organochlorine Compounds ♦ CHLORINATED HYDRO-CARBONS.

Organophosphorus Compounds. A group of synthetic non-persistent chemicals mainly used as insecticides (e.g. MALATHION), nowadays often in preference to the more persistent CHLORINATED HYDROCARBONS. They may have either a SYSTEMIC or a non-systemic action and mainly affect the nervous system.

Orpington. A general purpose breed of poultry combining both laying and table qualities. Black, buff or white in colour, with white legs and feet, a single comb, and hens laying tinted eggs.

Osier. Any willow grown as a COPPICE crop, the flexible twigs of which are used in basket making, particularly *Salex viminalis.* Also called withy.

Osmosis. The phenomenon by which the solvent of the weaker of two solutions of differing concentration passes through a semi-permeable membrane separating them, so that the concentrations become equalised. This is the process by which water in the SOIL SOLUTION passes into the cells of the root hairs of plants.

Out-crossing. The mating of unrelated animals of the same breed, normally in order to introduce a desired CHARACTER to the strain which is otherwise absent. (♦ IN-BREEDING, LINE-BREEDING, CROSS-BREEDING)

Outfall. The point where a DRAIN discharges into a ditch, or where a watercourse discharges to a river or the sea (sometimes associated with a sluice or flap valve).

Outfield, Out-lying Field. 1. A field situated some distance from the farm buildings usually on the farm boundary.
2. A Scottish term for a field under continuous arable cropping without being manured. (♦ INFIELD)

Outliers. Livestock kept outside during winter.

Out-run Land. The open unenclosed areas surrounding a HILL FARM, normally some distance from the farm buildings. (♦ IN-BYE LAND)

Outs ♦ INS.

Outside Bat ♦ STRAINING POLE.

Out-wintering. The practice of keeping cattle in the fields during the winter as distinct from keeping them under cover.

Ova ♦ OVUM.

Ovary. 1. The hollow organ at the base of the CARPEL of a flower, containing one or more OVULES (the future seeds).
2. The organ in a female animal which produce ova (♦ OVUM) and sex hormones, e.g. OESTROGEN.

Ovine. Of sheep, or sheep-like. Derived from *Ovis*, generic name for sheep.

Oviparous. Laying eggs which hatch to release the young. Birds, amphibians, most insects and some reptiles are oviparous. (♦ VIVIPAROUS)

Ovule. A small structure in the OVARY of a plant containing the egg-cell or female GAMETE which, after fertilization, develops into a seed.

Ovum. An egg-cell or female GAMETE. Usually restricted to those in animals which, after fertilization by a male gamete or sperm, develops into an EMBRYO. (Plural, ova)

Ox. A general term for a male or female of domestic CATTLE. Particularly applied to a castrated male or STEER, especially when used for draught purposes.

Oxford Down. The largest of the DOWN BREEDS of sheep derived from crosses between the HAMPSHIRE DOWN and the long-woolled COTSWOLD. Characterised by a dark brown face and legs and a high, boldly-held head, with a distinctive cap of wool over the forehead and on the cheeks. The breed was developed on the arable farms of the Oxfordshire and Gloucestershire DOWNS but is nowadays widely found and rams are used to produce cross-bred STORE lambs. A hardy breed suited to wet areas.

Oxidase. A type of ENZYME which acts as a catalyst in the OXIDATION of substances, removing hydrogen which then combines with oxygen to form water.

Oxidation. The combination of OXYGEN with a substance, or the removal of hydrogen from it ($\blacklozenge$ OXIDASE). In cellular RESPIRATION, oxygen reacts with CARBOHYDRATES such as GLUCOSE yielding carbon dioxide and water as by-products and liberating energy. This process is comparable to the oxidation of substances by burning, in which energy is liberated as heat. The term is also used to describe chemical reactions in which electrons are lost from atoms such as when ferrous iron (Fe^{2+}) is oxidised to ferric iron (Fe^{3+}).

Oxygen. (O). An odourless, colourless, gaseous element, forming about 2% of the atmosphere and a constituent of water. Essential to most living organisms for RESPIRATION and necessary for combustion.

Oxytocin. A hormone secreted by the posterior lobe of the pituitary gland. It is responsible for activating the LET-DOWN mechanism, releasing milk from a cow's udder. It also causes the contraction of smooth muscles in the uterus.

P

Packhouse. A building in which fruit, vegetables, eggs, etc., are packed into boxes or containers for despatch from a farm to a market, retail outlet, store, etc. CASUAL WORKERS are usually employed seasonally for packhouse work.

Paddle. A long-handled, spade-like tool used to remove mud from plough SHARES and MOULDBOARDS, etc. In some areas called spud.

Paddock. A relatively small enclosed pasture, usually near the farm buildings.

Paddock Grazing. A system of managing grassland in which grazing areas are divided into paddocks by permanent or semi-permanent fences. Each paddock is grazed in turn and then allowed to rest. (◊ GRAZING SYSTEMS)

Pale, Paling. A wooden stake driven upright into the ground for fencing. Also a wooden board nailed vertically with others to horizontal pieces to form pale fencing or paling.

Palea. The upper of the two BRACTS which enclose grass flowers. (◊ LEMMA)

Palm Kernel Cake. A CAKE derived from the residue of the kernels of the fruit of the African oil palm (*Elaeis guineensis*) after extraction of the oil. It has a high FIBRE content and is slightly gritty, and is therefore somewhat unpalatable. It is less rich in PROTEIN than other cakes and is slow to absorb water. Sometimes used to STEAM-UP dairy cows.

Palmate Leaf. One with several lobes or leaflets, like the fingers of a hand, spreading from a central point, e.g. horse-chestnut.

Pan. A hard layer in the soil that is (*a*) compacted, such as a plough pan caused by continual ploughing to the same depth, (*b*) indurated, such as an iron pan in which the soil particles have become cemented together by precipitated iron and silicates, or

(c) has a very high clay content. Pans often impede drainage and root growth, and are broken down using a SUBSOILER. Also called hardpan or sole.

Pancreatic Juice. An alkaline DIGESTIVE JUICE secreted by the pancreas containing various ENZYMES including AMYLOPSIN, STEAPSIN and TRYPSIN.

Panicle. A type of inflorescence comprising a vertical main stalk with divided branches or RACEMES each bearing stalked SPIKELETS. Typified in OATS. (♦ CEREALS)

Pannage. An old right to allow pigs to BROWSE in a wood or forest. Also the actual food eaten in such circumstances by pigs. (♦ MAST)

Paraquat. A CONTACT HERBICIDE which disrupts PHOTO-SYNTHESIS causing CHLOROSIS. Used to control a wide range of annual broad-leaved weeds and grasses. It is very poisonous to animals.

Parasite. An organism which lives in or on another (the HOST) from which it derives food, for all or part of its life. Often, though not always, harmful to the host. Animal parasites include internal worms (e.g. ROUNDWORMS, TAPEWORM, etc.) and external FLIES, LICE, TICKS, etc. Some are parasitic on a secondary host during part of their life cycle (e.g. tapeworm, LIVER FLUKE). Plant parasites include fungi and bacteria (causing such diseases as MILDEW, RUST, FIRE BLIGHT, etc.), various plants such as dodder and mistletoe, and numerous insects and their larvae, e.g. APHIDS, FRUIT FLY, EELWORMS, etc. A semi-parasite is one which feeds partly on its host but also manufactures some of its own food (e.g. mistletoe). (♦ SAPROPHYTE)

Parathion. An extremely poisonous ORGANOPHOSPHORUS COMPOUND (diethyl-para-nitrophenyl-thiophosphate) used to control a variety of insect pests including MITES, APHIDS, EELWORMS, etc.

Parent Material. Unconsolidated, chemically weathered mineral (sometimes organic) matter derived from rock (parent rock) form which soil develops.

Parent Rock ♦ PARENT MATERIAL.

Parent Stock. The individual plants or animals from which others have been bred, often in large numbers.

Park Cattle ◗ BRITISH WHITE.

Parkland. Park-like grassland dotted with mature, usually individual, trees.

Parlour ◗ MILKING PARLOUR.

Parthenogenesis. The growth of an OVUM into a new individual without FERTILIZATION. It occurs naturally in some plants (e.g. dandelion) and animals (e.g. APHIDS), and the offspring are normally all DIPLOID and genetically identical to the parent.

Partial Drought. A period (in the U.K.) of at least 29 consecutive days, during which the mean daily rainfall does not exceed 2.54 mm (0.1 in.). Not an internationally accepted definition. (◗ ABSOLUTE DROUGHT, DRY SPELL)

Pasteurised Milk. MILK heated to a temperature which kills the majority of the BACTERIA present, but without affecting its palatability. The treatment applied is that needed to destroy *Mycobacterium tuberculosis* (◗ TUBERCULOSIS), the most heat resistant PATHOGEN normally present in cow's milk. Most milk is pasteurised using the High Temperature, Short Time method (H.T.S.T.) by which it is heated to at least 72°C (161°F) for 15 seconds and then rapidly cooled to 10°C (50°F) or less. An alternative is the Holder method by which milk is heated to between 63°–66°C (140°–150°F) for at least 30 mins, and then cooled to 10°C (50°F) or less. The adequacy of the treatment and the keeping quality are determined, respectively, by two statutory tests, the phosphatase and methylene blue tests. Pasteurised milk should keep at least 2–3 days. The term is derived from the French chemist, Louis Pasteur, who pioneered the process. (◗ STERILISED MILK, ULTRA HEAT-TREATED MILK and UNTREATED MILK)

Pasture. An area of GRASSLAND used only for grazing as distinct from one cut to produce HAY or SILAGE (a MEADOW).

Pathogen. An organism or substance which causes disease.

Pattle. A Scottish form of PADDLE.

Paunch ♦ RUMEN.

Pears. Various high-quality varieties of pear (*Pyrus communis*), particularly Conference and Comice, are grown commercially in ORCHARDS in the East and South-East of England. Perry pears are grown in the West and South-West of England. The area grown (except Perry pears) has declined slightly in recent years and, according to the 1980 ANNUAL REVIEW White Paper, was estimated at 4500 hectares (*c.* 11100 acres) in 1979.

Peas. A leguminous crop (*Pisum sp.*) with a long trailing stem, compound leaves and leaf stalks terminating in climbing tendrils. Two species are grown:
P. sativum (with mainly white flowers) grown mainly in the drier eastern counties and harvested (*a*) mature as dried peas for canning or packeting as processed peas, (*b*) immature as vining peas for canning, dehydration or quick-freezing as garden peas (such peas are usually grown under contract) – the haulm being valuable as SILAGE, or (*c*) immature and picked fresh for sale in the pod by greengrocers and markets.
P. arvense (with mainly coloured flowers) harvested (*a*) mature as field peas for compounding and pigeon feed, or (*b*) immature as forage peas for direct grazing or ensilaging.
The dry. mature seeds of all types, including waste and stained dried peas, and seeded down vining peas, may be used for stock feed. Peas are often grown following a cereal crop in a ROTATION and enrich the soil with nitrogen. (♦ NITROGEN FIXATION)

Pease-straw. The HAULM of pea plants after removal of the PEAS.

Peat. An organic soil consisting of plant material accumulated in mainly anaerobic waterlogged conditions in which BACTERIA and earthworms, etc, are absent and decay is very slow. Fen peat develops when the water contains mineral salts and is slightly alkaline. It is black and well decomposed, and characteristic of the FENS of eastern England. If mineral salts are absent, acid bog peat forms, which is brownish in colour and contains distinguishable plant remains (mainly *Sphagnum* mosses). Such peat is characteristic of BLANKET PEAT bogs.

Peck. A measure of capacity for dry goods, e.g. grain, equal to 2 GALLONS or a quarter of a BUSHEL.

Peck Order. The social hierarchy in poultry and other birds. The term derives from the habit of dominant individuals pecking submissive ones on the head. (♦ BUNT ORDER)

Ped. A natural unit of soil structure as distinct from clods or soil fragments formed artificially by tillage.

Pedigree. The recorded line of ancestry of cultivated plant varieties or animals, much used in determining breeding programmes.

Pedigree Registered Animal. One whose forebears have been recorded with a BREED SOCIETY.

Pedigree Selection. The selection of animals in stock-breeding on the basis of desirable features or qualities in their ancestors. (♦ GENOTYPIC SELECTION, PHENOTYPIC SELECTION)

Pedology. The study of soils.

Peeler. A plant which has a high nutrient demand and in the absence of FERTILIZERS would impoverish the soil.

Pellet ♦ CUBER.

Pen. A small enclosure for animals or poultry. Also those animals kept in the pen and sufficient in number to fill it. Also called FOLD for sheep.

Pen Mating ♦ FLOCK MATING.

Pencil ♦ CAKE.

Pendro ♦ STURDY.

Penistone ♦ WHITEFACED WOODLAND.

Pepsin. An ENZYME occurring in GASTRIC JUICE which breaks down insoluble PROTEINS into more soluble PEPTONES.

Peptide. A product of the digestion of PROTEIN and PEPTONES. An organic compound of two or more AMINO ACIDS in the form of a chain or polypeptide. (♦ TRYPSIN)

Peptone. A product of the digestion of PROTEIN by PEPSIN, an ENZYME found in GASTRIC JUICE. Slightly more complex than PEPTIDES containing more AMINO ACIDS in the form of long polypeptide chains.

Perennial. A plant which continues growth from year to year. A herbaceous perennial is one in which the leaves and stems die away in autumn and new shoots develop in the spring from underground storage organs e.g. PERENNIAL RYEGRASS. Woody perennials include shrubs and trees, etc., in which each year's new growth arises from the woody stems, so that height and breadth are increased. (◆ ANNUAL, BIENNIAL)

Perennial Ryegrass. A grass (*Lolium perenne*) which in Britain forms the basis of most long LEYS and is the dominant grass in the best pastures. Extensively sown in SEED MIXTURES with other grasses and with CLOVER. Many varieties are now available, all of which grow rapidly, respond to fertilizers, set seed well, and stand grazing and cutting well. A loosely to densely tufted, smooth grass, with a flattened stem and green hairless leaves. The SPIKELETS are stalkless and are attached alternately on opposite sides of the RACHIS (*see p. 211*). (◆ ITALIAN RYEGRASS)

Performance Testing. The measurement of the growth rate of meat-producing animals by determining liveweight gain over a given period on a particular ration. Comparative tests between different strains or breeds are useful in the selection of breeding stock. (◆ PROGENY TESTING)

Perianth. The outermost, non-sexual part of a flower, which encloses the STAMENS and CARPELS, and usually comprises the CALYX of sepals and COROLLA of petals.

Permanent Wilting Percentage. The percentage water content of a soil in that condition when a crop has removed so much water from it, that that which remains is insufficient to meet the requirements of the crop, as a result of which it wilts. (◆ FIELD CAPACITY, READILY AVAILABLE MOISTURE, SOIL MOISTURE DEFICIT)

Persistent Chemicals. Those which remain active for long periods after they are applied as herbicides, insecticides, etc.

Pest(s). Animals, insects, fungi, weeds, etc., which are troublesome to crops or livestock, or cause harmful diseases, and result in considerable losses in agricultural or horticultural production in terms of both quantity and quality. (◆ PESTICIDES)

Pesticides. Various categories of poisonous chemicals used to kill PESTS, including fumigants, fungicides, HERBICIDES, INSECTICIDES and rodenticides. Most are synthetic organic chemicals, produced in concentrate form but diluted for application with various substances such as water, talc, clays, kerosene, etc., to ensure even distribution. They are available in a variety of forms including dusts, granules, solutions, suspensions and emulsions. Some are persistent in the environment (♦ CHLORINATED HYDROCARBONS) and are dangerous to wild-life populations. A considerable amount of information about pesticides is now available to users, and the introduction and use of pesticidal chemicals is now controlled under various voluntary schemes. (♦ AGRICULTURAL CHEMICALS APPROVAL SCHEME, PESTICIDES SAFETY PRECAUTIONS SCHEME)

Pesticides Safety Precautions Scheme. A voluntary scheme under which PESTICIDE manufacturers withhold marketing a new product for use in agriculture or food storage, or the introduction of a new use for an existing available chemical, before safe usage recommendations have been agreed with the appropriate Government Departments. Such recommendations are summarised on product labels and are published in full in Recommendations Sheets available from the MINISTRY OF AGRICULTURE, FISHERIES AND FOOD.

Petiole. A stalk by which a leaf is attached to the stem of a plant.

pH Value. A measure of the hydrogen ION concentration of a solution, and therefore an indication of its acidity or alkalinity. Expressed on a scale from 0 to 14. pH 7 is neutral, less than 7 is acid, more than 7 is alkaline. Also called reaction. (♦ ACID SOIL, ALKALINE SOIL, NEUTRAL SOIL)

Phage ♦ BACTERIOPHAGE.

Phagocytes. White blood corpuscles which engulf BACTERIA as part of the body defence mechanism of animals.

Phenotype. The visible characteristics of a plant or animal as distinct from its genetic constitution (GENOTYPE). A particular genotype may produce several different phenotypes according to the environment in which it develops. (♦ PHENOTYPIC SELECTION)

Phenotypic Selection. The selection of animals in stock breeding on account of their PHENOTYPE. Also called mass selection. (♦ GENOTYPIC SELECTION, PEDIGREE SELECTION)

Phloem. Tissue in vascular plants which carries synthesised food materials (e.g. SUGARS, PROTEINS, etc.) and some mineral IONS, from leaves to roots and vice versa. Located to the outer side of the CAMBIUM and also known as BAST in trees. (♦ XYLEM)

Phosphates. Various salts of phosphoric acid applied to the soil as phosphatic FERTILIZERS to supply PHOSPHORUS to crops.

Phosphatic Fertilizers ♦ FERTILIZERS.

Phosphorite ♦ ROCK PHOSPHATE.

Phosphorus. (P). A chemical element occurring naturally only in compounds, mainly as calcium phosphate in the mineral, apatite. Essential to life as a constituent of certain PROTEINS and for many metabolic (♦ METABOLISM) reactions. In animals it is important in the formation and maintenance of bones and teeth, and for efficient CARBOHYDRATE utilisation, and is mainly derived from BONE FLOUR, milk, cereals, and oil-seed CAKE. In plants phosphorus is necessary for root development. It also enhances crop ripening and counteracts the weakening effect of excessive applications of NITROGEN. It is applied to the soil in the form of various phosphatic FERTILIZERS.

Photoperiodism. The effect of the length of day and night on plant flowering. Some plants are long-day plants requiring 14–16 hours of sunlight per day to flower, e.g. WHEAT, lettuce, BEET, etc. Short-day plants require only 8–9 hours of sunlight to flower, e.g. kidney beans, rice, soya bean, chrysanthemum, etc. Some plants are unaffected by day length (day-neutral plants) e.g. tomato.

Photosynthesis. The process in green plants by which CARBOHYDRATES are synthesised from water and carbon dioxide using the energy of sunlight absorbed by CHLOROPHYLL. Also called carbon assimilation.

Phycomycetes. A class of fungi, mainly single-celled and found in damp situations, (e.g. powdery MILDEWS). Those possessing thread-like hyphae lack cross walls. Many attack crops (e.g. *Phytophthora infestans*, the cause of POTATO BLIGHT).

Physical Analysis. The determination of the composition of a substance by physical means. In SOIL ANALYSIS, for example, samples are dispersed in water or simple solutions and the particles allowed to settle. After oven drying the soil is shaken through sieves of varying mesh size to separate SAND, SILT and CLAY, and the proportion of each component then determined. This technique is used to accurately determine the TEXTURE of a soil sample.

Physiological Heat Value ♦ METABOLISABLE ENERGY.

Physiology. The study of the internal functioning of living organisms.

Picker Combine. A machine used to harvest MAIZE, the picking mechanism consisting of snapping rollers which remove only the cobs from the stems. 2-, 3-, or 4-row models are used.

Pickle. A grain of corn.

Pie ♦ CLAMP.

Piebald. A term applied to animals of two colours, mainly black and white, in patches. (♦ SKEWBALD)

Piecework. A basis of employment often used on farms by which seasonal workers such as fruit pickers are paid according to the amount of work done (e.g. per basketful picked), as distinct from according to an hourly rate.

Pietrain. A heavily-muscled, prick-eared breed of pig, mottled brown in colour, imported from Belgium and widely used in cross-breeding programmes.

Pig. The pig is a domesticated UNGULATE derived from the wild boar (*Sus scrofa*) with some crossing with the Chinese type (*Sus indicus*), and kept for meat production for human consumption. Pig production is carried on in most parts of Britain but is particularly important in eastern and southern England, in North-East Scotland and in Northern Ireland. There is an increasing concentration into specialist units and larger herds.

Whereas in 1963, 60% of breeding sows were in herds of less than 20, in 1979 about 72% of sows were in herds of 50 and over, although average herd size was 30. About 42% of total fattening pigs were in herds of 1 000 and over, although average herd size was 203. Total pig numbers in 1979 were almost 7.9 million (see table *Pig Numbers in the U.K.*). Pigs are now kept largely indoors in automated houses with controlled environments. (◗PIGGERY)

Pig Numbers in the U.K. (at June each year). ('000 head).

	Average of 1968–70	1975	1976	1977	1978	1979
Total pigs	7 753	7 532	7 947	7 736	7 708	7 873
of which: sows in pig and other sows for breeding	768	710	747	725	724	743
Gilts in pig	150	104	137	103	118	109

(Source: Annual Review of Agriculture, H.M.S.O. 1980).

The principal breeds now kept in the U.K. include the LARGE WHITE, LANDRACE, WELSH, BRITISH SADDLEBACK and HAMPSHIRE, the first three being the most widely used. A large number of cross-bred sows are also used in commercial pigmeat production to produce the following four classes of pigs (weights according to the MEAT AND LIVESTOCK COMMISSION):

Pork Pig: One bred for quick growth and maturity at light weight. Usually slaughtered at about 19 weeks, weighing 46–63 kg (100–140 lbs) liveweight (l.w.), 32–45 kg (70.5–100 lbs) deadweight (d.w.). Also called porker.

Bacon Pig: One specifically reared to produce bacon (green or smoked) and lean ham with a long carcase bearing minimum fat. Usually slaughtered at about 24 weeks, weighing 79–100 kg (174–220 lbs) l.w. and 59–77 kg (130–170 lbs) d.w. Carcases are not usually trimmed for fat. Also called Wiltshire bacon pig or baconer.

Cutter: A general purpose pig which, after slaughter, is cut into different parts each of which may be used differently for pork, bacon or processing. Usually slaughtered at about 23 weeks, cutters weighing 68.5–93 kg (151–205 lbs) l.w. or 50–67.5 kg (110–148.5 lbs) d.w., and heavy cutters weighing 92–109 kg (202–240 lbs) l.w. or 68–81 kg (150–178 lbs) d.w.

Heavy Hog: A general purpose pig noted for usually having carcases with a high d.w. to l.w. ratio, and reared to produce bacon (green or smoked), ham, pork and various by-products. Usually slaughtered at about 27–28 weeks, weighing 105 kg (231 lbs) or more l.w. or 81.5 kg (179 lbs) or more d.w. Carcases are usually stripped of excess fat.

Pig improvements are fostered through research by the Meat and Livestock Commission (◗ PIG IMPROVEMENT SCHEME) and by leading commercial firms. ARTIFICIAL INSEMINATION is available nationally.

In 1979 about 24% of the pigmeat produced was used for bacon and ham and 76% for pork and manufacturing into sausages and other processed forms.

(Sources: Annual Review of Agriculture, H.M.S.O. 1980; and 'Agriculture in Britain', Central Office of Information, R5961/1977).

Some of the common breeds of pigs kept in the U.K.

Welsh

Landrace

Hampshire

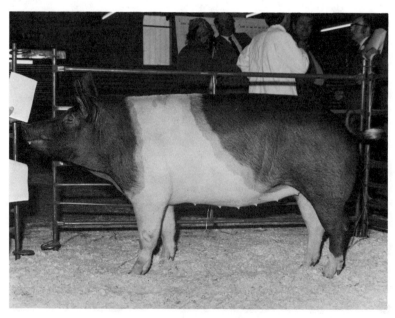

Large White

British Saddleback

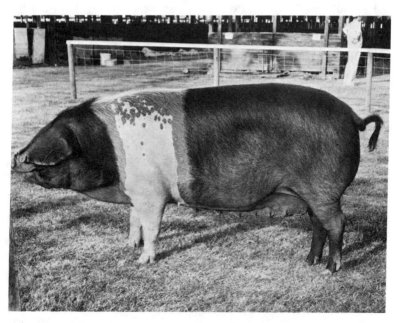

Pig Improvement Scheme. A scheme begun in 1971 by the MEAT AND LIVESTOCK COMMISSION for the genetic improvement of pigs. It is based on a first stage of intensive central performance testing and selection within a limited number of the best herds in Britain (about 65 Nucleus and 35 Reserve Nucleus herds), and a second stage of multiplication of improved nucleus stock by about 115 Nucleus Multiplier herds. The Nucleus and Reserve Nucleus herds submit about 5 000 boars a year to the M.L.C. for performance testing, by which important commercial characteristics are measured, and a reliable evaluation of a boar's ability to sire profitable progeny is provided. Tests are rigorous and competitive and only the best boars are subsequently offered for sale. About 2 000 tested and licensed boars are available annually, together with a large number of litter brothers and sons of tested boars. Almost half the boars in use in Britain are derived from herds in the scheme and the genetic quality of the national herd is steadily being improved. Under the scheme, pig breeding companies may also submit boars for testing and carcases for evaluation.

Pig Production Development Committee. A Committee constituted under the Pig Production Development Act (Northern Ireland) 1964, to provide services and facilities to improve the quality of pigs and thus enhance the profitability of pig production in Northern Ireland. It is financed, by Government Order, by a levy collected on all pigs sold to the PIGS MARKETING BOARD (N. IRELAND). There are 8 Committee members, 2 each nominated by the P.M.B., the Department of Agriculture, the Ulster Farmers' Union and Breeders' associations.

Pig Typhoid ♦ SWINE FEVER.

Piggery. A place where pigs are kept. A variety of forms of pig housing are in use. Breeding pigs are sometimes run outside on well-drained pastures and housed in movable sheds or arks, or in other temporary shelters. Store and fattening pigs are sometimes kept in covered or partly covered dunging yards, usually covered with straw bedding, and provided with low 'kennels' as warm sleeping quarters. Modern indoor systems, particularly for farrowing pigs, usually consist of a low, insulated building provided with a heat source and controlled ventilation, and pigs are often kept in pens arranged in rows with access passages and DUNGING CHANNELS. Design, particularly in fattening houses, varies according to whether dry or liquid feeding systems are used. In farrowing houses where FARROWING CRATES are used, piglets are kept warm by a covering hood housing an infra red heater. The main types of piggery are as follows:

Cottage Piggery: A type of piggery used for fattening pigs or sows and litters, consisting of a row of low-roofed pens containing straw bedding, each with access through a small door to an uncovered yard in which food and water troughs are sited, and in which the pigs usually dung. The McGuckian and Harper Adams are modifications of this simple design.

Danish Piggery: A type of totally enclosed, fully insulated piggery for use with fattening pigs. It consists of a central feeding passage flanked on each side by a range of pens, with dunging passages along both outside walls. Eight to twelve pigs are usually kept in each pen. Also known as the Scandinavian piggery.

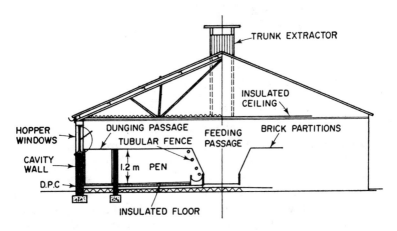

Danish-type piggery.

Flat Deck Rearing Piggery: A low, well insulated piggery for rearing weaned pigs from 2–3 weeks of age up to 7–8 weeks of age, with a linked heating and ventilating system which will maintain the temperature constantly at any desired level. There are several pens within each house, usually $1\,m \times 2\,m$ (3 ft 4 in. × 6 ft 8 in.) in size. Floors are made of perforated material, usually metal, to allow dung to fall through and collect in a slurry channel. The pigs are fed from self-feed hoppers against the central passage.

Harper Adams Piggery: A modification of the cottage piggery used for fattening pigs, consisting of a low linear building (sometimes a lean-to against an existing wall), with a sloping roof, containing a row of pens, with feeding troughs adjacent to a feeding passage at one end, and leading via a small door at the other end to an outside run, lined with straw litter, over which the roof partially extends.

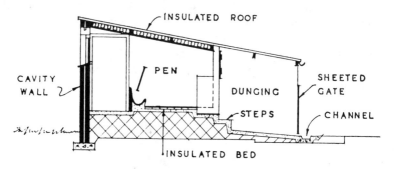

Harper Adams-type piggery.

McGuckian Piggery: A modification of the cottage piggery which is used for fattening pigs. The pens are arranged in pairs, each pair leading off an access passage from which they are separated by a wall and door. The lying areas of the pens are separated from each other by a feed passage along both sides of which are the feed troughs. Each individual pen has an outdoor un-roofed dunging area with access through a small door to the lying/feeding area.

Mono-pitch Fattening Piggery: A naturally ventilated piggery for fattening pigs with a roof which slopes from 0.9 m (3 ft) high at the back to 2 m (6 ft 8 in.) high at the front. Each entirely separate pen is 2.3 m × 5 m (7 ft 8 in. × 16 ft 8 in.) in size. Ventilation is controlled by flaps at the front and rear. Pigs are fed through a hatch at the rear of the roof, on the floor. There is either a slatted or solid dung channel near the gate at the highest point of the pen.

Solari Farrowing Pen: A mono-pitch building which has each pen entirely separate. The pens are 1.5 m (5 ft) wide by 5 m (17 ft) long and have removable rails which divide the pen into a 0.6 m × 1.5 m (2 × 5 ft) creep area at the lowest part of the pen, and a 2.4 m × 1.5 m (8 × 5 ft) farrowing bed and a 2.1 m × 1.5 m (7 × 5 ft) dunging area. Ventilation is controlled by a flap over the door.

Solari Piggery: A type of piggery for fattening pigs with double fattening pens arranged to each side of a central feeding passage, under an open-sided steel Dutch barn. Each pen comprises an inner sleeping section or 'kennel' with a hinged wooden roof which can be raised or lowered to control temperature and ventilation, and an open outer dunging pen, at a lower level, and lined with deep straw.

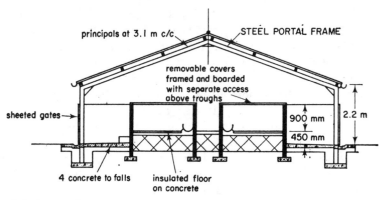

Solari-type piggery.

Suffolk Piggery: A very popular type of piggery for fattening pigs. It is housed under a Dutch barn clad with YORKSHIRE BOARDING, providing natural ventilation. It consists of two rows of pens, each comprising a kennel as a sleeping area leading to a scrape-through dunging area containing a feeding trough and abutting against a central feeding passage. Alternatively the rows of kennels may be adjacent centrally, with the feeding passages to the outside. This system is straw-based, usually with liquid feeding into the troughs, normally with about 10 pigs per pen.

Sweat-box House: A type of piggery for fattening purposes, comprising a row of insulated, totally closed houses each with room for about 20 pigs, with minimal ventilation, in which very high temperature and humidity conditions develop. Sloping floors lead to slat covered dunging channels. Water is delivered to automatic drinking bowls and pelleted CONCENTRATES are fed on the floor. Very rarely used in Britain.

Verandah House Piggery: A type of piggery for weaned pigs from 4–5 weeks of age up to 10–12 weeks, consisting of 2 rows of kennels with insulated lids to either side of a central passage, all of which is under an umbrella roof. Outside the roof on each side of the building, there is a slatted dung channel with a pop hole which gives access to the kennel. Pigs are fed either from self feed hoppers or on the floor.

Pig-sick ◗ SICK.

Pigs Marketing Board (Northern Ireland). A producer–controlled organisation first established in 1933 under the Pigs Marketing Scheme (Northern Ireland), to market pigs. Government controls on fatstock marketing introduced during the Second World War were dismantled in 1954 when the P.M.B. was reconstituted. Although pig numbers have declined in recent years pig production is still one of the most important agricultural activities in Ulster. Only registered producers in Ulster may sell bacon pigs (those weighing not less than 77 kg (169.8 lbs) liveweight or 56.5 kg (124.6 lbs) deadweight, and these must be sold to the P.M.B. Breeding stock, however, may be sold between producers. Sales of weaner pigs for fattening are not made through the P.M.B. Light pork pigs below bacon weight may be sold direct to bacon factories and abattoirs, although such sales are also made through the P.M.B. The P.M.B. also has powers to become more closely involved in production, e.g. to sell equipment to producers, promote co-operation and research, and to produce and sell pigmeat products.

During 1979 discussions involving the Board, the Department of Agriculture (Northern Ireland), the Ministry of Agriculture, Fisheries and Food, the Ulster Farmers' Union, the ULSTER CURERS' ASSOCIATION and the European Economic Commission continued with a view to arriving at a future framework for the Board, compatible with the E.E.C. Treaty and Regulations.

By the end of the year no conclusions had been announced but the Commission subsequently accepted that the non-enforcement of the provisions of the Pigs Marketing Scheme which required producers to market bacon pigs either through or as directed by the Board, coupled with the full information which had been given to producers of their rights to market independently had had the practical effect of transforming the Board into a voluntary organisation; the necessary amendments to legislation are being processed.

Producers in Ulster are paid a pooled price by the Board for their pigs, based on weight and grade, and the latter are scheduled to encourage pig production to the quality and weight required by the U.K. bacon market, the main outlet. A Pig Contract safeguards committed producers against unwarranted increases in marketings.

In the past all pigs were sold to local curers, distribution being according to a Government 'quota' whereby each curer received a fixed percentage of the total pigs marketed, the price being according to a formula negotiated annually between the P.M.B. and curers. In recent years this system has been modified to take account of changing circumstances and to introduce flexibility and greater competition. A certain proportion of pigs are now auctioned rather than being distributed under quota. Sales have also been made to curers in Eire, and sales of pork carcases made directly to buyers in Britain and overseas. Discussions took place in 1979 and 1980 between the Board and the Ulster Curers' Association with a view to phasing out the quota and replacing it with a contract system.

Through its wholly owned subsidiary, P.M.B. (Investments) Ltd, established in 1969, the P.M.B. now has a controlling interest in 3 curing factories, which together represent about 45% of the processing capacity of the pig industry in Ulster.

The P.M.B. comprises 11 members, 8 of whom are elected by registered producers, and 3 are appointed by the Department of Agriculture. (MARKETING BOARDS)

Pightle. A small enclosure or field. Also a CROFT.

Pike HAYCOCK.

Pin Bones. The two prominent pelvic projections to either side of a cow's tail.

Pin Fallow ◗ BASTARD FALLOW.

Pin Feather. A young undeveloped or unexpanded feather.

Pinder. A person who, in the past, officially impounded stray cattle in a parish in a pinfold or enclosure. Owners were required to pay a fine on retrieving them. Also called pounder.

Pines ◗ PINING.

Pinfold ◗ PINDER.

Pining. The loss of condition in sheep, and sometimes in cattle, due to cobalt difficiency. Also called pines, vinquish.

Pinnate Leaf. A type of COMPOUND LEAF with a feather-like appearance, with leaflets in pairs attached opposite each other to either side of a single main axis or rachis. In a bipinnate leaf the leaflets on the rachis are themselves divided into leaflets.

Pioneer Crop. A crop or mixture of crops (e.g. Italian ryegrass, rape and turnips) grown on ploughed-up grassland which has become worn out and the soil impoverished, prior to reseeding the grassland with a good seeds mixture. Pioneer crops are fertilized and grazed under heavy stocking (to accumulate dung and urine) and ploughed in, having the effect of improving the fertility of the soil for reseeding.

Pirls. Small knots of wool in a fleece, as found in the WENSLEYDALE.

Pistil. The OVARY of a flower together with the STYLE and STIGMA.

Pitch Pole. A type of HARROW with double-ended TINES mounted on central shafts, so that they are able to be turned through 180°. This enables clogged or obstructed ends to be turned and the 'clean' ends to be brought into play.

Placement Drill. A machine which places FERTILIZER close to the side of rows of seeds and slightly below them. Fertilizer placement improves the efficiency of usage by ensuring that nutrients are concentrated close to actively growing roots in young seedlings.

Placenta. 1. The membranous tissue uniting an unborn mammal to its mother's womb and acting as a site of nutrient exchange between the blood systems of mother and developing offspring. Discharged as afterbirth.
2. The part of a plant OVARY to which OVULES are attached.

Planker. An implement used to crush clods on land where a ROLLER cannot be used, consisting of a number of fixed overlapping planks, shod with iron bars along the working edges, which is pulled over the land. Also called float, rubber or scrubber.

Plant Breeders' Rights. Rights which act in a similar manner to patents and enable a breeder of a new plant variety to be rewarded for his work. A successful application for such rights gives a breeder the sole right to produce for sale and to sell reproductive material of the particular plant variety. In practice breeders license others to produce seed for sale and in the terms of the Licence the payment of SEED ROYALTIES is specified.

Plant Breeding. The development of new and improved varieties and strains of agricultural crops and other plants by selection and hybridisation of individuals or varieties possessing desirable characteristics or qualities.

Plant Food Ratio. The ratio of NITROGEN to PHOSPHATE and POTASH in a FERTILIZER. Thus a COMPOUND FERTILIZER containing 20% N, 10% P, and 10% K has a ratio of $2:1:1$. Such ratios are a guide to the type of compound. A number of fertilizers may have the same ratio and therefore be usable for the same purpose, but the quantity needing to be applied will vary in relation to the actual percentage nutrient content.

Plant Louse ◗ APHID.

Plashing ◗ LAYERING.

Plat. A plot of ground.

Plate Mill. A machine which coarse grinds grain, consisting of two circular plates of equal diameter positioned one above the other, with an adjustable gap between them to control the fineness of grinding. The lower plate is fixed whilst the upper one rotates against it. Plate diameter varies between 15 and 45 cm (6–18 in.). (◗ HAMMER MILL, CRUSHING MILL)

Platyhelminthes ◗ FLATWORMS.

Pleaching ◗ LAYERING.

Plough. One of the oldest types of tillage implement which breaks up the land, turning over the soil into ridges and furrows, burying surface vegetation and manure and exposing the new soil surface to the air in preparation for sowing seed. The earliest types were wooden and of simple design. The principal parts of modern ploughs comprise a steel frame to which are fixed a number of bodies, each consisting of a COULTER which cuts a vertical slice, and a SHARE which makes a horizontal cut under the furrow slice which is then turned by the MOULDBOARD. The lateral accuracy and stability of each body is maintained by a LANDSIDE. Three main categories of tractor-drawn ploughs are in common use on farms, viz, (*a*) fully mounted types, attached to the three-point linkage and raised or lowered hydraulically, (*b*) semi-mounted types, in which a rear wheel partially supports the weight, and (*c*) trailed types, pulled by a drawbar. The following types are described alphabetically elsewhere; DIGGER, SEMI-DIGGER, DISC, LEA, REVERSIBLE and RIDGER.

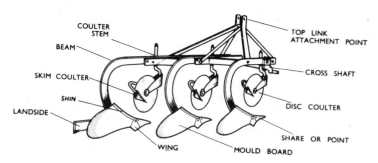

A three-furrow mounted plough.

Plough Pan ◗ PAN.

Plough-in. To bury a crop, stubble, weeds, etc., by turning them under the soil with a plough.

Ploughing. The three main methods of ploughing fields commonly practiced include (*a*) one-way ploughing with a REVERSIBLE PLOUGH, in which ploughing is begun at one side of a field and successive furrows turned until the other side is reached, (*b*) systematic ploughing, in which fields are ploughed in LANDS, and (*c*) round and round ploughing, in which fields are ploughed from the centre to the outside or vice versa.

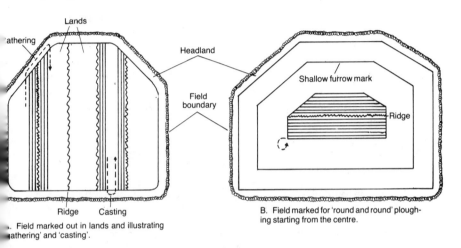

A. Field marked out in lands and illustrating 'gathering' and 'casting'.

B. Field marked for 'round and round' ploughing starting from the centre.

Note: Dotted arrows indicate movement of tractor and plough.

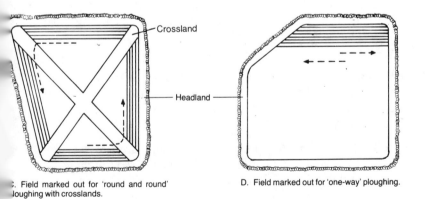

C. Field marked out for 'round and round' ploughing with crosslands.

D. Field marked out for 'one-way' ploughing.

PLOUGHING METHODS

Ploughing Iron. A term for any of the iron parts of a plough, e.g. COULTER, SHARE, etc.

Ploughing Rate. This can be approximately calculated by the following formulae which assume a field efficiency factor of 70% to take account of headland turning and non-productive time.

$$\text{Hectares per hour} = \frac{\text{working width (m)} \times \text{speed (km/h)}}{14.28}$$

$$\text{Acres per hour} = \frac{\text{working width (yds)} \times \text{speed (m.p.h.)}}{3.93}$$

Ploughland. 1. Arable land.
2. The amount of land that could be tilled in a year by a plough drawn by a team of eight oxen, together with a proportionate amount of pasture, used historically as a basis for determining taxes.

Ploughshare ◖ SHARE.

Plough-sick ◖ SICK.

Plucker Finger. A wire loop in a hop picking machine which severs the cones from the BINE.

Plumule. 1. The young shoot of an embryo seed plant which develops into the stem and produces true leaves.
2. A down feather.

Plymouth Rock. A heavy, dual-purpose breed of poultry which originated in the U.S.A. Variously coloured including barred, white, black, blue, buff, etc., with yellow legs and a medium-sized single comb. Hens lay brown eggs.

Poaching. The trampling of land when wet, mainly by cattle, so that the soil becomes churned and muddy. Poaching is a particular problem of heavy land, and leads to the deterioration of soil structure.

Pocket. 1. An elongate sack into which hops are pressed, about 2 m (6½ ft) tall and 60 cm (2 ft) in diameter. Also the quantity of hops contained in such a sack, about 76–89 kg (1½–1¾ cwt).
2. A measure of wool equal to 76 kg (168 lbs).

Podsol. A type of soil in which minerals have been leached from the surface layer and deposited in the subsoil, often forming a hard pan impenetrable by roots. The upper leached layer is usually extremely poor in mineral salts and consists of bleached sand. It is often covered with very acid, decomposing plant material. Such soils are common on quick draining sandy HEATHS, and under conifer forest, particularly in the cool, wet areas of northern and western Britain.

Point. A SHARE or TINE.

Point of Lay. The state of a PULLET when about to lay its first egg.

Points. The important features or body characteristics of animals which are considered when judging livestock.

Poke. A loosely woven sack for green hops, able to contain 8–10 BUSHELS (0.29–0.36 cubic m). Also called greenbag, greensack.

Pole. An old measure of length equal to $5\frac{1}{2}$ yds (4.6 m), or of area equal to $30\frac{1}{4}$ sq. yds (25.3 sq. m). Also called rod or perch.

Poll. 1. The top of the head.
2. To dehorn an animal.
3. To cut off the top of a tree.

Poll Dorset. A breed of sheep developed in Australia and identical to the DORSET HORN except that it is hornless.

Pollard. 1. An animal from which the horns have been removed.
2. A tree having had its crown cut off, and with new branches growing from the top of the stem.

Pollards. A term once given to a category of offal from wheat milling, consisting of fine BRAN and other products. All wheat offals, apart from bran, are now sold as WEATINGS, and are used as a FEEDINGSTUFF.

Polled. A descriptive term for animals naturally lacking horns.

Pollen. The minute spores, often yellow, produced by the anthers of flowers and carried by the wind or insects, and sometimes by water, to the stigmas (pollination). Each pollen grain is a male GAMETE.

Pollination. The transfer of POLLEN from anther to stigma either (*a*) of the same flower or of another flower on the same plant (self-pollination), or (*b*) of a flower on another plant of the same species (cross-pollination).

Pollinator. A tree planted in an ORCHARD to provide POLLEN for the FERTILIZATION of surrounding trees. Pollinators are usually of a different variety to the rest of the trees in the orchard.

Polyploid. Having more than twice the normal HAPLOID number of CHROMOSOMES, e.g. triploid, tetraploid.

Polysaccharide ♦ CARBOHYDRATES.

Pooking Fork. A fork bearing three tines, used to push corn swaths together into sheaves for tying.

Pore. A sweat gland in the skin of an animal.

Pore Space. The space not occupied by soil particles in the soil.

Pork Pig, Porker ♦ PIG.

Potash. A term applied to potassium carbonate (K_2CO_3), potassium hydroxide (KOH), potassium oxide (K_2O), and to potassium salts in general. (♦ POTASSIUM)

Potash Fertilizers ♦ FERTILIZERS.

Potash Nitrate. A FERTILIZER containing about 15% NITROGEN and 10% POTASH, derived from natural deposits in Chile, and mainly used in market gardening as a top dressing. (♦ NITRATE OF POTASH)

Potassium. (K). A metallic element found naturally in various salts and present in the soil in CLAY MINERALS. It is essential to life, influencing both CARBOHYDRATE and PROTEIN metabolism, and the movement and utilisation of water in living organisms. In plants it ensures healthy growth, improving both resistance to drought and fungal disease, and winter hardiness. It assists in the transportation of the products of PHOTOSYNTHESIS and counteracts the effects of excessive applications of NITROGEN. In animals, it contributes to the maintenance of the osmotic balance of the body fluids, particularly intracellular fluid, and influences muscular and nerve activity. Continuous cropping, lack of manuring, excessive dressings of nitrogen, and certain

342

crops with a high potassium requirement (e.g. mangolds, lucerne, potatoes, sugar-beet, etc.) may deplete the soil of potassium. It may also be deficient in sandy, peaty or chalk soils. It is applied in the form of various potash FERTILIZERS.

Potato. A perennial herb (*Solanum tuberosum*) introduced to Britain from South America in the late sixteenth century and grown for its CARBOHYDRATE-rich edible tuber, mainly for human consumption. Characterised by compound pinnate leaves with large and small leaflets in alternate pairs. Four categories are produced:

Early potatoes, varieties of which can be harvested for consumption early in the summer. The earliest crops of new potatoes are called first earlies, and have foliage of a dwarf nature. If left to mature they produce lower yields than maincrop potatoes. These are followed by second earlies which are slightly slower to bulk.

Maincrop potatoes, varieties of which bulk later than earlies, develop larger foliage, produce greater yields, and are usually harvested in October.

Processing potatoes, normally grown under contract, for chips, dehydrated powders, crisps, canning, etc.

Seed potatoes, produced under a statutory certification scheme and mainly grown in cool areas in Scotland, N. Ireland and North-West England. (◗ SEED POTATOES)

Potatoes for sale are mainly grown in Britain by registered producers on a quota system, operated by the POTATO MARKETING BOARD (◗ WARE POTATOES). They are grown in most parts of Britain. Early potatoes for immediate summer marketing are grown in most areas, though the earliest growing districts are concentrated in the south and on the west coast. The production of maincrop potatoes (which may be stored until the following spring) is centred on the eastern side of England and Scotland and in N. Ireland.

First supplies of early potatoes normally become available at the end of May, and the lifting of maincrop varieties usually starts in September. Seed potato production for Britain and for export is chiefly centred in Scotland and N. Ireland. (Source: 'Agriculture in Britain', Central Office of Information, R5961/1977).

Potatoes are frequently grown as a grass break crop following cereals or in ROTATIONS following PEAS, BEANS or a grass

343

LEY, and usually on the ridge system. They are particularly susceptible to virus diseases (e.g. LEAF ROLL) transmitted by APHIDS, to POTATO BLIGHT and to frost. Potatoes are fed raw to mature cattle, and after cooking or steaming (which facilitates starch digestion) to pigs and poultry. Surplus potatoes are sometimes used for SILAGE making with grass, or, if cooked, for ensilaging alone.

Potato Apple. The small green fruit of the POTATO, similar to a tomato.

Potato Blight. A fungal disease (*Phytophthora infestans*) of potatoes which devastated the British crop in 1845 and caused the Irish famine in 1846. Characterised initially by dark green, turning to dark brown, leaf patches, resulting in death of the haulm and loss of yield. Tubers, which become separately infected, show brown staining and usually rot. (◖ BEAUMONT PERIOD)

Potato Harvester ◖ ELEVATOR DIGGER.

Potato Marketing Board. Established in 1934 under the Potato Marketing Scheme to promote the orderly marketing of potatoes in Britain. The Board was suspended in 1939, after which potato production was directly controlled by the Ministry of Food. It was reconstituted by Parliament in 1955 following a Public Inquiry. The Board systematically controls the total area of potatoes planted, in an attempt to match production with estimated demand. All producers planting 0.4 ha (1 ac) or more of potatoes must register with the Board if it is intended to sell any part of their crops. Each new producer is allocated a basic area which determines both the area of potatoes he may plant in any year without incurring liability for excess production, and the level of his annual contribution to the Board's operating costs. In some years quotas may be imposed which restrict plantings to a proportion of each producer's basic area. Registered producers may only sell potatoes for human consumption through merchants licensed by the Board. Quality standards and minimum sizes for such potatoes (◖ WARE POTATOES) are also prescribed. The Board is responsible for implementing GUARANTEED PRICE arrangements and SUPPORT BUYING programmes for potatoes in Britain, when necessary. It also undertakes and encourages research, organises working machinery demonstrations, and conducts publicity campaigns. Apart from four Ministerial appointments the Board comprises elected registered potato producers.

Potato Pit. A CLAMP of potatoes.

Potato Planter. A tractor-drawn machine, similar to a DRILL, which plants SEED POTATOES in widely spaced rows by feeding them down tubes and covering them with a ridge of soil. Various types exist including (*a*) hand-fed planters, in which an operator feeds the tubers into a rotating seed wheel which drops them down the tubes, (*b*) semi-automatic planters, in which the operator channels the potatoes into cups which more rapidly feed the tubes, and (*c*) automatic planters, which require no operator to feed the cups.

Potato Spinner ◗ SPINNER.

Potential Evapotranspiration ◗ EVAPOTRANSPIRATION.

Poulard. A fattened or spayed hen.

Poult. A young turkey less than 8 weeks old.

Poultry. A term usually restricted to domestic fowl or chickens kept for both egg and meat production, but also applied to ducks, geese and turkeys kept for their meat.

Most breeds of fowl are believed to be descended from the Red Jungle Fowl, *Gallus gallus*. Hybrids have largely replaced the pure-breds and crossbreds in commercial production, which is now mainly carried out in intensive, controlled-environment buildings (◗ BATTERY HENS, DEEP LITTERING, BROODER). Fowl may be classified as follows:

Table birds (◗ BROILER), resulting from crosses between White Cornish Game (derived from INDIAN GAME) males and White PLYMOUTH ROCK females.

Laying birds, most commercial ones being hybrids produced from inbred strains. Light types are derived from the White LEGHORN. Medium to heavy types are often derived from the RHODE ISLAND RED. Other breeds used in hybrid production include Rhode Island White, Barred and White Plymouth Rock, NEW HAMPSHIRE Red, and LIGHT SUSSEX. A synthetic strain is also used, derived from several breeds. There is a limited demand for pure strains of Rhode Island Red, Light Sussex and Marans.

Dual-purpose birds, which are now less popular, the Rhode Island Red cross Light Sussex being frequently used.

Fancy breeds, kept mainly for showing.

The British poultry industry has expanded rapidly in recent years, aided by the application of improved husbandry and management techniques in intensive production units and by genetic improvements in stock. Only a small proportion of total production is in the hands of the smallest producers – in 1979 about 85% of the laying birds on farms were in flocks of 5 000 or more, and just over half were in flocks of 20 000 or more, although average flock size was 732. About 74% of the broilers were in flocks of 50 000 or more, and just over half in flocks of 100 000 or more, with average flock size about 26 000. The average yield of eggs per bird in 1979 was about 245 a year and nearly all eggs consumed in Britain are home-produced. In 1979 poultry numbers totalled about 133 million birds of which over 55 million were table fowls (including broilers), about 48 million were laying fowls, and about 15 million were growing PULLETS (see table *Poultry Numbers in the U.K.*).

Poultry Numbers in the U.K. (at June of each year). ('000 head).

	Average of 1968–70	1975	1976	1977	1978	1979
Total poultry of which:	132 467	136 572	142 222	134 286	137 329	132 997
Table fowls (incl. broilers)	42 971	56 708	61 325	56 153	56 319	55 445
Laying fowls	53 473	49 359	49 085	49 119	50 488	48 076
Growing pullets	23 020	18 195	18 383	16 341	17 273	15 173

(Source: Annual Review of Agriculture, H.M.S.O. 1980).

Pounder ♦ PINDER.

Poussin. A type of poultry killed for the table at about 7 weeks of age weighing about 0.9–1.0 kg (2–2.25 lbs) liveweight.

Power Harrow. A type of HARROW consisting of a number of p.t.o driven tranverse bars bearing TINES which reciprocate in a rotary manner. Used mainly to produce a fine tilth in a single operation, particularly for potatoes.

Pre-basic Seed ♦ SEED CERTIFICATION.

Precision Chop Harvester ♦ METERED-CHOP HARVESTER.

Pre-emergence Herbicide. A HERBICIDE applied following the sowing of a crop but before it emerges from the ground, either before or after the weeds emerge.

Prepotent. A term used in breeding for a parent which has the ability to transmit more characteristics to its offspring than the other parent.

Press ♦ FURROW PRESS.

Pressed Honey. Honey obtained by pressing broodless honey-combs with or without the application of moderate heat.

Price Review ♦ ANNUAL REVIEW.

Prilled Fertilizer. A FERTILIZER in the form of small spherical granules, about 2–3 mm in diameter.

Primary Host ♦ HOST.

Proboscis. The sucking mouth parts of some insects.

Producer-Retailer. A person who produces a crop or commodity and sells it direct to the public as distinct from to a Marketing Board or through a market. Particularly applied to milk. (♦ UNTREATED MILK)

Production Ration. Food supplied to an animal in excess of a MAINTENANCE RATION so that it is able to gain weight, produce milk, etc.

Proficiency Test. A test set and carried out under the authority of the National Proficiency Tests Council, to determine the practical skill of an agricultural worker in a particular skill, e.g. tractor driving and handling, sheep dog handling, etc. A CRAFT CERTIFICATE may be issued to a worker who has passed the specified number of tests appropriate to the particular CRAFT, after which the worker is classified as a CRAFTSMAN for wages purposes. Tests are organised by County-based Proficiency Tests Committees which comprise representatives from all sides of the agricultural industry, and are carried out by a panel of experts appointed by the Committee and drawn from the industry.

Profile ▶ SOIL HORIZONS.

Progeny Testing. The evaluation of the performance of the progeny of an animal (e.g. in milk production) or plant variety, used to provide a guide for its breeding value. (▶ PERFORMANCE TESTING)

Progesterone. A HORMONE secreted by the ovaries of mammals which stimulates the uterus to prepare for pregnancy and to nourish the developing embryo. With OESTROGEN it controls the development of the MAMMARY GLANDS.

Propagation. The reproduction of a species. The term is used in horticulture to mean the artificial multiplication of plants by vegetative means, including the taking of cuttings, LAYERING, BUDDING, and grafting (▶ GRAFT), etc.

Protein. A class of complex organic compounds, containing Carbon, Oxygen, Nitrogen (about 16%), Hydrogen, and some also containing Phosphorus and Sulphur. They consist of thousands of AMINO ACIDS linked in polypeptide chains (▶ PEPTIDE) and sometimes folded in a complicated manner. About 21 different amino acids are found in proteins, the sequence of their arrangement in the chains being specific for each protein and giving it its characteristic properties. Proteins may be simple, containing only amino acids, or conjugated, being combined with other compounds, e.g. CARBOHYDRATES (muco- or glycoproteins), FATS (lipoproteins), nucleo-proteins (phosphorus-containing proteins combined with DNA and forming the CHROMOSOMES of cell nuclei), etc. With water they form the basic constituents of PROTOPLASM, and play an important role in the structure of organisms. Plant proteins are often stored in seeds and other organs and include globulins, albumins, gluteins and prolamins. They are digested by animals into their component amino acids which are then used to synthesise animal proteins. Examples include ALBUMEN in eggs, osein in bones, COLLAGEN in connective tissue, CASEIN in milk, globin in HAEMOGLOBIN, creatine in muscle, etc. Protein may be provided to livestock in feedingstuffs of animal origin (e.g. FISH MEAL, MEAT AND BONE MEAL, etc.), or of vegetable origin (e.g. oilseed CAKE, peas, beans, etc.), although the latter tend to be deficient in one or two 'essential' amino acids. ENZYMES are also proteins. (◗ CRUDE PROTEIN, DIGESTIBLE CRUDE PROTEIN and TRUE PROTEIN)

Protein Equivalent. An indication of the value of a FEEDINGSTUFF expressed as the percentage of TRUE PROTEIN plus half the percentage of non-protein nitrogen, on the assumption that the latter is capable of conversion to protein by animals. (◗ CRUDE PROTEIN, DIGESTIBLE CRUDE PROTEIN and STARCH EQUIVALENT)

Protoplasm. The contents of the cells of living organisms, including the nucleus.

Proud. A descriptive term meaning advanced or excessively developed. For example, a winter proud crop is one that has put on more growth than usual during the winter. Proud flesh is a growth of tissue around a wound or ulcer, etc.

Proven Sire. A bull, boar or ram, etc., which has been shown to throw progeny which produce meat, milk, or wool, etc., of good quality and/or quantity, and which can therefore be recommended for use in breeding programmes.

Provender. Dry food for livestock such as hay and corn. Particularly applied to cut straw or hay mixed with MEAL.

Proximate Analysis. A method of analysing large numbers of animal FEEDINGSTUFF samples, involving the measurement of the relative proportions of moisture, ASH, CRUDE PROTEIN, ETHER EXTRACT, CRUDE FIBRE and NITROGEN-FREE EXTRACT. These fractions are not precise and the chemical compounds in, for instance, the crude protein fraction of cereals and potatoes will differ.

Psalterium ◗ OMASUM.

Pseudomonas. A genus of BACTERIA present in the soil which convert amino compounds to AMMONIA and ammonium compounds. (♦ NITROGEN CYCLE)

Pseudo-Rabies ♦ AUJESZKY'S DISEASE.

Ptyalin. An ENZYME present in saliva which converts STARCH to SUGARS.

Pug. 1. CHAFF, particularly when derived from clover.
2. A term for a HOGG.

Pullet. A female FOWL between 20 weeks and 18 months old, in its first laying year.

Pullorum Disease ♦ BACILLARY WHITE DIARRHOEA.

Pulsator. A mechanical device in a MILKING MACHINE which provides an intermittent suction action and stimulates the release of the milk from a cow's udder.

Pulse Crops. Leguminous crops such as PEAS and BEANS yielding edible seeds (pulse).

Pund. A sheep FOLD.

Pupa. A resting stage in insect METAMORPHOSIS between LARVA and young adult, during which movement and feeding ceases and a great change in form occurs. Also called chrysalis.

Pur Lamb ♦ SHEEP NAMES.

Pure-bred. A term for animals or plants bred from parents of the same breed or variety. (♦ CROSS-BREEDING)

Purple Melick-grass, Moor Grass ♦ FLYING BENT.

Pykel. A HAYCOCK.

Pyrethrum. A genus of composite plants, the powdered flowerheads of which are used as an insecticide. Now less commonly used due to the introduction of synthetically produced insecticides.

Q

Quarter. 1. One of the four divisions of an UDDER, including one teat.

2. A butcher's term for the limb of an animal and adjacent body parts, detached from a carcase.

3. A measure of weight, one quarter of a cwt, equal to 28 lbs.

4. A measure of capacity, equal to 8 BUSHELS.

Quartz. A natural crystalline form of silica. The commonest mineral and the main constituent of sand and sandstone.

Queen Excluder. A perforated metal sheet which prevents a queen bee from passing from the brood chamber of a BEEHIVE to the honey chamber, but which allows the small workers to pass through.

Quey. A Scottish term for a HEIFER.

Quick ◗ COUCH GRASS.

Quick Hedge. A hedge of living plants, particularly hawthorn. Also called quickset.

Quicklime ◗ CALCIUM OXIDE.

Quickset ◗ QUICK HEDGE.

Quiet Cab ◗ SAFETY CAB.

Quilt. A thick, protective cover, placed over the frames of a BEEHIVE.

R

Rabbit. A burrowing rodent (*Oryctolagus cuniculus*) of the hare family. It was introduced into Britain from the Mediterranean countries in the twelfth century and escaped and spread throughout the countryside. The myxomatosis epidemic of 1953–56 severely reduced numbers but the population has increased again considerably in recent years and they are again a severe pest in many areas, grazing cultivated crops. The stock-carrying capacity of grassland with a heavy rabbit population can be considerably lowered. Control measures include ferreting, gassing, trapping and shooting. (◗ RABBIT CLEARANCE SOCIETIES)

Rabbit Clearance Societies. Corporate bodies comprising land occupiers, established under a scheme introduced in 1958 providing Government grants, aimed at systematically controlling and eradicating wild rabbits in local districts. Each Society was required to adopt a plan of control, to have a membership of about two-thirds of the occupiers in its area, and received a grant of 50% of its operating costs. In the late 1960s many Societies amalgamated to reduce costs, sharing secretaries and operating staff. The grant was withdrawn in 1971 and many Societies gradually ceased to function, though a few continue.

Rabies. A NOTIFIABLE DISEASE caused by a virus which infects the central nervous system. All mammals are susceptible. It is transmitted in the saliva mainly by biting. Rabies is not present in Britain, but in recent years it has been spreading in the Continental European fox population, its main reservoir and vector. Affected cattle, sheep, pigs and goats show abnormal changes in behaviour and develop any of the following symptoms; anxiety, aggressiveness, frequent bellowing, excessive saliva with choking, difficulty in eating, grinding of teeth and constipation or diarrhoea with straining and tail swishing. Progressive paralysis develops with death occurring on the 3rd–7th day. Also called hydrophobia.

Race. 1. A term used by breeders for the descendants of a common ancestor which exhibit common characteristics, but which differ slightly from other individuals or races within the species. Also used generally to mean a breed, variety, stock, group, class or genus of plant or animal.

2. A white streak on an animal's face.

3. A passageway along which livestock may be moved (e.g. from pen to DIPPING TANK), normally made of gates set in two parallel lines sufficiently apart to take one animal at a time.

Sheep undergoing routine treatment in a race.

Raceme. A type of inflorescence with a main stem bearing stalked flowers, e.g. lupin.

Rachis. 1. The main stem of an inflorescence.
2. The main axis of a PINNATE LEAF.

Rachitis ◆ RICKETS.

Rack. 1. A frame from which livestock may pull down FODDER.
2. A term sometimes used for the neck and spine of a forequarter of a carcase.

Rack Up. To fill a RACK with hay or straw.

Raddle. 1. A flexible piece or strip of wood used for weaving between uprights in the making of fences or HURDLES.
2. ◆ RUDDLE.

Radicle. The young root which develops from a germinating seed.

Radnor. A small breed of hill sheep, intermediate between the KERRY HILL and the WELSH MOUNTAIN. It has a clear tan face, aquiline nose, and short legs. Although ewes are polled, rams usually possess small, closely set horns which curl forwards, then backwards. The kempy wool is used in Wales in the production of woollen materials.

Raett. A sheep FOLD.

Rafter. To plough land so that each furrow is overturned onto an unploughed strip.

Ragwort. Composite plants of the genus *Senecio*, particularly the coarse yellow-flowered *S. jacobaea* which is a common weed of pastures, and is poisonous to livestock. It is scheduled as an INJURIOUS WEED and must be controlled either by cutting or treatment with weedkillers.

Raik ◆ RAKE.

Raiting ◆ RETTING.

Rake. 1. A hand implement with a long handle attached to a crossbar bearing several prongs, used for scraping, gathering material (e.g. cut grass) together, smoothing a seed bed, etc.
2. A tractor-drawn wheeled implement with long, curved, spring tines, used for gathering hay, scraping weeds into heaps, etc.
3. A very thin horse.
4. A northern term for a track, particularly one on a pasture or hillside, or in a gully.
5. A pasture. Also called raik.
6. To keep a flock of sheep moving from pasture to pasture.

Ram. An uncastrated adult male sheep. Also called a tup. (◗ SHEEP NAMES)

Rape. A plant (*Brassica napus*) related to the turnip and grown for FORAGE, particularly for sheep, during the autumn and early winter. There are giant and dwarf varieties which produce rapid, succulent, leafy growth, smothering weeds. It is sometimes included in grass SEEDS MIXTURES as a COVER CROP and grazed early. It is also often grown as a CATCH CROP after PEAS or early POTATOES, or in cereal stubble. Rape is very susceptible to CLUB ROOT and MILDEW. Also called forage rape, cole or coleseed. Certain varieties possess seeds rich in oil. (◗ OILSEED RAPE)

Rape Kale. A hybrid variety of KALE with a short, thick stem and wrinkled leaves, grown mainly in southern England in cereal stubbles to produce FORAGE in the spring, mainly from side shoots. The plant is frost resistant but is susceptible to CLUB ROOT and MILDEW. (◗ HUNGRY GAP KALE)

Rating ◗ RETTING.

Ration ◗ MAINTENANCE RATION, PRODUCTION RATION.

Ray Fungus. A bacterium (*Actinomyces bovis*) which forms radiating threads in grasses and cereals and may cause ACTINOMYCOSIS if eaten by livestock.

Reaction ◗ pH VALUE.

Reactor. An animal which is shown to possess a certain disease by a positive response or reaction to a test for the particular disease.

Readily Available Moisture. The volume of water held in the soil between FIELD CAPACITY and PERMANENT WILTING PERCENTAGE, which is capable of being removed by plants. As a soil dries between these two reference levels the growth rate and yield of a crop are reduced. (⬥ SOIL MOISTURE DEFICIT)

Reap. To cut a crop, particularly a cereal, to harvest the grain.

Reaper-binder ⬥ BINDER.

Recessive Gene ⬥ DOMINANT GENE.

Rectum. The terminal part of the large intestine of animals from which faeces are voided via the anus.

Red Clover. A deep-rooting, short-lived, species of CLOVER (*Trifolium pratense*), with an erect hairy stem, untoothed leaflets, short, broad, sharp-pointed stipules, and pale rose flowers. The following types are important agriculturally:

Red Clover.

(a) Broad red clover, varieties of which have broader leaves than late-flowering types and flower about two weeks earlier (about mid-June). They produce good HAY or SILAGE and good AFTERMATH, but then recover poorly. Mainly used in one-year LEYS. Also called early-flowering red clover or double-cut red clover.

(b) Late-flowering red clover, varieties of which are more persistent and hardy than broad red clover, tiller more abundantly and are less susceptible to CLOVER ROT. They produce good hay but relatively little aftermath. Normally sown in SEED MIXTURES for long leys. Also called single-cut red clover or cow grass.

(c) Extra-late-flowering red clover, varieties of which tiller very abundantly and produce persistent growth late in spring, and are sown for supporting intensive grazing, particularly by sheep.

Red Fescue. A species of grass (*Festuca rubra*) normally undersown in thinly sown spring cereal crops. Best suited to relatively poor upland and marginal soils.

Red Mould ▶ HOP MILDEW.

Red Poll. A hornless, dual-purpose breed of cattle derived from crosses between the now extinct Suffolk Dun breed and red cattle from Norfolk. Deep red in colour, sometimes with white on the end of the tail and udder, with a long, deep, muscular body and short legs. A hardy breed which has declined in popularity. The annual average MILK YIELD in 1978/79 was 3 880 kg (3 779 litres; 830 gal.) with an average BUTTERFAT content of 3.80%.

Red Rubies. A popular name for DEVON cattle due to their deep cherry red colour.

Reddle ▶ RUDDLE.

Redwater. A parasitic disease of cattle caused by a microscopic protozoan, *Babesia (Piroplasma) bovis,* which attacks red blood cells releasing HAEMAGLOBIN, and causes listlessness, salivation, fever, diarrhoea, reddish urine, and eventual death if untreated. The disease is transmitted by the common sheep tick, *Ixodes ricinus*, the intermediate HOST.

Reed. 1. A tall grass (*Phragmites communis*) which grows extensively in thick reed-beds in swampy conditions. The dried stems are used in East Anglia for thatching.
2. The uterus of a EWE.

Reed Stomach ▶ ABOMASUM.

Reel. 1. A rotating cylinder on a pick-up BALER, bearing spring tines, which picks up hay or straw from a swath and feeds it into the machine.
2. The rotating pick-up mechanism of a COMBINE HARVESTER or BINDER, often fitted with spring tines to pull laid crops into the machine. Bat reels are not fitted with tines and are used when the crop is in good standing condition.

Reference Price. 1. The minimum price at which certain fruit and vegetables can be imported into the E.E.C. under the COMMON AGRICULTURAL POLICY. These prices are broadly equal to average market prices on representative markets in the E.E.C. during the three preceding seasons. Produce imported at below the Reference Price may be subject to a levy (countervailing duty).
2. The weighted average market prices for livestock in the E.E.C. (▶ SLUICEGATE PRICE)

Reforestation. The planting of trees on land where a forest has previously stood, but has been destroyed, e.g. by a forest fire. (▶ AFFORESTATION)

Regime. 1. ▶ COMMON PRICES.
2. The flow and behaviour pattern of a river through the seasons and in all conditions.

Regional Panels. Advisory panels established in 1972 in each of the M.A.F.F. regions of England and Wales following the abolition of COUNTY AGRICULTURAL EXECUTIVE COMMITTEES. Each comprises 12–13 members appointed personally by the Minister as representing the main farming systems and areas within each region. One third of the appointments are reviewed each year. The Panels exercise, with the Ministry, a two-way communications role, so that both the Minister and the agricultural and horticultural industries are fully informed of important developments in their respective spheres. The Panels also advise the Minister of specific occasions where an independent non-Ministry view is desirable, such as the consideration of representations against decisions by officials in GRANT AID and subsidy cases.

Relative Weed. One that may have some use to a farmer in certain circumstances. (♦ ABSOLUTE WEED)

Rendzina. A shallow type of soil developed over calcareous rocks, with a brown to black horizon of MULL HUMUS, often as the only SOIL HORIZON, over the parent rock. A neutral to alkaline soil with a variable content of free CALCIUM CARBONATE.

Rennet ♦ RENNIN.

Rennet Stomach ♦ ABOMASUM.

Rennin. An enzyme occurring in the GASTRIC JUICE of young mammals which causes milk to clot and curdle by converting the soluble protein caseinogen into casein which then forms an insoluble calcium-casein compound. This is the principle of cheese-making and impure rennin (rennet) is extracted for this purpose from the rennet stomach or ABOMASUM of calves.

Rent (Agriculture) Act, 1976. An Act giving certain agricultural workers occupying service houses (tied cottages) security of tenure on ceasing to be employed in agriculture, and establishing AGRICULTURAL DWELLING HOUSE ADVISORY COMMITTEES.

Resazurin Test. A test used to determine the keeping quality of milk. (♦ MILK QUALITY SCHEMES)

Re-seed. To re-establish a LEY by sowing seeds of GRASSLAND SPECIES, usually in the form of a SEEDS MIXTURE. Reseeding may be carried out after ploughing up a previous ley which has deteriorated, or as part of a ROTATION. (♦ DIRECT RESEEDING, UNDERSOWING)

Reserve Nucleus Herd ♦ PIG IMPROVEMENT SCHEME.

Residual Herbicide. A HERBICIDE which, when applied to the soil, persists for some time, killing weeds on germination.

Residual Value ♦ MANURIAL VALUE.

Respiration. 1. The action of breathing, involving gaseous exchange between the lungs and the atmosphere, in which oxygen is taken in and carbon dioxide given out.
2. A metabolic process occurring in the cells of all living organisms by which CARBOHYDRATES are chemically broken down, releasing energy. Respiration is mostly aerobic, involving a complex series of ENZYME controlled oxidation-reduction reactions, with the consumption of oxygen and production of carbon dioxide. The energy released is equivalent to that produced by burning carbohydrates (e.g. glucose) in air. Respiration may be anaerobic, occurring in the absence of oxygen, with carbohydrate breakdown releasing less energy than in aerobic respiration. Examples include the degradation of glucose to yield carbon dioxide and ethyl alcohol in yeast, and the breakdown of GLYCOGEN (a storage carbohydrate) to lactic acid in muscles. In animals oxygen is delivered to the body's cells from the lungs, dissolved in the blood. Plants derive oxygen mainly by gaseous exchange through leaf pores (STOMATA).

Resting Pasture. One from which grazing animals have been removed to allow it to recover and put on new growth.

Restitutions. Export subsidies paid under the COMMON AGRICULTURAL POLICY to assist the disposal abroad of E.E.C. surpluses of agricultural commodities in order to prevent such surpluses upsetting the E.E.C. market.

Restorative Crops. Crops which are grazed whilst growing so that livestock return ORGANIC MATTER and nutrients to the soil in the form of faeces and urine, e.g. a LEY, folded roots (◗ FOLD), KALE, etc.

Reticulum. The second stomach of a RUMINANT which receives harder food material passed on from the first stomach or rumen and in which accidentally eaten small stones normally accumulate. Also called the honeycomb stomach after its internal mucous membrane lining which has a honeycomb-like structure.

Retting. The exposure of harvested FLAX to moisture so that it partially rots and the fibres are more easily separated from the rest of the stem. Also called rating or raiting.

Return ◗ TURN and JUNE RETURN.

Reversible Plough. A popular type of plough having two sets of bodies (◗ BODY), used for one-way ploughing (◗ PLOUGHING). One set of bodies, turning furrows to the right, are used for ploughing in one direction; the other left-handed set for ploughing in the other direction. Automatic, hydraulic or mechanical mechanisms are used for turning over the sets of bodies. The reversible plough has the advantage that headland use is much reduced; no ridges, depressions or finishes are left in the field and the surface is left level. They are mainly used in Britain for deep ploughing before preparing seed beds for root and vegetable crops. Also called one-way plough.

A four-furrow reversible plough.

Reversion. 1. The process by which soluble monocalcium phosphate, a constituent of SUPERPHOSPHATE, is rapidly converted in the soil to the less soluble di-calcium salt, producing a finely divided precipitate which is easily available to plants.

2. ◗ ATAVISM.

3. A viral disease affecting blackcurrants causing the development of narrower and darker leaves than normal and reducing the amount of fruit produced.

Rex Rabbits. Varieties of fur breeds of rabbit with very fine, densely textured coats. They can be bred in any colour.

Rhine. A large, broad ditch or open water-course in flat marshy areas, particularly in the South of England.

Rhinitis ♦ ATROPHIC RHINITIS.

Rhizobium ♦ NITROGEN FIXATION.

Rhizome. A horizontal creeping underground stem growing just below the soil surface, from which axillary buds produce new stems and roots giving rise to new separate plants, acting as a foodstore and means of vegetative reproduction, e.g. COUCH GRASS, raspberry.

Rhode Island Red. A hardy, medium-sized, laying breed of FOWL, rich reddish-brown in colour with black or greenish-black tail feathers, yellow legs and a single comb. Hens lay large brown eggs.

Ribbing ♦ BREAK-FURROWING.

Ribbon-Footed Corn Fly ♦ GOUT FLY.

Rice. 1. Small branches or twigs. Also pea straw.
2. A cereal (*Oryza sativa*) grown in the tropics mainly for human consumption, poor in fat, protein and minerals. Sometimes fed to poultry in Britain.

Rick. A STACK.

Rickets. A disease of young animals due to a deficiency of either Vitamin D or Phosphorus. Bones are unable to absorb Calcium from food and fail to harden or ossify properly. The long limbs tend to bend and develop swellings on the ends at the joints.

Rickyard ♦ STACKYARD.

Riddle. A large coarse sieve. Also called screen. Riddles of specified mesh size are used to sort WARE POTATOES from a potato crop.

Ridge. 1. The soil thrown up by a plough between two furrows.
2. A raised strip ploughed to mark out LANDS in a field before it is systematically ploughed, consisting of a balanced number of furrow slices leaning from opposite directions towards the centre. There are two types, (*a*) the arable ridge, used mainly in

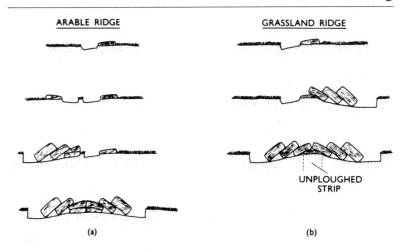

ARABLE RIDGE   GRASSLAND RIDGE

UNPLOUGHED
STRIP

(a)   (b)

The various stages in ploughing ridges.

fields in which an arable crop was previously grown, in which all the land beneath the ridge is ploughed, and (b) the grassland ridge, used for ploughing grass fields in which a central strip under the ridge is left unploughed. Also called an opening, rig, split or stitch.

3. A raised strip in which potatoes are planted. (◗ RIDGER)

Ridger. 1. A type of plough bearing one or more double MOULDBOARDS, each throwing two furrows simultaneously, one to each side. Once commonly used (before the advent of the POTATO PLANTER) to form ridges for potatoes and other root crops. Still used for earthing up purposes. Also known by various local names including booter, bowter, baulking plough, and double-breasted plough.

2. ◗ BACK CHAIN.

Ridging Plough ◗ RIDGER.

Rig. 1. ◗ RIDGE.

2. A male animal in which one or both testicles have not completely descended from the abdomen to the scrotum at the normal time.

Right-of-way. The right to pass over land. Also the track or path along which such a right exists. A right-of-way must be over a defined strip of land and may be either 'private' (e.g. for a particular person or group of persons generally, or for a specific purpose, such as access to a church, with or without vehicles) or 'public' (i.e. a public footpath, bridleway or public vehicular highway), depending upon the nature of the rights which have been dedicated to, or acquired by, the person or persons to which they apply. Private rights-of-way are a matter as between the grantor of the particular right and those entitled to use it, whilst public rights-of-way are subject to public highway law.

Rind Graft ◗ CROWN GRAFT.

Rinderpest ◗ CATTLE PLAGUE.

Ring. 1. A metal circle fixed through the nose of an animal, e.g. in pigs to prevent grubbing in the ground with the snout, or in bulls for roping and controlling movement (also called BULLDOG). Also to fix such a ring through an animal's nose.
2. A combination of buyers at a market or auction who arrange not to bid against one another in order to keep prices down.
3. A segment of a CATERPILLAR or WORM.
4. ◗ RINGBARK.
5. ◗ ANNUAL RING.
6. ◗ FAIRY RING.

Ring Fence. A continuous fence encircling a farm or estate.

Ring Roll. ◗ CAMBRIDGE ROLL.

Ringbark. To remove a circular strip of BARK from a tree to prevent growth. Also called ring.

Ringbone. A callus of bony tissue which sometimes develops on the pastern bone of a horse's hoof.

Ringworm. A contagious skin disease characterised by ring-shaped patches caused by certain fungi growing on the skin surface or in hairs growing in the affected skin. In livestock young store cattle are most commonly affected.

Riparian. Land adjacent to a river bank. Also the owner of such land.

Rippling. A North Country term for the practice of ploughing a shallow furrow to mark HEADLANDS and SIDELANDS.

River Basin. The total area or CATCHMENT drained by a river and its tributaries.

Roan. The colour of an animal's coat, usually a mixture of white and other coloured hairs, e.g. red, brown, black, chestnut, etc.

Rock Phosphate. A natural mineral found in the form of a sedimentary rock containing various calcium phosphates. After grinding it is used as a phosphatic FERTILIZER. Also called phosphorite or mineral phosphate.

Rod ♦ POLE.

Roding. A term used in the FENS for the practice of cutting and cleaning out dyke vegetation.

Rogue. A sub-standard plant or one of a different type, particularly in a seed crop. Also to remove such plants.

Roll ♦ ROLLER.

Roll Bar ♦ SAFETY CAB.

Roll Over Protection Structure ♦ SAFETY CAB.

Rolled Grain ♦ CRUSHED GRAIN.

Roller. A tractor-drawn implement consisting of heavy rotating cylinders. Used to crush clods, consolidate soil, flatten and smooth the surface, break crusts, or prepare fields for sowing seeds. Many types exist, the two main ones being the CAMBRIDGE ROLL and FLAT ROLL. Also called roll.

Roller Crusher. A machine used in HAYMAKING which passes freshly cut grass between steel or rubber faced rollers rotating at high speed, bruising the grass and allowing sap to be removed subsequently by the action of sun and air. The rollers may be cleated or grooved to provide more effective bruising. (♦ CRIMPER)

Roller Mill ♦ CRUSHING MILL.

Roman. A moderate-sized, white-coloured breed of goose, producing between 50–60 eggs per year.

365

Romney. A hardy, polled breed of sheep, well suited to the rigorous conditions and lush wet pastures of the Romney Marsh in Kent, where it was developed and is kept under dense stocking. Characterised by long, close, fine wool, a white face and legs, and thick tufted forelock. Some claim the breed to be resistant to LIVER FLUKE and FOOT ROT. It is commonly crossed with the SOUTHDOWN to produce lambs which are particularly in demand in South-East England. Ewes are also crossed with North Country CHEVIOT rams, producing the ROMNEY HALFBRED. The breed has been widely exported, particularly to New Zealand (*see p. 392*). Also called Kent.

Romney Halfbred. A crossbred type of sheep produced by crossing ROMNEY ewes with North Country CHEVIOT rams. The MULE and SCOTCH HALFBRED are more prolific and profitable, and more popular.

Rood. An old measure of land equal to ¼ acre (1 210 sq. yds) or 40 sq. POLES.

Rookworm ♦ COCKCHAFER.

Roost. 1. A perch or resting place for a bird. Also a number of birds nesting together.
2. A henhouse.

Rooster. A domestic COCK.

Root. 1. The part of a vascular plant which anchors it to the ground and through which it absorbs water and nutrients from the soil via minute tubular outgrowths (root hairs) from epidermal cells just behind the root tip. Root tips comprise a mass of cells (the root cap) which are produced by and protect the growing point, and which wear away and are replaced. There are two main systems of root growth, (*a*) tap roots, which develop from the radicle or primary root of a young seedling (e.g. carrots), and (*b*) fibrous roots, which are numerous and replace the radicle, growing from the base of the stem (e.g. grasses) (♦ also ADVENTITIOUS ROOTS). Roots are frequently modified for food storage and are valuable to man, e.g. beet, carrots, turnips. (♦ ROOT CROPS)
2. ♦ ROOTLE.

Root Cap ♦ ROOT.

Root Crops. Certain biennial plants grown for their edible swollen roots which act as food reserves, commonly in the form of sugar. The MANGEL, TURNIP, SWEDE and FODDER BEET are mainly grown as stock feed. The total area grown has decreased during this century due to high labour demands, but they are increasing again in popularity with mechanisation. SUGAR BEET and CARROTS and also the POTATO (usually classified as a root crop) are mainly grown as cash crops. The 'tops' or foliage of root crops (except carrots and mangels) are valuable fodder, and sheep or pigs are often folded (◗ FOLD) on them. Root crops are often grown following or between cereal crops in ROTATIONS. Also called roots.

Root Hair ◗ ROOT.

Root Graft. A GRAFT which lies beneath the soil surface, in which the scion is joined to the root of the stock.

Root Nodule ◗ NODULES.

Rootle. To turn up earth with the snout. Also called root or rout.

Roots ◗ ROOT CROPS.

Rootstock ◗ STOCK.

Rotary Cultivator. A type of tractor-mounted or trailed CULTIVATOR consisting of a horizontal shaft driven by p.t.o. from the tractor, and bearing a series of tines or blades of various designs which rotate with a 'stirring' action. Used for a variety of purposes including (a) shattering large clods, (b) producing a tilth rapidly, combining ploughing and cultivating in one operation, and (c) burying surface trash (e.g. straw) which can clog a plough, or (d) eradicating bracken. Small models are popular on horticultural holdings for simultaneously burying previous crops and preparing seed beds.

Rotary Parlour ◗ MILKING PARLOUR.

Rotation. A cropping system in which two or more crops are grown in a field in a fixed sequence. If a LEY is included it is known as 'alternate husbandry' or 'ley farming'. One of the earliest systems practiced was the four-course NORFOLK ROTATION which comprised the sequence root crop – cereal – ley – cereal. This has been modified since

367

the start of the century to meet the changing needs of agriculture. In Norfolk sugar beet replaced other rootcrops, and in some cases potatoes were introduced in a five-course rotation, still widely practiced, which comprises sugar beet – barley – 1 year ley – potatoes – wheat. In some areas six-course rotations developed (◗ EAST LOTHIAN ROTATION), whilst in others three course systems evolved (◗ FENLAND ROTATION). The benefits of rotation include reduced accumulation of disease and pests which accompany monoculture, weed control, the maintenance and improvement of fertility, spreading the risk of specific crop failure, and the even distribution of labour requirements over the year. In recent years farming has moved away from rigid traditional cropping programmes to more simplified systems due to various developments. These include the production of pesticides and ARTIFICIAL FERTILIZERS, improved and increased mechanisation, and guaranteed crop prices. (◗ ANNUAL REVIEW)

Rotational Grazing. A method of managing grassland in which successive areas are intensively grazed for a period, in which defoliation is rapid and complete, followed by a rest period during which regrowth occurs. This method (also called the on-off principle) is used in a number of GRAZING SYSTEMS as a means of matching livestock requirements with grass growth.

Rothamsted Experimental Station. Said to be the oldest agricultural experimental station in the world. Founded in 1843 by John Bennet Lawes at Rothamsted Manor, Hertfordshire, it now undertakes research into all aspects of crop production, except PLANT BREEDING. Early work on FERTILIZERS and the usage of agricultural statistics were pioneered at Rothamsted.

Rough Fell. A horned moorland breed of sheep found locally in Cumbria and North-West Yorkshire. Square-shaped and characterised by a dark coloured face with a tinge of brown, a grey muzzle, and a fleece of very coarse wool.

Rough Grazing. A term generally applied to grassland of poor quality with a low stock-carrying capacity, particularly that covering rough land, HEATH, MOORLAND and hill areas, where cultivation and the use of FERTILIZERS is impractical, and which is characterised by coarse grasses, sedges, heather, bracken or mosses, etc. Such land is usually grazed by hardy hill breeds of sheep and some cattle.

Rough Leaf. One of the pair of leaves which develop after the COTYLEDONS or smooth leaves in BRASSICAS.

Rough Stalked Meadow Grass. A close-growing perennial grass (*Poa trivialis*) with short creeping STOLONS, a rough, flat stem, and an erect panicled flower head, reddish, purplish or green in colour. It is common in lowland meadows and pastures, particularly on rich moist soils. It is included in SEEDS MIXTURES for permanent pastures, particularly those on wet heavy soils, and grows vigorously producing a low closely knit sward which remains green through the winter. It is very palatable to livestock and is also useful for hay. (◗ SMOOTH STALKED MEADOW GRASS)

Roughage. Bulky FEEDINGSTUFFS containing FIBRE in significant amounts, e.g. hay and straw, which stimulate intestinal muscular activity.

Round. A circular wall in which sheep shelter from snow drifts.

Roundel. A circular kiln used for hop drying.

Roundworms. A class of worms (Nematoda), mainly parasites but also including free-living forms in the soil and freshwater. The parasitic forms are found in both plants (e.g. EELWORMS) and animals, where they infect the small intestine, stomach, lungs, etc., of livestock causing various diseases, e.g. gastro-enteritis, GAPES, HUSK.

Roup. A chronic respiratory disease of poultry causing loss of condition and reduced egg production. Also called croup.

Rout ◗ ROOTLE.

Row Crops. Those crops planted in widely-spaced rows facilitating cultivations between the rows, e.g. ROOTCROPS, potatoes, cabbages, etc.

Row-crop Tractors. Tractors designed to work in growing ROW CROPS. They usually have high, narrow wheels giving good ground clearance, an adjustable track width, and a narrow turning circle. They are used with special tools for cultivating between the rows. Most models are four-wheeled although some have only three wheels.

Rowen ◗ AFTERMATH.

Royalty ◗ SEED ROYALTY.

Ruakura Farrowing Crate ◗ FARROWING CRATE.

Rub. A measure of the quality of a sample of hops obtained by rubbing the sample with the hands and estimating the amount of sticky resinous material left on the hands, and the 'feel' or silkiness of the hops.

Rubbed Seed. Sugar beet seeds used for sowing which have been separated from the clusters in which they are naturally fused together. Also called graded seed.

Rubber ◗ PLANKER.

Rubbers. A term for SHEEP SCAB.

Ruddle. A colouring material (often red ochre) applied to the chest of a ram so that those ewes which have been covered (mated with) are marked. Also called keel, raddle or reddle.

Rumen. The first stomach of a RUMINANT, a large sac lined with a mucous membrane in which coarse, partly chewed food is churned, partially digested and stored until it is regurgitated and chewed again (chewing the cud), before being re-swallowed and passed to the second stomach or RETICULUM. Also called paunch.

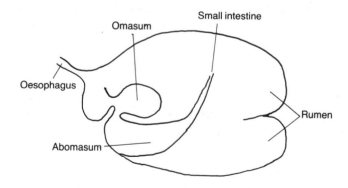

The Ruminant Stomach.

Rumen Tympany ♦ BLOAT.

Ruminant. An animal which chews the cud and possesses a complex digestive system including a four-part stomach comprising the first stomach or RUMEN, second stomach or RETICULUM, third stomach or OMASUM and fourth stomach or ABOMASUM. Ruminants include cattle, sheep and goats which lack upper incisor teeth. They consume large amounts of bulky fibrous foods such as grass, which their complex stomach is designed to store and digest.

Ruminate. To chew the cud. (♦ RUMEN)

Run. An enclosure for animals, particularly poultry, in which they are free to roam and feed.

Runner ♦ STOLON.

Runt. 1. A small, undersized or stunted animal, particularly the smallest piglet in a litter. Also called sharger, dolly or Anthony.
2. An old cow or ox.
3. A STONE animal.
4. A dead tree trunk or stump.
5. A cabbage stem.

Rusts. An Order of small parasitic BASIDIOMYCETE fungi (Uredinales), the spores of which form spots, often rust coloured, on the leaves and stems of infected plants, e.g. BLACK RUST of cereals, YELLOW RUST of barley.

Rut. A seasonal period of sexual excitement in male animals, particularly deer, when there is an urge to mate.

Rye. The hardiest of the CEREAL crops (*Secale cereale*). It has a similar structure to WHEAT, with single spikelets attached to each notch of the RACHIS, but also has small narrow GLUMES and conspicuous large outer pales (♦ PALEA) tapering to long whiskery AWNS. Rye is of relatively minor importance in the U.K. It is grown mainly on poor or acid light soils where other cereals would be uneconomic and is used mainly for stock feeding. It also provides an early bite for sheep or dairy cattle after which it recovers well to set grain. Alternatively it may be ploughed in after folding (♦ FOLD) often in preparation for a root crop. Some rye is grown under contract for rye crispbread and the straw is used for thatching (*see p. 95*).

Ryegrass. One of the most important genera of grasses (*Lolium sp.*) used in SEEDS MIXTURES. (◗ ITALIAN RYEGRASS, PERENNIAL RYEGRASS and WESTERWOLDS RYEGRASS)

Ryeland. A minor, short-woolled breed of docile sheep developed in Herefordshire. It is small in size, with a dull white face and legs, a well-woolled poll, forehead and cheeks, and produces a uniform fleece of good quality. The breed provides high quality, lightweight lambs.

S

S.19. A strain of vaccine injected into heifer calves to suppress BRUCELLOSIS.

'S' Strains ◗ ABERYSTWYTH STRAINS.

Saanen. A breed of GOAT imported from Switzerland, usually white, sometimes cream, with short hair, a dished face and erect ears. Females weigh about 50–55 kg (110–121 lbs) and produce milk containing about 4.3% BUTTERFAT. Adult males weigh about 75 kg (165 lbs).

Sack. 1. A measure of capacity, particularly for grain, equal to 4 BUSHELS. Also called coomb. Now in disuse in the grain trade since most grain is now transported in bulk by lorry.
2. A large bag of coarse material.

Saddle. 1. A rider's leather seat strapped to a horse's back. Also a similar apparatus for draught horses from which the shafts of a cart or towed implement are suspended.
2. A butcher's cut including some of the backbone and ribs.

Saddleback ◗ BRITISH SADDLEBACK.

Safety Cab, Safety Frame. A structure fitted to a tractor designed to protect the driver in event of an over-turning accident. Fitment of these structures is subject of the Agriculture (Tractor Cabs) Regulations. According to the Regulations the term 'Safety Cab' embraces the safety frame or roll bar. The Safety Cab is a fully clad structure with windows and doors and in some circumstances this will be a 'Quiet Cab', which is so designed that the noise level reaching the tractor driver's ear from the operation of the tractor is reduced to less than 90 dbA under specified test conditions. The Safety Frame is a protective structure usually consisting of four posts supporting a top framework. In some cases a roof is added, although it is otherwise unclad. The Roll Bar is a two post structure in the form of a hoop, usually sited just behind the driving position. These structures are sometimes referred to as Roll Over Protection Structures (R.O.P.S.).

Sainfoin. A perennial leguminous fodder plant (*Onobrychis viciifolia*), once extensively grown for both HAY and SILAGE, with AFTERMATH grazing, mainly in chalky areas in eastern and southern England, usually on its own, but sometimes included in SEED MIXTURES. It has a similar nutritive value as LUCERNE which it is sometimes substituted for on potash-deficient soils, being somewhat drought-resistant. It has a strong, woody, long root, a branching stem with PINNATE leaves, pink flowers in a conical RACEME, and single-seeded pods. Also called Cock's Head, Holy Grass and St Foin.

Salmonella. A genus of BACTERIA. Many types exist some of which cause infectious diseases (Salmonellosis) in livestock. Important types are *S. dublin* and *S. typhimurium*, which often affect cattle, causing various symptons including diarrhoea (sometimes bloody), fever, loss of appetite, lower milk yield, abortion, sometimes pneumonia, and death, etc. Infection can be transmitted in faeces, in slurry spread on grassland, in contaminated milk or animal feeds, by healthier carriers, by cross-infection at markets, etc. *S. typhimurium* and certain other types are the cause of food poisoning in man. (◆ BACILLARY WHITE DISEASE)

Salt Fertilizer. (Na Cl). Common salt, sodium chloride, long used as a FERTILIZER for mangels, sugar beet and carrots, being applied 2–3 weeks prior to sowing, and producing increased yields.

Salt Lick ◆ MINERAL MIXTURES.

Saltpetre ◆ NITRATE OF POTASH.

Salving. The protecting of sheep against cold or parasites by smearing them with a mixture, usually of tar and butter.

Sand. A constituent of SOIL comprising coarse mineral particles (e.g. QUARTZ), intermediate in size between silt and gravel, being 0.02–2.0 mm in diameter (according to the international particle size system). Derived mainly from weathered, particularly siliceous, rocks (e.g. sandstone, granite). (◆ CLAY)

Sandy Soils. Soils containing a high proportion of SAND particles and very little CLAY, usually less than 5%. Such soils have a low water holding capacity, drain freely, and due to LEACHING are usually poor in plant nutrients. They are also well aerated and ORGANIC MATTER quickly decomposes. They therefore require adequate applications of FERTILIZERS and MANURES and are consequently often called 'hungry soils'. LIMING is also needed to correct acidity. They tend to dry out quickly allowing them to be worked throughout the year. They are often used for MARKET GARDENS using irrigation. Crops suited to sandy soils include barley, carrots, peas, potatoes, sugar beet, etc. Sometimes called 'light soils' because they are relatively easy to cultivate compared with heavy or CLAY SOILS.

Sapling. A young tree.

Saprophyte. An organism which feeds on decaying organic material, e.g. many bacteria and most fungi. (♦ PARASITE)

Savoy. A type of winter CABBAGE with wrinkled leaves and a large close head.

Sawflies. A family of insects which derive their name from the saw-like ovipositor which is used to lay eggs in twigs, leaves or blossom. The caterpillars of the Apple Sawfly, *Hoplocampa testudinea*, feed on the core and seed of the growing apple. Several other species are also noted pests: Gooseberry Sawfly, *Nomatus ribesii*; Turnip Sawfly, *Athalia spinarum*; Pear Sawfly, *Eriocampa limacina*; Rose Slug, *E. rosae* (the larvae of *Eriocampa spp.* are called slugworms), Rose Sawfly, *Argeodiropus*.

Scab. 1. A skin disease of animals characterised by scales or pustules in patches, especially one caused by MITES (♦ SHEEP SCAB). Also called MANGE.
2. A fungal disease of various kinds of fruit and vegetables, causing the growth of scaly crusts, e.g. Common Scab of potatoes (*Streptomyces scabies*) and apple scab (*Venturia inequalis*).

Scabies. An infectious skin disease caused by a parasitic MITE (*Sarcoptes scabei*) which affects farm livestock and man and results in severe scratching. Commonly called itch.

Scaly Leg. A disease of poultry caused by a MITE (*Sarcoptes mutans*, var. *gallinae*), in which hard scales develop on the unfeathered parts of the legs, accompanied by irritating itchiness.

Scandinavian Piggery ♦ PIGGERY.

Scarecrow. Any object, often a dummy resembling the human form, set up in a field to scare birds.

Scarifier. A CULTIVATOR, the action of which breaks up and stirs the soil without turning it over. Also called scuffler.

Scarp Slope. A short steep slope to one side of a line of hills (e.g. North Downs), as distinct from the long, gentle DIP SLOPE to the other side.

Scheduled Diseases. Those diseases listed under the DISEASES OF ANIMALS ACT, 1950, some of which are NOTIFIABLE DISEASES.

Scheduled Weeds. Certain WEEDS listed under the Weeds Act, 1959. (♦ INJURIOUS WEEDS)

Scion ♦ GRAFT.

Scolt ♦ SOLE FURROW.

Scottish Blackface. A breed of mountain sheep, small but very hardy. Characterised by a black, or mottled black and white, face and legs, from which wool is absent. The breed is well horned, those of rams having a distinctive double spiral. The wool is long, coarse and hairy, and is used in carpet making. Some is exported to Italy for filling mattresses. It is a very important breed in Britain dominating the heather moors of Scotland and Northern England. Ewes are drafted to the lowland after about 3 years, and commonly crossed with DOWN-BREED or BORDER LEICESTER rams, the latter cross called the GREYFACE (*see p. 391*).

Scottish Half-bred. A uniform crossbred type of sheep with a white face and long ears resulting from a BORDER LEICESTER ram crossed with a CHEVIOT ewe. Rams of DOWN BREEDS are mated with such crossbed ewes to provide fat lambs.

Scours. Diarrhoea in livestock, a symptom of a variety of diseases.

Scrag. 1. The neck of a sheep.
2. A lean or skinny animal.

Scratch. A term for two light FURROWS ploughed to mark the setting out or opening of a plot, which form the base of the first two furrows (the crown furrows). Also called scribesod.

Screen. 1. A WINDBREAK. (♦ LEWING)
2. A RIDDLE.

Screw. A BROKEN-WINDED horse.

Scribesod ♦ SCRATCH.

Scrub. 1. An undersized or inferior animal.
2. A stunted tree. Also an area of such trees, bushes, low shrubs and BRUSHWOOD.

Scrubber ♦ PLANKER.

Scuffler. A SCARIFIER.

Scuppet. A type of large shovel, the scoop consisting of a wooden frame covered with hessian or ticking. Used for moving hops in an OAST house, either from the drying to cooling floor, or into POCKETS for pressing.

Scur. A loose, horny knob that sometimes develops in horned cattle at the site where a horn would normally grow. Also called snag.

Scythe. A hand implement with a long wooden handle with two hand grips (nibs), bearing at its lower end, fixed at right-angles, a large gently curved blade. Used for mowing by sweeping it through the vegetation. Sometimes spelt sithe.

Seams. The angles or shaped sections of the SLICES turned over in ploughing where they touch each other.

Season. 1. One of the four climatic divisions of the year, e.g. Spring.
2. The oestrus or 'heat' period of a female animal. (♦ OESTROGEN)
3. To prepare a SEED BED.

Second. To hoe between rows of ROOTCROPS which have previously been subject to singling. (♦ SINGLE)

Second Cut. The second crop or cut of grass taken from a field in a season for HAY or SILAGE.

Second Early Potatoes ▶ POTATO.

Secondary Host ▶ HOST.

Secondary Thickening. The growth of additional (secondary) vascular and structural tissues (i.e. PHLOEM and XYLEM) in a plant by cell division in the CAMBIUM, accompanied by increasing diameter in stems and roots.

Seconds. Those grains of a cereal crop intermediate in size between the largest ones (head corn) and the smallest ones (tail corn) threshed out during harvesting with a COMBINE HARVESTER.

Sedges. A large family of grass-like plants (Cyperaceae), particularly the genus *Carex*. They are distinguished from grasses by having a solid, usually triangular, stem, and only one scale beneath each flower (grasses have two). Flowering heads are greenish, brownish or purplish. Sedges are common in wet or badly drained areas and are regarded as weeds.

Seed. 1. A reproductive structure of flowering plants, conifers and various other plants. It develops from a fertilized OVULE and comprises an EMBRYO and a food reserve contained in a protective coat or testa.
2. To sow seeds of a crop in a field. (▶ DRILL)

Seed Barrow. A hand-pushed or tractor-mounted implement used for sowing seeds, consisting of a long narrow box with a perforated base, mounted laterally on a wheelbarrow-like frame with a single wheel. The seeds were brushed through the holes. Also called shandy barrow.

Seed Bed. An area of land or a field cultivated to a level, fine tilth, in which seeds are sown.

Seed-borne Diseases. Those plant diseases which may be transmitted from one generation to the next, either within or on the surface of seeds, e.g. many fungal diseases.

Seed Certification. All seeds offered for sale (except standard seeds of vegetables or uncertified Pre-Basic Seeds) must be certified. Such seed is produced under various certification schemes technically supervised in England and Wales by the Seed Production Branch of the NATIONAL INSTITUTE OF AGRICULTURAL BOTANY and operated in N. Ireland and Scotland by the respective Departments of Agriculture. All crops entered for certification (i.e. grasses, herbage legumes, cereals, field beans, beets, oil and fibre plants) must be inspected, whilst growing, for trueness to cultivar, absence of contamination, etc. Samples of the seed produced are laboratory tested according to rigid standards for health, purity, germination and weed seed content. The aim is to produce clean healthy seeds for sale which germinate well and produce crops of the stated variety and true to type.

Under the various schemes a plant breeder or maintenance breeder may enter for certification, seed of selected varieties (Pre-Basic Seed) from which it is intended to produce, by multiplication, Basic Seed or subsequently Certified Seed. Following crop inspection and later laboratory testing of samples, the seed may be certified as Basic Seed if it complies with the rigid standards. Basic Seed is usually used by seed companies or seed growers to produce Certified Seed. The growing crop must again be inspected and seed samples laboratory tested. Certified Seed is usually sold commercially to farmers and growers for crop production. For cereal seed it may be sold as first or second generation Certified Seed.

Crops of cereal seed being grown for Basic Seed or Certified Seed may be entered for certification at a higher level – Higher Voluntary Standard (H.V.S.) – for which they must meet higher standards for crop conditions, varietal and species purity, and wild oats content. As far as grasses and herbage legumes are concerned, only certain species being grown for Certified Seed may be entered for H.V.S. certification, which applies only to seed purity and weed seed content. (◖ COMMERCIAL SEED, SEED POTATOES)

Seed Cleaning. The removal from seeds, usually by machine, of impurities such as weed seeds, broken leaves, insects, etc.

Seed Coat ◖ TESTA.

Seed Cord ◖ SEED-LIP.

Seed Corn. A cereal grown to provide grain for seed.

Seed Dressing. The chemical treatment of seeds, particularly cereals, with fungicides and sometimes insecticides, to protect them against soil and seed-borne diseases and pests. A dye may be added which enables treated seed to be distinguished.

Seed Drill. A tractor-mounted or trailed machine which sows seeds in rows. It consists of a seed hopper incorporating seed metering units which are generally driven by gearing or chains from the land wheels. These units, which are adjustable to suit various seed sizes and sowing rates, transfer the seed from the hopper down coulter tubes to a groove cut in the soil by COULTERS.

Drills for cereals and grasses normally sow the seed at random, while Precision Drills, used mainly for sugar beet and vegetable crops, place individual seeds at specific pre-set intervals in the rows. The feeding mechanisms that are available include:

Random Spacing Drills	*Precision Drills*
Internal force feed	Belt feed
External force feed (fluted or studded roller)	Cell-wheel feed
	Pneumatic or vacuum feed
Brush feed	Cup feed
Sponge roller feed	

A narrow-row, random spacing seed drill, bearing a double row of coulter tubes and trailing a spring-fingered following harrow.

The common drill coulters in use include the tine type (useful on stony or heavy land), the single-disc (used on heavy ground, but damaged by stones) and the V-shaped Suffolk hoe type (used on light soils). (♦ COMBINE DRILL)

Seed House. A commercial company providing seed varieties for crop production.

Seed Inoculation. The dressing of seeds of leguminous plants with a culture of nitrogen-fixing (♦ NITROGEN FIXATION) bacteria, usually when the soil is deficient in the bacterial strain appropriate to the legume concerned and would otherwise restrict root NODULE formation.

Seed Leaf. A COTYLEDON.

Seed Potato. Relatively small tubers, usually 32–57 mm (1¼–2¼ in.) in diameter. Produced from approved parent material under the statutory certification scheme administered by the respective Agricultural Departments in the U.K. Growing crops of 'seed' are officially inspected for health, purity and vigour prior to certification. Two categories of 'seed' are certified, viz, (*a*) Basic Seed, used mainly for multiplying further as 'seed,' and (*b*) Certified Seed, used for the production of crops of WARE POTATOES, graded within each category as follows:

Basic Seed: V.T.S.C. Virus Tested Stem Cuttings. (Produced in the Protected Region, i.e. Scotland, N. Ireland and certain parts of N. England.)

F.S. 1–4. Foundation Seed (Produced in the Protected Region).

S.S. Stock Seed. (Produced in England and Wales outside the Protected Region.)

A.A.1. (Produced in Scotland.) ⎤ First quality com-
⎟ mercial seed with
⎬ nil tolerance for
⎟ tobacco veinal
A. (Produced in N. Ireland)⎦ necrosis virus.

A.A. First quality commercial seed (not produced in N. Ireland).

Certified Seed: C.C. Healthy commercial seed (not produced in N. Ireland or the Protected Region).

Both categories must produce crops which reach specified standards including trueness to varietal type stated, freedom from certain diseases (particularly virus), and freedom from undesirable variations and rogues.

Basic Seed potatoes are mainly grown in relatively cool areas of Scotland, N. Ireland and certain areas of North and North-West England, where APHID attack is less likely, whilst Certified Seed can be produced anywhere else in England and Wales.

Seed Potato Marketing Board for Northern Ireland. A MARKETING BOARD set up in 1961 with the objective of regulating the marketing of SEED POTATOES in Northern Ireland. In the past it has operated by intervening between growers and merchants to stabilise and guarantee prices to growers and by promoting the sale of seed potatoes in markets in Britain and abroad.

The Board comprises 9 members, 6 of whom are elected by registered potato growers and 3 are appointed by the Minister responsible for Agriculture in the Northern Ireland Office.

The Board is responsible for purchasing all certified seed potatoes (currently within the riddle sizes 35 × 60 mm) for shipment from registered growers in Northern Ireland who contract their production to the Board prior to the commencement of the season, and for selling this seed. Seed potatoes purchased from contract growers are bought at a fixed price, according to the variety, throughout the season, but the price at which they are sold by the Board varies from time to time according to market circumstances. (◆ POTATO MARKETING BOARD)

Seed Quality. The quality of crop seeds has two principal components. Firstly, the value of different varieties for crop production, assessed by yield trials and tests for disease reaction and produce quality. Secondly, the value of different 'seed lots' for sowing, assessed by laboratory analysis of samples for purity, germination, weed content and seed-borne diseases. (◆ SEED CERTIFICATION)

Seed Royalties. Fees paid under licence for the use of seed of protected (patented) plant varieties, and usually included in or added to the sale price of seed. Royalties are paid to breeders and are used to finance continued PLANT BREEDING work. (◆ PLANT BREEDERS' RIGHTS)

Seed Testing ♦ NATIONAL INSTITUTE OF AGRICULTURAL BOTANY.

Seeding Year. The year in which a grass (SEEDS MIXTURE) is sown.

Seedless Hay. HAY derived from the remains of a grass seed crop after threshing out the seed heads with a COMBINE HARVESTER.

Seedless Hops. Hops, the CONES of which lack seeds, not having been pollinated.

Seedling. A young plant, grown from seed as distinct from one grown from a cutting or by a GRAFT.

Seedling Crab ♦ SEEDLING ROOTSTOCK.

Seedling Rootstock. A STOCK grown from a pip or seed of a wild crab apple. Also called seedling crab.

Seed-lip. A basket carried against the side of the body and supported by a strap round the neck, from which seed is sometimes broadcast by hand. Also called seed cord or sidlip.

Seeds. A LEY, usually a mixture of grasses and clover. (♦ SEEDS MIXTURE)

Seeds Executive ♦ UNITED KINGDOM SEEDS EXECUTIVE.

Seeds Harrow. A light type of ZIG-ZAG HARROW, with short, straight TINES, usually comprising several sections attached to a frame. Used to prepare SEED BEDS and cover seeds after sowing, and often hitched behind the DRILL. Sometimes also used for light weeding.

Seeds Hay ♦ HAY.

Seeds Ley. A short-term LEY, sown with a SEEDS MIXTURE of grasses and clover.

Seeds Mixtures. Mixtures mainly of grass and clover seeds, sometimes including other herbage legumes, sown to produce both short-term and long-term LEYS. Mixtures are precisely compounded with varieties suitable to the conditions and needs for which the ley is required. They may be prepared to order by a merchant or standard mixtures can be purchased from seeds firms. Also called small seeds.

Seedy-cut. A blackish blemish sometimes found in pigmeat.

Seepage. The emergence of GROUNDWATER from the soil surface in an 'oozing' manner, often along an extensive line, as distinct from a continuous flow from a spring at a particular spot.

Seg ▶ STAG.

Segmented Seed. The 'seeds' of certain rootcrops (e.g. MANGEL, SUGAR BEET) naturally borne in clusters of fruits each containing a single true seed, which are often separated for sowing. If the cluster itself is sown, several seedlings may develop which makes SINGLING a slow operation.

Selection. The basis of animal and PLANT BREEDING in which individuals are selected for breeding on the basis of specific desired characteristics or qualities. (▶ GENOTYPIC SELECTION, PEDIGREE SELECTION and PHENOTYPIC SELECTION)

Selective Herbicide. A HERBICIDE capable of killing or stunting the growth of weeds growing in a crop but having little or no harmful effect on the crop itself. (▶ CONTACT HERBICIDES, TRANSLOCATED HERBICIDES, TOTAL HERBICIDES)

Self-binder ▶ BINDER.

Self-feed Silage. A management system whereby stock are allowed to graze SILAGE *in situ*, the amount taken being controlled by an electric fence or movable barrier set a short distance from the silage.

Self-fertilization. The process by which certain plants are able to fertilize themselves by the transference of pollen from the stamens to the stigma of the same flower. (▶ CROSS FERTILIZATION)

Self-mulching Soil. A soil, the surface of which is so well aggregated that it does not crust over under rainy conditions and is able to act as a surface mulch when the soil dries.

Self-pollination ▶ POLLINATION.

Self-sown. A term applied to a plant which grows from wild as distinct from sown seed.

Dairy cows self-feeding from silage.

Self-sterile Plant. A plant which is not able to fertilize itself with its own pollen. (♦ SELF-FERTILIZATION)

Semen. A liquid produced by the male reproductive organs of animals containing SPERM. (♦ ARTIFICIAL INSEMINATION)

Semi-digger. A type of MOULDBOARD with a gently concave curvature and twisted along its length, sometimes with a renewable leading edge or shin, and producing a broken furrow

385

slice. It is in general use for various operations including moderately deep ploughing (e.g. for ROOTCROPS), ploughing-in FARMYARD MANURE and surface trash, rapid seed bed preparation in the spring, and winter ploughing on well drained soils. It ploughs to a depth of 30–35 cm (12–14 in.) and is therefore intermediate between the DIGGER and GENERAL PURPOSE MOULDBOARD.

Semi-parasite ♦ PARASITE.

Semolina. Particles of fine, hard wheat which do not pass into flour during milling.

Sepal. One of the green leaf-like parts of a flower which form the CALYX and surround the petals.

Separated Milk. Milk from which the CREAM has been removed by centrifuging (♦ CREAM SEPARATOR). It has about half the energy value of whole milk, but a relatively higher content of PROTEIN and minerals and is used as an animal FEEDINGSTUFF. It is sometimes fed in the liquid form to pigs combined with barley MEAL. More usually it is reduced to a dried powder, with a protein content exceeding 30%. Such powder has various uses such as the production of substitute milks for calves and orphan lambs (when vegetable oils are incorporated before drying), and inclusion in chick rations and in meals for CREEP FEEDING to piglets. Also called skim milk.

Separator ♦ CREAM SEPARATOR.

Series ♦ SOIL SERIES.

Serum. A colourless watery liquid which separates from blood as it clots and which constitutes LYMPH. Essentially it is blood less the CORPUSCLES.

Serum Agglutination Test ♦ AGGLUTINATION TEST.

Serve. To copulate with. Also called cover.

Service. The act of a male animal serving, covering or copulating with a female.

Service Crate ♦ BREEDING CRATE.

Service House. A dwelling house on a farm provided usually as part of the terms of employment to an agricultural worker, either free or at a statutory nominal deduction from wages. The nature of many farm jobs (e.g. cowmen, stockmen) require them to live as close as possible to the farm. Loosely called a tied cottage. (◗ RENT (AGRICULTURE) ACT 1976)

Service Pen ◗ BREEDING CRATE.

Set. 1. To form fruit or seed.
2. To put a hen on eggs to hatch them or to put eggs under a broody hen.
3. To plant young plants, bulbs and potatoes.
4. A SEED POTATO or a piece of a potato used for planting.
5. A plant cutting or shoot used as a scion in a GRAFT, or placed in a rooting medium or the ground to grow roots.
6. A BADGER'S burrow.
7. A breeding group of geese, usually comprising 1 gander with 3–5 geese.
8. A young HOP plant.
9. To adjust a PLOUGH for work.

Set Stocking. A GRAZING SYSTEM whereby a fixed number of livestock are allowed to graze a given area for the entire growing season. Both animal and grass productivity are low under this system, and the grassland deteriorates due to both over- and under-grazing.

Set-to. An orphan lamb fostered by another ewe.

Set Up. To mark out the HEADLAND in a field with a marker furrow, and the various LANDS with RIDGES, before commencing ploughing.

Sewage Sludge. A partially dried residue from sewage treatment works sometimes used as a fertilizer. Compared with FARMYARD MANURE it is poor in POTASH and less beneficial on soil texture.

Sex-linkage. The distribution of certain GENES on the same CHROMOSOME as that determining sex. For example, in certain breeds of poultry the colour of the feathers is sex-linked so that all males are of one particular colour whilst the females are differently coloured.

Shab. A term for SHEEP SCAB.

Shamble. A slaughterhouse or a meat market.

Shandy Barrow ♦ SEED BARROW.

Shank. The lower part of the foreleg, from the knee to the foot.

Share. A pointed steel or cast iron blade attached to the front of a MOULDBOARD on the BODY of a plough which makes a horizontal cut under the furrow slice which is then turned by the mouldboard (*see p. 338*). Also called ploughshare or point. (♦ BAR-POINT SHARE)

Share Face. The underside of a SLICE cut by a SHARE during ploughing, and turned to become the back of the furrow slice. (♦ COULTER FACE)

Sharger ♦ RUNT.

Sharp. 1. A MEAL in which the particles are clearly visible.
2. A descriptive term for light, gravelly land which tends to wear implements.

Sharps. A term once given to a category of wheat offal resulting from milling. All wheat offals, apart from BRAN, are now sold as WEATINGS for use in FEEDINGSTUFFS.

Shaw. 1. A small wood or COPPICE.
2. The aerial parts of ROOTCROPS.

Sheaf. A bound bundle of unthreshed corn stalks, Plural, sheaves.

Shear. 1. To cut or clip a fleece from a sheep. (♦ SHEARING)
2. A term used in loosely describing a sheep's age. For instance, a three-shear ewe is between 3 and 4 years old, having been shorn three times.
3. A Scottish term meaning to reap with a SICKLE.

Shearing. 1. The clipping of a fleece from a sheep, nowadays almost always by machine shearing as distinct from the old practice of hand shearing. This normally takes place when the warm summer weather has caused the yolk (grease) to rise in the wool. In Britain shearing usually commences in southern England in May, with flocks on semi-arable upland being shorn in June, and hill and mountain sheep in July or August. A skilled shearer can handle 100–300 sheep per day. The fleeces

Sheep shearing, using electric shears.

are rolled and tied and are usually packed in WOOL SHEETS for delivery to wool merchants' warehouses. The BRITISH WOOL MARKETING BOARD now runs a shearing training scheme.
2. ◗ SHEEP NAMES.

Shearling. A young sheep between its first and second SHEARING. (♦ SHEEP NAMES)

Sheaves ♦ SHEAF.

Shed. 1. To separate one or more animals from a flock or herd. **2.** Cereal crops are said to shed grain when a strong wind or heavy rain, etc., causes the grain to drop from the EARS.

Shedder. A RACE leading to a two-way swing gate, along which sheep may be driven in order to separate or shed some individuals from a flock.

Sheeder Ewe ♦ SHEEP NAMES.

Sheep. A RUMINANT (*Ovis aries*) kept in the U.K. mainly for the production of meat and also for its wool. In some areas of southern Europe sheep are kept for their milk used in cheese making, and this practice is being introduced into the U.K. They are found in a wide range of environmental conditions under numerous management systems, from lowland areas (even on mainly arable farms) to uplands, moorlands and mountains. The majority of the British sheep population is nowadays found on the higher land, being able to make use of the large areas of ROUGH GRAZING land unsuitable for other agricultural usage. About 50 recognised breeds exist in the U.K. at present together with a variety of local types and numerous crossbreds.

In the hills, and especially in the north of England, in Scotland and in Wales, winter conditions are usually severe. In this environment, hardy native breeds are utilised for the production of lambs, the majority of which are transferred each autumn to lowland farms for fattening or growing-on (in the case of ewe lambs). The old or 'draft' ewes from these hill flocks are brought to farms at intermediate altitudes, where they are mated with rams which introduce increased prolificacy and milking ability to female offspring; then the cross-bred progeny pass on to the milder lowland conditions, where they become crossed again with Down rams for fat lamb production. The hill breeds bring in hardiness and mothering ability and the Down rams introduce early maturity and fleshing ability. Lamb production provides the main source of income for sheep farmers but wool is also of considerable importance, especially to farmers in hill areas. (Source: National Sheep Association).

**Some of the common breeds of sheep
kept in the U.K.**

Derbyshire Gritstone.

Scottish Blackface.

North Country Cheviot.

Romney.

Suffolk Ram.

Welsh Mountain.

Border Leicester.

Jacob.

Sheep Numbers in the U.K. (at June each year). ('000 head).

	Average of 1968–70	1975	1976	1977	1978	1979
Total sheep and lambs	26 896	28 270	28 265	28 104	29 686	29 967
of which: Ewes	10 968	11 279	11 298	11 215	11 444	11 692
Shearlings	2 362	2 471	2 369	2 487	2 717	2 877

(Source: Annual Review of Agriculture, H.M.S.O. 1980.)

Total sheep and lambs in the U.K. in 1979 numbered almost 30 million. Average flock size was 175 and about 40% of total breeding sheep were in flocks of 500 and over. The recognised breeds of sheep, each of which is dealt with separately in the text, may be classified as follows:

Long-woolled Breeds

Blue Faced Leicester*	Lincoln Longwool
Border Leicester	Romney
Devon and Cornwall Longwool	Teeswater
Dartmoor	Wensleydale
English Leicester	

Short-Woolled Breeds (Close-woolled)

Down Breeds:	Dorset Down	Shropshire
	Hampshire Down	Southdown
	Oxford Down	Suffolk
Horned-Whitefaced:	Dorset Horn	
	Wiltshire Horn (no wool)	
Polled-Whitefaced:	Devon Closewool	Ryeland

Mountain, Moorland and Hill Breeds

Cheviot	Lonk
Clun Forest	Radnor
Derbyshire Gritstone	Rough Fell
Exmoor Horn	Scottish Blackface
Herdwick	Shetland
Jacob	Swaledale
Kerry Hill	Welsh Mountain
Limestone	Whitefaced Woodland

Important crossbreds include Greyface, Halfbred, Masham, Mule, Romney Halfbred, Scottish Halfbred and Welsh Halfbred.

* This breed has very short, curly wool, but for convenience is usually classified with the long-wools.

Sheep Cote. An enclosure for sheep.

Sheep Counting Systems. Shepherds' numerals still in use in some parts of Great Britain for counting sheep, take several forms. Two of these run as follows:

(a) *Onetherum, twotherum, cocktherum, qutherum, setherum, shatherum, wineberry, wigtail, tarrydiddle, den.*

(b) *Yan, tan, tethera, pethera, pimp, sethera, lethera, hovera, covera, dik.*

Of higher figures, 15 was bumfit and 20 figgit. A remote connection with Latin is traceable. Qutherum, for instance, suggests quatuor; pimp and wigtail are phonetically reminiscent of quinque and octo; it is also instructive to compare the set with the Welsh numerals:

Un, dau, tri, pedwar, pum, chwe, saith, wyth, naw, deg.

(The above information is reprinted, by kind permission, from an article in *The Field*, 2 December 1965.)

Sheep Dip. An approved chemical used in a DIPPING BATH diluted in water, to disinfect sheep to control parasitic diseases such as SHEEP SCAB. (◗ DIP)

Sheep Louse ◗ KED.

Sheep Maggot Fly. A type of BLOW FLY (*Lucilia sericata*), also known as a green bottle, metallic blue-green in colour and of similar appearance to the house-fly. It lays its eggs in wounds and in the fleece, and on hatching the maggots bore into the flesh, causing the condition known as STRIKE.

Sheep Names. The names given to sheep are probably more numerous than for any other of the domesticated animals, and a great many local terms and variations exist. The table on p. 397 gives a list of the commoner terms.

Sheep Nostril Fly. A type of BOT FLY (*Oestrus ovis*), greyish yellow-brown in colour, which either lays its eggs or deposits the hatched maggots in the nostrils of sheep. The maggot crawls upwards towards the sinuses causing a nasal discharge and the condition known as STAGGERS. When fully grown it is discharged by sneezing and forms a PUPA in the ground.

Common Sheep Names.

Periods	Male		Female	Remarks
	Uncastrated	Castrated		
Birth to weaning	Tup lamb Ram lamb Pur lamb Heeder	Hogg lamb	Ewe lamb Gimmer lamb	A sheep until weaning is a lamb
Weaning to shearing	Hogg (also used for the female) Hogget (also used for the female) Haggerel or Hoggerel Tup teg Ram hogg Tup hogg	Wether hogg Wedder hogg He teg	Gimmer hogg Ewe hogg Sheeder ewe Ewe teg	Hogget wool is wool of the first shearing
First to second shearing	Shearing, or Shearling, or Shear hogg Diamond ram Dinmont ram tup One-shear tup	Shearing wether Shear hogg Wether hogg Wedder hogg Two-toothed wether	Shearing ewe Shearling gimmer Theave Double-toothed ewe Double-toothed gimmer Gimmer	'Ewe', if in-lamb or with lamb; if not a 'barren gimmer'; if not put to a ram is a 'yield gimmer' (Scotland)
Second to third shearing	Two-shear ram Two-shear tup	Four-toothed wether Two-shear wether	Two-shear ewe	A ewe which has ceased to give milk is a 'yeld ewe'; taken from the breeding flock she is a 'draft ewe' or a 'draft gimmer'
Third to fourth shearing	Three-shear ram Three-shear tup	Six-toothed wether Three-shear wether	Three-shear ewe Winter ewe (Scotland)	
Afterwards	Aged tup or ram	Full-mouthed, full-marked or aged wether or wedder	Ewe	After fourth shearing 'aged', or 'three-winter'.

(This table is reproduced, with permission, from *Black's Veterinary Dictionary* Thirteenth Edition, 1979, edited by Geoffrey P. West, and published by A. & C. Black Ltd.)

Sheep Pox. A very contagious viral disease (*Variola ovina*) of sheep causing fever, cessation of feeding, difficulty in breathing, depression and skin eruptions, with a severe loss of condition if badly affected. In Britain it is a NOTIFIABLE DISEASE with compulsory slaughter of all affected sheep.

Sheep Run. An area of open grassland used for keeping sheep on.

Sheep Scab. A NOTIFIABLE DISEASE of sheep caused by the MITE *Psoroptes communis*, which lives on the skin surface and feeds on SERUM emanating from puncture wounds which it inflicts and which become inflamed, forming scabs. An irritant poison secreted by the mites causes the sheep to scratch and rub itself, with the fleece becoming detached in patches revealing the scabs, which are themselves rubbed off leaving ulcerations. Prevention is by compulsory dipping (◗ DIP). The disease reappeared in 1973 having been eradicated from the U.K. in 1952. Also called belt, shab, tag and rubbers.

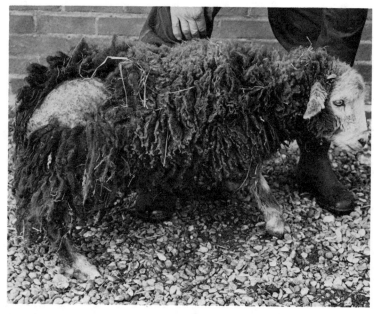

An advanced case of sheep scab, showing loss of wool.

Sheep's Fescue ⬦ FESCUE.

Sheep Sick ⬦ SICK.

Sheep Tick ⬦ TICK and KED.

Sheep Walk. A range of pasture on which sheep are allowed to graze.

Sheepdog. A dog trained by and under the control of a SHEPHERD, to assist him to watch sheep, and particularly used when rounding up sheep or driving them from place to place.

Sheep-gate. 1. A HURDLE used for penning sheep.
2. A right to graze sheep on land. Also the land on which such a right exists.

Sheet ⬦ WOOL SHEET.

Sheet Erosion. The erosion or removal of surface soil in a uniform manner under the influence of surface run-off water.

Shell. To remove the grain from the EARS of cereals.

Shell Board ⬦ MOULDBOARD.

Shelter Belt. A line or 'belt' of trees purposely planted to act as a shield and provide shelter against the weather, particularly the prevailing winds. Soil erosion is checked and crop yields show an improvement.

Shepherd. A person who tends or looks after sheep. A full-time shepherd may have as many as 1 000 ewes under his control.

Shetland. 1. The hardy and thrifty native breed of cattle of the Shetland Islands. Black and white in colour, with soft skin (similar to the JERSEY), short legs, a large body, and slender, slightly curved horns. A dual-purpose breed with low feed requirements and able to be fattened on poor quality pasture, and cows having a large udder.
2. A small, very hardy breed of mountain or moorland sheep, now almost confined to its native Shetland Islands. Characterised by a fleece of dense, fine soft wool used for making shawls and scarves in the famous cottage industry of the islands. The wool is mainly white although other shades exist. The breed has erect pointed ears and a short 'fluke' tail, broad at its base. Rams have light horns but ewes are normally polled.

Shieling. A Scottish term for rough summer pasture, usually in the hills. Also a hut or shelter for a SHEPHERD tending a flock on such pasture.

Shift. One of the crops in a ROTATION (e.g. the barley shift), or the entire sequence or rotation itself (e.g. a four-course shift).

Shim. A horse shoe.

Shin. The leading edge of a MOULDBOARD which in some types (e.g. DIGGER, SEMI-DIGGER) may be replaced when it becomes worn. Sometimes called a cutter.

Shippen, Shippon. A COWHOUSE.

Shire. The largest and heaviest breed of draught horse, with well feathered legs, a heavy head, short arched neck, and a shortish, wide, very strong, muscular body. Males can stand over 17 HANDS high and may weigh over 1 000 kg (1 ton). Variously coloured, often with white markings on the head and feet.

Shock ◗ STOCK.

Shoddy. A waste product of the woollen industry consisting of discarded shreds and fragments of material, used as a FERTILIZER, particularly by horticulturalists. Its quality depends on the amount of pure wool (which is entirely protein) in it. The presence of other waste products (e.g. cotton) results in the nitrogen content varying considerably in the range 3–12%.

Shoot the Red. The stage in growth of a young turkey when the face reddens and the WATTLE develops.

Short Dung ◗ FARMYARD MANURE.

Short Fallow ◗ BASTARD FALLOW.

Short Land ◗ SHORT WORK.

Short Manure. Another term for short dung. (◗ FARMYARD MANURE)

Short Sheep. Short-woolled sheep. (◗ SHEEP)

Short Work. The ploughing of those odd corners left unploughed (short land) during the main systematic ploughing of a field.

Shorthorn. In the eighteenth century the Colling brothers used the inbreeding methods of BAKEWELL to improve the race of short-horned cattle (the Teeswater or DURHAM) which had bred in the Tees valley in the North-East of England. This improved Teeswater, or Shorthorn as it came to be called, spread throughout the British Isles and has given rise to four separate breeds, the BEEF SHORTHORN, DAIRY SHORTHORN, NORTHERN DAIRY SHORTHORN and the WHITEBRED SHORTHORN. The Shorthorn was also crossed with local cattle in Lincolnshire to produce the LINCOLN RED.

Shots. Small lambs which are sometimes culled.

Shropshire. A medium sized Down breed of sheep, with a characteristic black face and woolled poll, and fleece of close growing wool. It was derived in the early nineteenth century from crosses between the SOUTHDOWN and old breeds in the West Midlands and Welsh Borders, particularly the Longmynd and Morfe Common breeds. It was mainly bred for export, particularly to North America, where it is still found in considerable numbers. Relatively few flocks remain in England.

Shrub. A low woody plant or bush, with little or no trunk, but with several stout stems arising from a single rootstock.

Shut Up. To prevent stock from having access to a meadow or pasture so that the grass is able to grow and subsequently be cut for HAY or SILAGE.

Siblings. Progeny with the same parents, i.e. brothers and sisters.

Sick. Diseased. Sick land (sometimes called stale land) is land which has been continuously used in the same manner over a long period of years (e.g. barley grown year after year, or grazed by the same stock over many years) so that pests and diseases have built up in the soil. Examples include fowl sick, pig sick and sheep sick land, etc. Such land should be put down to another crop or rested from the particular stock. Plough-sick land is land which needs sowing to a LEY.

Sickle. 1. A hand implement with a short handle and a hooked-shaped, toothed blade, used mainly for cutting cereals.
2. One of the long, curved feathers of a cock's tail.

Side. A butcher's term for a half of a CARCASE, divided along the backbone.

Side Delivery Rake. A type of SWATH TURNER which turns two SWATHS to one side of the machine combining them both into one windrow. Usually used prior to baling (◗ BALER) or sweepings (◗ SWEEP).

Side Graft. A type of GRAFT in which the SCION is attached to the side of a branch rather than into the cut end of a STOCK.

Side String ◗ SPRIG.

Sideland. A strip of land left at the side of a field to be ploughed out with the HEADLANDS.

Sideling. Land which slopes sideways in relation to the line of work, e.g. when ploughing along the contour.

Sidlip ◗ SEED-LIP.

Sieve. 1. A frame containing a mesh or perforated base for sifting material. Normally finer than a RIDDLE.
2. A basket for fruit with a BUSHEL capacity.

Silage. A feedingstuff consisting of forage crops (e.g. GRASSLAND SPECIES, kale, beet tops, pea haulm, etc.) cut or harvested in the green state and preserved in a SILO in a succulent condition for later use. The principle of silage making is the fermentation, by bacteria, of CARBOHYDRATES in the plant material to organic acids, and of PROTEINS to amino acids, which act as preservatives. Well made silage is yellowish-brown in colour and contains mainly lactic acid (as much as 2% of its fresh weight) produced by *Lactobacilli*, with some acetic acid and with a pH of 4.0 or less. Crops contaminated with faeces and soil tend to produce a dull olive green silage containing undesirable butyric acid, produced by *Chlostridia*. *Lactobacilli* grow best when the silage material has a high sugar content. Crops are ideally cut after dry sunny weather which stimulates PHOTO-SYNTHESIS, and therefore sugar production, and wilts the plants, increasing sugar concentration. Cut crops may also be crimped or lacerated to increase wilting. When sugar content is low various additives are sometimes included, e.g. molasses, which provides supplementary sugar and formic acid, which increases acidity and suppresses *Chlostridia*.

Silica. (SiO₂). Silicon dioxide, occurring in crystalline forms such as quartz, the main constituent of SAND, and in combination with various metal oxides as silicates, e.g. hydrated aluminium and magnesium silicates, the principal constituents of CLAY MINERALS.

Silk. The STYLES or 'tassels' of the female flowers of MAIZE.

Silo. A container in which SILAGE is made and stored. A number of types are in use including the various forms of CLAMP on the surface, in a pit, or in a large sealed polythene 'bag' evacuated of air (♦ VACUUM SILAGE). Silage may also be prepared in wooden, concrete or steel towers (sometimes under airtight conditions) in which carbon dioxide production restricts RESPIRATION and allows good FERMENTATION. The term silo is sometimes applied to a moist-grain storage tower.

Silo System. A technical term used to describe the mechanism of the E.E.C. cereals regime (♦ COMMON PRICES) which, when

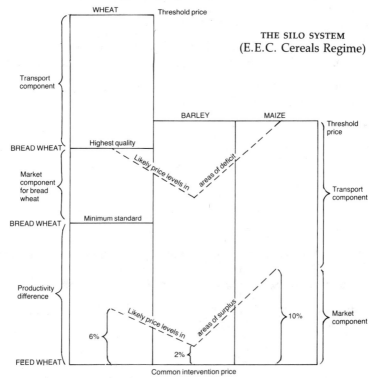

403

shown graphically, resembles a silo. The base of the silo represents the common INTERVENTION PRICE for all Community grown cereals (i.e. wheat, barley and maize), whilst the roof of the tower block represents the THRESHOLD PRICE of competing imported wheat and the main roof of the silo represents the threshold price of competing imported feed cereals (i.e. barley and maize). The body of the silo represents the Community's internal cereals market price structure, and reflects both the relative feed values of the three main cereals and the transport costs from areas of surplus to areas of deficit.

Silt. A constituent of SOIL comprising particles intermediate in size between CLAY and SAND, being 0.002–0.02 mm in diameter (according to the international particle-size system). Silt is often carried suspended in running water and is deposited on river beds and banks or in lakes as alluvial sediment.

Simmental. A heavy dual-purpose breed of cattle originating in the Simme valley in Switzerland, now the most widely distributed breed in Europe. Yellowish-brown to red in colour with a white head, belly, legs and tail tassel. Also characterised by shortish legs, a long, well-muscled body, well developed dewlap, and horns which curve outwards and forwards (*see p. 86*).

Simple Leaf. A leaf which is not divided into separate leaflets, as distinct from a COMPOUND LEAF.

Sinew ▶ TENDON.

Single. To reduce the number of plants in a ROOTCROP row by separating and removing all except one of the individual plants in each clump.

Single Comb. A type of narrow, serrated, erect COMB in poultry.

Single-cut Red Clover ▶ RED CLOVER.

Single-suckling. A method of feeding beef cattle in which calves are allowed to suckle their own mothers. (**▶** DOUBLE-SUCKLING, MULTIPLE-SUCKLING)

Singletons. Single lambs born to EWES, as distinct from twins, triplets, etc.

Sire. The father of an animal. Also to beget offspring as the father. (♦ DAM)

Sisal. A species of plant, *Agave sisalana*, native to Mexico and Central America, grown for its hard fibre which is used for manufacturing binder twine and rope.

Sit, Sitter. To BROOD, a BROODY HEN.

Site of Special Scientific Interest. An area notified by the Nature Conservancy to a local Planning Authority as being of special interest in respect of its geology, physiography, flora or fauna. Such notification requires the Planning Authority to consult the Conservancy prior to granting planning permission for any development which could affect the site. Notification does not impose any obligations on landowners or occupiers in the area, or provide public rights of access or entry. Often abbreviated to S.S.S.I.

Sithe ♦ SCYTHE.

Six-tooth Sheep. Sheep of about 24–27 months of age. (♦ TWO-TOOTH, FOUR-TOOTH, FULL MOUTH and SHEEP NAMES)

Skep. An old type of straw or wicker beehive.

Skewbald. Coloured white and another colour (except black) in irregular patches. (♦ PIEBALD)

Skim Coulter ♦ COULTER.

Skim Milk ♦ SEPARATED MILK.

Skip-jack ♦ CLICK BEETLE.

Slack Bine. Hop BINE with relatively few branches and a shortage of foliage.

Slack Dried Hops. Insufficiently cured hops, with a moisture content exceeding about 12%.

Slade ♦ SOLE.

Slag ♦ BASIC SLAG.

Slaked Lime ♦ CALCIUM HYDROXIDE.

Slaughterhouse. A place where animals are hygienically slaughtered and carcases are prepared for sale to the public for human consumption. All carcases are examined by qualified inspectors employed by the Local Authority and only those free from disease are allowed to leave the premises for sale to the public.

Most slaughterhouses in large cities are owned by, but not necessarily operated by, Local Authorities. Many large slaughterhouses in rural areas are owned and operated by meat trade companies (e.g. FATSTOCK MARKETING CORPORATION). Small owner-operated slaughterhouses are becoming increasingly uneconomic and only a few remain, mainly in rural areas combined with retail outlets. Also called abattoir.

Sledge ◗ BALE SLEDGE.

Slender Foxtail. An annual grass (*Alopecurus myosuroides*) commonly found on heavy and poorly drained land. An arable weed, particularly in early-sown winter cereal crops. A tufted, smooth-sheathed, green or purplish grass, with a slender, cylindrical, tapering PANICLE. Also called black grass.

Slice. The strip of soil turned over in ploughing to leave a furrow. Also called sod.

Sling Gear ◗ CHAIN HARNESS.

Slink, Slink Calf. A prematurely born animal, particularly a calf; or a very young calf, or one removed from its mother's womb when the latter has been slaughtered. The meat, sometimes called slink veal, is used mainly for such products as meat paste.

Slip. 1. To miscarry or abort.
2. A SCION or CUTTING.

Slob. Land in a muddy condition.

Slough. A muddy hollow or a marsh.

Slubbing Out. A term used in the FENS for cleaning DYKES.

Sludge ◗ SEWAGE SLUDGE.

Sluice. A structure in a watercourse used to control the flow of water, usually by means of a door, flap or hatch. Also a drainage channel.

Sluicegate Price. A price which, with the addition of a 'basic levy', forms the minimum import price into the E.E.C. for eggs, poultry and pigmeat. Fixed quarterly (in advance) as the total cost of producing and marketing 1 kg of the particular product outside the E.E.C. A supplementary levy is added to the basic levy on imports brought into the E.E.C. at below the sluicegate price. (◊ REFERENCE PRICE)

Slurry. A semi-fluid mixture of faeces and urine, often also containing rain water and washing-down water from livestock buildings. It is sometimes mixed with litter, mainly straw, to produce FARMYARD MANURE. It may also be stored in a lagoon or tank, where it is diluted with water and is then piped onto fields (when fibre-free), or it may be sucked, under vacuum, direct into a tanker and directly distributed on fields (◊ MANURE SPREADERS, GULLE SYSTEM). Machines are also available which pass the slurry through a hydraulic press providing a 'solid' friable residue which can be composted, the liquid being spread on the land. Slurry contains valuable NITROGEN, PHOSPHATE and POTASH but its composition varies according to the type of livestock, diet, dilution, etc. (◊ LIQUID MANURE)

Small Seeds ◊ SEEDS MIXTURES.

Smallholding. A general term for a small farm. In the privately-owned sector the term is not subject to any further definition. Under various Acts of Parliament during this century statutory smallholdings estates have been established. Some are owned by Smallholdings Authorities (County Councils in England and Wales, and the Greater London Council) for letting to persons with adequate agricultural experience who want to become farmers. Other statutory smallholdings are owned by the Ministry of Agriculture, Fisheries and Food and are managed for the Minister by the LAND SETTLEMENT ASSOCIATION. The latter type are provided, with organised services, for horticulture. Loans of working capital are available to tenants of statutory smallholdings, although not to smallholders generally. Statutory smallholdings are at present subject to controls exercised by the Minister through the Agriculture Act, 1970 (Part III). They are restricted in size to holdings over 0.4 ha (1 ac.) and under

900 Standard Man Days (i.e. less than the amount of land that would provide a full year's employment for 3 workers calculated with reference to the Smallholdings (Full-time Employment) Regulations, 1970). Holdings may exceed this upper limit only with the Minister's approval.

Smooth Leaves. The COTYLEDONS of BRASSICAS. (◗ also ROUGH LEAF)

Smooth Stalked Meadow Grass. A tufted, rhizomatous, perennial grass (*Poa pratensis*), with smooth, greyish-green leaves, with either an abruptly pointed or a blunt hooded tip and a smooth-sheathed, folded stem. The purplish, green or greyish flowers are in the form of an ovate PANICLE. It is an important hay and pasture grass in North America and on continental Europe. In Britain it is mainly sown on roadside banks, playing fields, etc., and is sometimes substituted for ROUGH STALKED MEADOW GRASS in SEEDS MIXTURES to be used on light soils.

Smother Crop. A crop which grows vigorously when well fertilized (e.g. Kale, arable silage crops, etc.) and occupies most of the growing space, tending to retard or smother the growth of weeds.

Smout. A hole in a wall through which only one sheep at a time may pass. Also called lunky.

Smudging. The practice of making smoke to produce an artificial cloud, in an attempt to reduce heat loss from the ground and prevent frost damage to crops.

Smuts. An order of BASIDIOMYCETE fungi which produce black spores, many species causing plant diseases. The main species of agricultural importance attack cereals, the spores developing in the grains which are destroyed, e.g. BUNT.

Snag ◗ SCUR.

Snath, Sned, Sneath. The curved wooden handle or shaft of a SCYTHE. Also called snead.

Sned. 1. To cut branches from trees or to prune.
2. ◗ SNATH.

Soay. A small 'wild' breed of sheep found in restricted numbers on Soay and a few other islands in the Outer Hebrides. Ewes may be horned or polled, but rams bear heavy horns and have a darkish hairy mane. The fleece contains a mixture of wool and hair and is either dark brown or fawn, except on the rump, chest and belly which are lighter in colour. A few small breeding flocks exist on mainland Britain and some have been exported to Europe and North America.

Sock. A plough SHARE.

Sock Lamb ♦ COSSET LAMB.

Sod. A SLICE.

Sodium. (Na). A metallic element, an essential constituent of the body fluids in animals. It is mainly present as various salts, mainly sodium chloride (NaCl) or common salt, its concentration in the blood being finely controlled with excesses excreted. Reduced performance results from sodium deficiency and livestock diets are usually supplemented with salt in MINERAL MIXTURES. Sodium chloride is also used as a SALT FERTILIZER.

Soft Fruit. Loosely berried FRUIT with relatively soft flesh, divided into bush fruit (e.g. black, red and white currants, gooseberries, etc.) and cane fruit (e.g. raspberries, blackberries and hybrid berries), and also strawberries. The latter are the most widely grown soft fruit, the principal maincrop areas being in Kent and East Anglia. Blackcurrants are also widely grown in Britain, mainly for the preparation of juice for the manufacture of soft drinks and flavours for confectionery. The largest concentration of raspberry plantations in the world is located around Perth in Scotland. (♦ TOP FRUIT)

Softwoods ♦ CONIFERS.

Soil. The unconsolidated material covering the surface of the earth in which plants grow, and in which many animals (e.g. insects, worms, beetles, bacteria, etc.) live and derive food. It consists mainly of particles of SAND, SILT and CLAY, closely associated with ORGANIC MATTER, the relative proportions of each determining the soil type (♦ SANDY SOILS, CLAY SOILS, LOAM). Soil mineral matter is derived from the weathering and erosion of rock. Water percolating through the soil depletes the surface layers of both soluble and fine insoluble substances (◆ LEACHING), and has an effect on the hydrogen and hydroxyl ION content of the soil, thus determining its acidity (◆ ACID SOIL, ALKALINE SOIL, NEUTRAL SOIL). From the agriculture viewpoint the top horizons (◆ SOIL HORIZONS), which crop roots penetrate to obtain water and nutrients (◆ CATION EXCHANGE), are the most important.

Soil Air. The gaseous content of the soil (i.e. that part of the volume of the soil other than solids or liquid) occupying part of the capillary spaces between soil particles and the soil pores. Soil air differs from atmospheric air, having a higher concentration of carbon dioxide due to RESPIRATION by plant roots and micro-organisms, and is normally saturated with water vapour. The carbon dioxide dissolves in the soil moisture forming carbonic acid, which assists in dissolving mineral matter to release plant nutrients.

Soil Analysis. The determination of the composition of soil by various laboratory chemical and physical methods. Mechanical analysis is used to accurately determine SOIL TEXTURE, an important factor in soil and land use classification. The analysis of soil in terms of its chemical composition is important in revealing its state of fertility and in recommending the amount and type of FERTILIZER treatment required.

Soil Association. A group of two or more SOIL SERIES (or other defined taxonomic units) distributed together in a specific pattern throughout an area or region, and sometimes mapped on a soil map as one mapping unit, either because the individual soils cannot be delineated at the scale used or because delineation is not required.

Soil Classification. The systematic arrangement of soils into various categories according to their characteristics. This may be (*a*) according to texture, e.g. sandy loams, clays, etc., (*b*) on a geological basis, e.g. chalk soils, sandstone soils, etc., or (*c*) on a pedological basis. The basic pedological unit of classification now used for soil mapping is the SOIL SERIES. The five most important soil groups in Britain are BROWN EARTHS, calcareous soils (e.g. RENDZINAS, BROWN CALCAREOUS SOILS), GLEY soils, PEAT soils and PODSOLS.

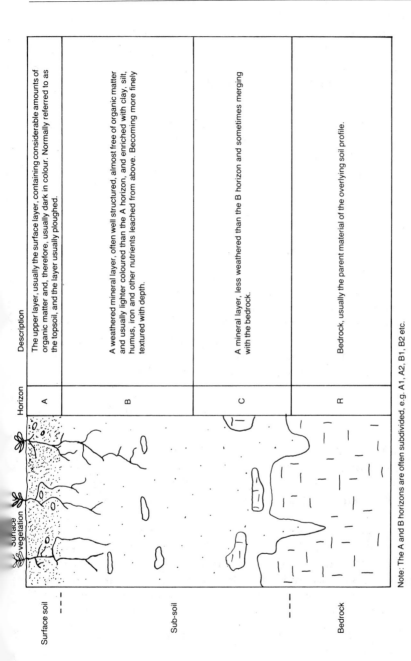

Horizon	Description
A	The upper layer, usually the surface layer, containing considerable amounts of organic matter and, therefore, usually dark in colour. Normally referred to as the topsoil, and the layer usually ploughed.
B	A weathered mineral layer, often well structured, almost free of organic matter and usually lighter coloured than the A horizon, and enriched with clay, silt, humus, iron and other nutrients leached from above. Becoming more finely textured with depth.
C	A mineral layer, less weathered than the B horizon and sometimes merging with the bedrock.
R	Bedrock, usually the parent material of the overlying soil profile.

Note: The A and B horizons are often subdivided, e.g. A1, A2, B1, B2 etc.

Generalised soil profile illustrating horizon development from underlying bedrock.

Soil Horizon. A layer of soil approximately parallel to the surface with fairly distinctive properties such as texture, colour, structure, mineral or chemical composition, etc. A vertical section of soil (a soil profile) will reveal the various horizons present from the surface to the parent material (if soil development has been *in situ*). The main horizons are illustrated (*see p. 411*).

Soil Moisture Deficit. The amount of water which a soil requires to be added to return it to FIELD CAPACITY. The size of this deficit, together with the condition of the crop, provide an indication of whether irrigation is required. (◖ PERMANENT WILTING PERCENTAGE and READILY AVAILABLE MOISTURE)

Soil Pan ◖ PAN.

Soil Profile ◖ SOIL HORIZONS.

Soil Series. The basic unit of soil classification and mapping consisting of soils which are identical in all the main profile characteristics, except surface texture, and predominantly confined to one parent material. The soil series can be considered as comparable to the GENUS in the classification of living organisms. It can be subdivided into lower categories (comparable to SPECIES) on surface texture, depth, slope, etc. (◖ SOIL ASSOCIATION)

Soil Sterilant. A chemical applied to the soil to kill pests, diseases or weeds.

Soil Texture. 1. The 'feel' of a soil, e.g. soapy, gritty, sticky, etc. Soils containing small CLAY particles, with small interparticle spaces are called close-textured, as distinct from open-textured soils containing larger particles (e.g. SAND) with larger air spaces.
2. A soil sample may be described as belonging to a particle class of soil texture (e.g. sandy loam, silty clay, sandy clay loam, etc.) following PHYSICAL ANALYSIS to determine the relative proportions of clay, sand and silt. The various textural classes have strict compositional limits, e.g. sandy clay loam containing 20–35% clay, less than 28% silt, and 45% or more sand.

Soil Water ◖ CAPILLARY WATER, GRAVITATIONAL WATER and IMBIBITIONAL WATER.

Soilage. FODDER CROPS, cut and fed to livestock in the fresh state. Also called ZERO GRAZING.

Soiling Crop. A GREEN CROP cut and fed to stock elsewhere, i.e. not at the field where grown. Such cut FORAGE is sometimes called green meat.

Solari Farrowing Pen, Solari Piggery ♦ PIGGERY.

Sole. 1. The base of the BODY of a plough which rests in the furrow bottom. Also called slade.
2. ♦ PAN.

Sole Furrow. The last SLICE cut during ploughing. Also called scolt.

Solids-not-fat. The various constituents of MILK other than BUTTERFAT and water. On average they represent about 8.7% of milk and consist of PROTEINS and other nitrogenous substances, LACTOSE or milk sugar, and various mineral salts and VITAMINS, etc. (♦ ASH)

Soot. Soot from domestic chimneys and boilers containing ammonium sulphate is sometimes used as a nitrogenous FERTILIZER.

Sorghum. A tropical cereal (*Sorghum sp.*) the grain of which is imported for feeding to poultry.

Sour Soil. An infertile ACID SOIL, as distinct from a sweet or fertile ALKALINE SOIL.

South Devon. 1. The largest native breed of cattle in Britain. A long-lived, docile breed, pale red in colour, with a long broad head, a long deep body, and short curved horns. At one time the South Devon was kept as a dual-purpose breed, but is now mainly regarded as a beef breed. Records of MILK YIELD are still kept, however, and in 1978/79 the average annual yield was 2 955 kg (2 878 litres; 616 gal.) with a 3.84% BUTTERFAT content. Also called South Hams.
2. Until 1977 a breed of long-woolled sheep, somewhat larger than the DEVON LONGWOOL, and producing slightly coarser wool. Mainly found in South Devon and Cornwall, and now amalgamated with the Devon Longwool to form the DEVON AND CORNWALL LONGWOOL.

South Hams ♦ SOUTH DEVON.

Southdown. The smallest of the Down breeds of sheep developed by John Ellman in Sussex from a local type found on the South Downs. Characterised by a distinct compact body, mouse-coloured short legs and face, and small, roundish, wool-covered ears. It produces a very high quality carcase, and a dense fleece of high-quality short wool. The breed has contributed to the development of other Down breeds (e.g. DORSET DOWN, HAMPSHIRE DOWN, RYELAND, SHROPSHIRE DOWN and SUFFOLK).

Sow. 1. An adult female pig, after having produced her first litter. (♦ GILT)
2. To scatter seed on the ground by broadcasting or to place it in the ground, usually with a DRILL.

Soya Bean. A leguminous plant (*Glycine max*) grown in warm temperate climates, producing BEANS rich in oil and protein. Soya bean CAKE and MEAL are imported mainly from Canada and the U.S.A. and are used to balance the cereal content of livestock rations. They are a particularly rich source of the AMINO ACID Lysine.

Spacing Drill. Another term for a precision SEED DRILL.

Spade Husbandry. The cultivation of land with a spade rather than a plough.

Spado. A castrated or impotent animal.

'Spaghetti' Wheat. A term sometimes applied to a variety of wheat used for the production of spaghetti and other pasta products. (♦ DURUM)

Span. A team of oxen or a pair of horses.

Spangled. A term for poultry feathers with a shiny dark spot at their tips.

Spay. To remove the ovaries (♦ OVARY) of a female animal to prevent it from breeding.

Spayn. To wean.

Spean. A teat.

Species. The smallest group or unit used in the classification of living organisms, consisting of individuals having a great number of common characteristics, but distinct from other individuals of different species within the same GENUS, and able to breed with each other and produce fertile progeny, but usually unable to breed with individuals of other species. In the binomial Latin nomenclature used in classification two names are given to a living organism, the first is its genus, the second denotes the species, e.g. *Hordeum sativum* (barley).

Speckled Yellows. A disease of sugar beet due to MANGANESE deficiency, in which the leaves become yellow between the veins, giving a streaked appearance. Some parts may turn brown. The disease is common on soils with a high LIME content.

Spelt. An inferior species of wheat (*Triticum spelta*) in which the inner husks stick to the grain, as in oats. Grown in mountain areas in Europe.

Sperm. A male GAMETE. They are produced by male animals in millions in a liquid medium called semen.

Spike. 1. An unbranched flower-head with individual, stalkless SPIKELETS attached to a long axis or common stem. Characteristic of grasses.
2. An ear of corn.

Spiked Link Harrow. A type of CHAIN HARROW with a spike on each link.

Spikelet. An individual unit on a SPIKE, generally consisting of 2 GLUMES and one or more flowers each borne between a LEMMA and PALEA.

Spile. A cleft piece of wood used for fencing, either linked by wire or nailed to a horizontal batten.

Spine. 1. The backbone of an animal. Also called spinal column.
2. A stiff thorn on part of a plant, often a modified branch or leaf.
3. The heart wood of a tree or other woody plant.

Spinner. A tractor-drawn implement used for harvesting potatoes. Trailed and mounted types are in use. The main components are a digging share which lifts the row of potatoes and loosens the soil, and a revolving wheel set at right-angles to the row and bearing rotating tines. The tines strike the lifted row flinging the potatoes and soil to the side often against a rope net, so that much of the soil passes through and the potatoes fall to the ground in a row. The spinner is particularly useful in wet soil conditions, but can cause damage to the tubers.

Spinney ◗ COPSE.

Spit ◗ GRAFT.

Spiv Ewe. A ewe in poor condition and which will not fatten.

Splitting ◗ CASTING.

Spoil. Excavated material from ditches or rivers.

Spore. A reproductive structure, either single- or several-celled, set free by various plants (e.g. ferns, mosses, fungi), bacteria and protozoa. They give rise to new individuals and, unlike seeds, may be formed asexually.

Spraing. A disease of potatoes caused either by Tobacco Rattle Virus (T.R.V.) carried by a free-living NEMATODE worm (*Trichodorus sp.*), or by Potato Mop-Top Virus (P.M.T.V.) transmitted by the powdery scab fungus (*Spongospora subterranea*). It is characterised by rust-coloured arcs in the flesh of the tubers.

Spray. A liquid applied under pressure, via a nozzle, in the form of a mist of fine droplets. Insecticides, herbicides, liquid fertilizers and disinfectants, etc., are often applied as sprays. (◗ SPRAY DRIFT, SPRAYER)

Spray Drift. The tendency of a SPRAY of fine droplets, produced by a low-volume nozzle, to drift in the air from the field of application to other fields. Such drifts may cause damage to other crops and can poison grazing animals or bees if the spray contains harmful chemicals. Drift damage is minimised by spraying in suitable weather conditions and by using sprays of sufficient dilution.

Spray Irrigation. The IRRIGATION of crops (*a*) by pumping water along distribution pipes to various types of distributor (e.g. large nozzle rain guns, small nozzle rotary sprinklers, fixed spray lines), or (*b*) by applying water from a tank carried on either a water-operated vehicle fitted with a spray boom or oscillating rain gun, or on a tractor or other vehicle fitted with a sprinkler boom.

Spray Race. A type of RACE along which nozzles are fitted at intervals. Used to treat cattle and sheep with liquid parasiticides.

Sprayer. A machine used to apply a SPRAY by forcing the liquid through a nozzle under pressure. Field crop sprayers are classified in terms of their capacity for application, as follows:

Volume	litres/ha	gal.ac.
Low	less than 220	less than 20
Medium	220–660	20–60
High	more than 660	more than 60

Sprayers essentially consist of a tank from which the liquid is pumped to the nozzles which are fitted at regular intervals along a boom. Nozzles may be of the fan type (used for low and medium volume spraying) or cone type (used for high volume spraying). Sprayers with tanks up to 450 l (c. 100 gal.) capacity may be tractor mounted. Those of larger capacity, up to 1 800 l (400 gal.), are either trailed or mounted on 4-wheel drive vehicles. Spraying may also be carried out by aeroplane or helicopter. (♦ AERIAL SPRAYING)

Sprig. 1. A shoot or twig.
2. A SCION.
3. A tiny spike sometimes present on the side of a single-comb in poultry. Also called side-sprig.

Spring Cleaning. The ploughing of a field in the spring as the soil dries, following ploughing during the winter, to speed the drying process and to raise weeds (e.g. COUCH GRASS) to the surface. Further cultivation and harrowing reduces the soil to a tilth allowing the weeds to be removed, usually by CHAIN HARROW, heaped and burnt. If they are fully dry they may alternatively be ploughed in. Usually carried out prior to sowing late-planted ROOTCROPS. (♦ STUBBLE CLEANING)

Spring Cultivations. Those cultivations carried out on the land in the spring when the frosts are over, in preparing a SEED BED, e.g. rolling, discing, harrowing, etc.

Spring Varieties ◗ VERNALISATION.

Springer. A cow almost ready to give birth to a calf. Also called a down-calver.

Spring-tined Harrow. A versatile cultivation implement, the TINES of which are sickle-shaped and made of spring steel, so that they are able to vibrate and shatter the soil. The angle of the tines may be varied giving different working depths suited to the type of cultivation required, so that the implement can be used either as a light HARROW or a light CULTIVATOR (*see* p. 380).

Sprout. 1. A young shoot, root, bud, etc. Also to put forth such new growth, or to cause seeds or plants to do so before planting.
2. ◗ BRUSSEL SPROUTS.

Spud. 1. A common term for a potato tuber.
2. ◗ PADDLE.
3. ◗ GRUBBER.

Spur. 1. A horny, claw-like growth on the back of the leg of a COCK or other bird.
2. A short side branch on a tree or other woody plant, usually bearing flowers or fruit. Also a lateral root.
3. A hollow pouch-like projection at the base of a petal, often containing nectar.

Square Ploughing. A method of round and round PLOUGHING, sometimes used for deep work in fields of more or less regular shape. A LAND with a similar configuration to the field boundaries is ploughed in the centre of the field. The field is then ploughed in a clockwise manner round this land. The plough is raised out of the soil at corners. This method obviates the need for OPEN FURROWS.

Squeaker. A piglet.

'S' Strains. ◗ ABERYSTWYTH STRAINS.

Stable Fly. A greyish fly (*Stomoxys calcitrans*) which breeds in faeces and rotting vegetation. They are a serious pest of horses, cattle and pigs in the warm summer months when they bite their legs to obtain blood, causing severe irritation and restlessness. They are often found in buildings housing livestock, e.g. stables, cowsheds, etc.

Stack. A large pile of hay, straw, or corn. (♦ HAY BALE STACK, HAYSTACK)

Stackyard. An enclosed area or farmyard in which a stack of hay or straw is built. Also called a rickyard, and in the case of hay a hayguard. (♦ HAYSTACK)

Staddle. 1. A support structure on which STACKS are built. Also called steddall or steddle.
2. A small tree left unfelled. Also a stump left to produce COPPICE.
3. A rootstock tall enough to produce a STANDARD fruit tree when a SCION has been grafted on.

Stag. 1. An adult male deer or male turkey.
2. An animal castrated when mature, particularly a boar (also called brawner, seg or steg).

Stag Set. A set or young hop plant which has remained in a nursery bed for more than one year.

Staggers. A condition of animals in which they are caused to stagger about. (♦ GRASS STAGGERS, LOUPING ILL)

Stale Furrow. Land which has been left for some time after ploughing so that the soil has had time to settle down and consolidate.

Stale Land ♦ SICK.

Stall. A partitioned compartment in a cowshed or stable in which cattle or horses stand or lie down.

Stallion. An uncastrated male horse, of 5 years of age or more. Mainly kept for breeding. Also called an entire.

Stamen. The male part of a flower consisting of a stalk or filament and a pollen-producing anther.

Standard. 1. A fruit tree, the lower branches of which are about 1.8 m (6 ft) above the ground. A half-standard is one in which the branches are about 1.2 m (4 ft) above the ground.
2. A tree left to grow in COPPICE woodland. (◆ COPPICE WITH STANDARDS)
3. An upright post supporting fencing wire.

Standing Crop. A crop growing in a field.

Standing Furrows. A term for those FURROWS still unploughed. Also called green furrows.

Standings. The raised parts of the floor of a COWHOUSE on which the cows are kept in pairs in stalls, each normally about 1.5 m (5 ft) deep and 2.1 m (7 ft) wide.

Staple. Wool fibres naturally binding together in a fleece forming a lock. The staple length of wool is an indication of the average length of the wool fibres. (◆ WOOL GRADES)

Starch. A polysaccharide CARBOHYDRATE, one of the main energy storage substances of plants, formed as a product of PHOTOSYNTHESIS. It is stored in granule form in seeds (particularly cereal grains), tubers (e.g. potato) and roots, which are consequently important in the diet of livestock and humans as sources of carbohydrate. Starch occurs in two component forms, amylose and amylopectin. Each form consists of 10–25 chains of GLUCOSE units, each chain containing about 20 units. The chains are linked end to end in amylose, forming one unbranched long chain, whilst they are joined by cross-links in amylopectin forming a very branched structure. Cereal and potato starches contain 20–30% amylose and 70–80% amylopectin. On digestion starch is broken down, first to MALTOSE and then to glucose. Carbohydrate is not stored in animals in the form of starch but as GLYCOGEN, sometimes called animal starch. (◆ STARCH EQUIVALENT)

Starch Equivalent. A term used in calculating livestock rations to indicate the feeding value of a FEEDINGSTUFF. The S.E. system, developed by the German nutritionist Kellner, gives values expressed as percentages which indicate the number of lbs of pure STARCH having the same energy value as 100 lbs of a particular feedingstuff. S.E. values vary considerably, e.g. Barley 71%, Soya Bean Meal 64%, Kale about 9%, whilst grass silage can vary from 0.1% to 12%. The use of S. E. values is gradually being replaced by metric METABOLISABLE ENERGY values.

Steading. Farm buildings, with or without a farmhouse.

Steamed Bone Meal, Bone Flour. Steamed bone meal is an organic phosphatic FERTILIZER consisting of bones which have been gently steamed to remove fats and gelatin for use in glue manufacturing. Steamed bone flour is the product of more intensive steaming followed by finely grinding the bones.

Steaming Up. The practice of feeding CONCENTRATES in addition to normal MAINTENANCE RATIONS to pregnant cows or heifers during the two months prior to calving, in order to build up their condition for the commencement of LACTATION. (◗ CHALLENGE FEEDING)

Steapsin. An ENZYME found in PANCREATIC JUICE which splits FATS into FATTY ACIDS and glycerol.

Stecklings. Young MANGEL or SUGAR BEET plants grown in a seed bed for later transplanting to produce seed crops, or alternatively undersown in barley and allowed to develop into a seed crop *in situ.*

Steddal, Steddle ◗ STADDLE.

Steer. A castrated male ox over 1 year old. Also called a bullock or stott.

Steerage Hoe. A type of HOE, rear-mounted on a tractor, requiring an additional operator to the tractor driver, who is able to steer or guide the implement to avoid crop damage. Steerage is effected either by a direct linkage mechanism from the tractor or by controlling the depth of ground wheels.

Steg ◗ STAG.

Stell. An open stone or corrugated metal enclosure or shelter for sheep and cattle, usually circular, built on hills and moorland.

Stem Eelworm. An EELWORM (*Ditylenchus dipsaci*) which is a pest of oats and red clover, causing seedlings to become distorted and swollen, and in oats causing the condition known as 'tulip root'.

Stepped Feeding ◗ FLAT RATE FEEDING.

Sterile. 1. Infertile or barren, and unable to produce offspring, fruit, seeds, spores, etc.
2. A term for land unable to grow crops.
3. Lacking micro-organisms, particularly those causing disease, e.g. sterile surgical instruments.

Sterilised Milk. One of the types of liquid MILK sold for public consumption. Such milk must pass the TURBIDITY TEST after prolonged heating, sometimes for up to 30 minutes, to between 110° and 115°C. (230°–240°F). This heating, carried out in sealed airtight containers in which the milk is later sold, kills all the BACTERIA present and further contamination is prevented by the airtight conditions. Such milk has a storage life of at least 7 days but the sterilising process affects its flavour and colour. (◗ PASTEURISED MILK, ULTRA HEAT-TREATED MILK and UNTREATED MILK)

Stetch. A type of RIDGE, about 2.4 m (8 ft) wide, separated from the next by an OPEN FURROW. Such ridges and furrows have been ploughed in low-laying parts of Suffolk and Essex for many generations to remove surface water.

St Foin ◗ SAINFOIN.

Stibble. A Scottish form of STUBBLE.

Stibbler. A horse put out to graze on STUBBLE.

Sticking. A term used at abattoirs for bleeding livestock, by cutting their throats, after they have stunned and hoisted.

Stigma. The expanded end of the STYLE, part of the CARPEL of a flower, forming a surface which receives POLLEN.

Stile. A series of steps to either side of a wall or fence, to facilitate climbing over.

Stilts. The handles of a horse-drawn plough.

Stint. 1. To successfully impregnate, i.e. to get in-calf, in-lamb, etc.
2. A fixed allowance, e.g. of pasture on COMMON LAND.

Stipule. One of a pair of leaf-like or spiny outgrowths at the base of a leaf-stalk.

Stirk. A term used in Scotland for male or female cattle less than 2 years old, although in England it is often restricted to females with the males termed STEERS.

Stitch. 1. ◗ RIDGE.
2. ◗ STOCK.

Stock. 1. The main stem of a plant, particularly the trunk of a tree.
2. ◗ GRAFT.
3. The perennial parts of herbaceous plants.
4. The race, family type or source material from which a plant or animal has been bred.
5. The animals on a farm (livestock) or the various stores, implements and other equipment (deadstock). Also to purchase such livestock and deadstock for a farm.
6. To sow a LEY.
7. To graze animals in a field.
8. A colony of bees.
9. To impregnate a female animal.
10. A HILL.

Stock Nitrogen. The NITROGEN present in animal faeces as distinct from in ARTIFICIAL FERTILIZERS.

Stock Seed. Seed supplied by a merchant to a seed grower from which further supplies of seed may be grown. (◗ SEED CERTI-FICATION, SEED POTATO)

Stocking. Distinctive coloured hair or FEATHER on an animal's leg.

Stocking Rate. The number of grazing animals that are allowed to graze a given area of pasture, normally expressed in terms of LIVESTOCK UNITS per acre or hectare.

Stockman. A man in charge of livestock.

Stolon. A modified type of plant stem which creeps horizontally above the ground, forming roots and buds at the NODES to produce new plants which may subsequently become indepen-dent, e.g. a strawberry runner.

Stoma. One of numerous pores in the surface tissue of plants, particularly in the leaves, through which gases and water vapour are exchanged with the atmosphere. Plural stomata.

Stone Fruit ♦ DRUPE.

Stook. A number of sheaves (♦ SHEAF) of corn set up in a field in 'pyramid' form to support each other, whilst drying out. Also called hile, shock, stitch or trave.

Stool. A tree stump from which shoots (stool shoots) sprout. (♦ STOOLING)

Stooling. A method of propagation in which the rootstocks of apple and pear trees are cut off at ground level to produce a STOOL from which shoots sprout. These stool shoots, which produce their own roots, are cut off and planted to produce new rootstocks.

Storage Drying. A method of drying BALES of HAY in a barn over a ventilated floor using a flow of unheated air. Initially about four layers are dried, and subsequently further layers are added and dried until the barn is filled. The bales are then stored *in situ*. Also called deep storage drying and barn hay drying. (♦ HAYMAKING, BATCH DRYING, TUNNEL DRYING)

Store. An animal kept at a steady rate of growth prior to later fattening for market.

Stot, Stott. A STEER.

Stover. A term for various types of FODDER such as CLOVER HAY, cereal stubble, and the broken pieces of straw from threshing.

Straight. 1. An animal feed composed of only one type of FEEDINGSTUFF, which may or may not have been processed before sale. (♦ COMPOUND FEED)
2. ♦ CAKE.

Straight Fertilizer. A FERTILIZER containing only one substance, usually providing only one, but sometimes two, of the major plant nutrients (i.e. nitrogen, phosphorus and potassium). (♦ COMPOUND FERTILIZER)

Strain. A breed, race, stock or type within a species of plants or animals, the individuals of which have specific characteristics distinguishing them from those of other strains.

Straining Pole. A sturdy pole anchored to the ground in a HOP GARDEN at the margin of a block of wire-work, and supporting a taut top wire to which strings are attached for training BINES. Also called outside bat.

Straw. 1. A term mainly used for the dry stalks of cereals but sometimes applied to the HAULM of peas and beans. It is used for litter, thatching and as a FEEDINGSTUFF. It is less digestible and has a lower feeding value than HAY, having a high FIBRE content and a lower content of PROTEIN, minerals and VITAMINS. It is usually fed in the long form, but sometimes chopped or ground, with other feedingstuffs such as CONCENTRATES. Wheat and rye straws have the lowest digestibility and energy value and are best suited for litter. Straw is mainly stored in BALE form.
2. A container holding one dose of processed SEMEN for use in ARTIFICIAL INSEMINATION.

Straw Spreader. A mechanism (often a rotating disc) sometimes fitted behind a COMBINE HARVESTER at the straw outlet to spread the straw on the field.

Straw Yard. An enclosed area covered with STRAW as litter, in which cattle are kept during winter. (◗ DEEP LITTERING)

Strawberry Comb ◗ WALNUT COMB.

Stretch. A set of 5 or 6 HURDLES together with stakes and shackles.

Strig. The central stalk of a hop CONE to which the BRACTS and seeds are attached.

Strike. 1. The infestation of the flesh of sheep by maggots hatched from the eggs of the BLOW FLY laid in the fleece. This condition, in which sheep are said to have been 'struck', causes intense irritation and death can occur within a week. Protection is afforded by DIPPING.
2. To take root.

Strip. To remove the last few drops of milk from a cow's udder.

Strip Cup. A small container into which the first few squirts of milk are drawn by hand from a cow's teats before milking starts. This foremilk is discarded as it has a high bacterial content.

Strip Grazing. A GRAZING SYSTEM whereby cattle are allowed access to a limited area of fresh pasture up to twice daily by means of a movable ELECTRIC FENCE. As each strip is grazed a 'back fence' is also moved forwards to protect the grazed area to allow it to recover. This is an efficient method which provides for accurate rationing, and ensures that the cattle eat the entire plant (i.e. stems as well as leaves) and limits damage by trampling and fouling. When grazed down the field is allowed to rest for 3–5 weeks before being grazed again.

Strippings. The last drops of milk drawn from a cow's udder during milking.

Stroke. To till a field once with a HARROW.

Strong Land. Land covered with a HEAVY SOIL.

Strong Pasture. Grassland which is too rich (i.e. has a high PROTEIN content) for young animals to graze, e.g. the EARLY BITE.

Strong Wheat. Varieties of wheat, the grain of which produces good quality baking flour, rich in GLUTEN. The grains are of a steely appearance as distinct from the white, mealy grains of the inferior quality 'weak' wheats which are usually used for biscuit making or as livestock feed.

Struck. 1. A form of entero-toxaemia in sheep caused by the bacterium *Chlostridium welchii*, restricted to certain valleys in North Wales and to Romney Marsh. It causes deaths in sheep of 1–2 years of age normally in the spring.
2. ♦ STRIKE.

Stubble. That part of a crop left in the ground after harvesting, including the roots and short stem remains, particularly the dry stalks of cereals. Also called gratten.

Stubble-breaker ♦ BROADSHARE.

Stubble Cleaning. The cultivation of the land in the autumn after harvest, either with a PLOUGH or heavy CULTIVATOR, followed by harrowing (♦ HARROW) to form a tilth and free the weeds from the soil so that they may be removed, usually by a CHAIN HARROW, heaped and burnt. Further cultivating destroys any seedlings which germinate. Also called autumn cleaning. (♦ SPRING CLEANING)

Stud. An establishment where horses are kept for breeding. Also a collection of horses or other animals, e.g turkeys. The term 'stud animal' is usually applied to a male used to sire progeny.

Stud Book. A record of the pedigree of thoroughbred horses, equivalent to a FLOCK BOOK or HERD BOOK.

Sturdy. A disease of sheep and sometimes cattle, characterised by dizziness or staggering, caused by tapeworm cysts (also called blebs) in the brain membranes. The cysts are formed by embryos of the dog tapeworm, *Taenia caenurus*, hatched from eggs deposited in grass in the faeces of infected dogs and subsequently eaten during grazing. Also called dunt, gid, gig, goggles, turn, turnsick, thorter-ill and pendro.

Sty. A pen in which pigs are housed, usually consisting of a 'kennel' leading to a walled run or dunging area.

Style. Part of the CARPEL of a flower consisting of the slender stalk growing from the OVARY with an expanded end or stigma for receiving POLLEN.

Subsoil. The layer of soil below the topsoil, of lower ORGANIC MATTER content. (◗ SOIL HORIZON)

Subsoiler. A type of heavy CULTIVATOR with long tines or 'chisels' drawn through the soil at between 30 and 60 cm (12 and 24 in.), and used to break up a compact SUBSOIL without inverting it. Deep cracks and fissures are caused, improving drainage, air and root penetration, and soil structure.

Succession. 1. A ROTATION of crops.
2. The sequence in which plants colonise a new habitat or bare ground, in which each plant COMMUNITY is superseded until a climax vegetation develops.

Successional Cropping. 1. The growing of successive but different crops in the same field in the same season.
2. The planting of a crop in small areas over a period so that a little ripens at a time, ensuring a gradual continuous supply over a long period.

Succulent Foods. Those FEEDINGSTUFFS with a relatively high water content, approximately 70–90%. They are highly digestive but are low in oil, PROTEIN and FIBRE. They are valued

427

particularly for winter feeding to balance dry fodders which tend to cause constipation, and especially for pregnant, sick or high yielding animals. Examples include the CARBOHYDRATE-containing tubers and root crops such as potatoes, carrots, mangels, swedes, and turnips, etc. Various by-products with a low dietetic value are sometimes regarded as succulent foods, such as wet brewers' grains, wet sugar beet pulp, liquid skim milk, etc.

Succus Entericus ◗ INTESTINAL JUICE.

Sucker. 1. An unweaned animal.
2. A new shoot which grows from a root or underground stem, and sometimes from the branches, which develops its own roots and develops into a separate plant.

Sucking Pig. A young milk-fed piglet.

Suckler Cow. A cow which is allowed to rear its own calf and is then used for beef production, as distinct from being used as a dairy cow for milk production. (◗ SINGLE SUCKLING, DOUBLE SUCKLING, MULTIPLE SUCKLING)

Sucklers. The flowering heads of CLOVER.

Suckling ◗ SINGLE SUCKLING, DOUBLE SUCKLING and MULTIPLE SUCKLING.

Sucrose ◗ CANE SUGAR.

Suffolk, Suffolk Down. A hardy, prolific breed of sheep characterised by a long body with a close, short fleece, and by a black face with legs which lack wool in contrast to other DOWN BREEDS. The breed was developed in the arable areas of East Anglia from crosses between SOUTHDOWN rams and ewes of the now extinct Norfolk Horn breed. It now has the widest distribution of the various Down breeds and rams are commonly used for crossbreeding, producing quick growing, well-muscled lambs (*see p. 393*).

Suffolk Coulter ◗ SEED DRILL.

Suffolk Punch. A heavy breed of horse used for draught work, chestnut coloured, sometimes with a white, star-like mark on the forehead. The legs have very little FEATHER.

Sugar. A general term for any of the sweet, soluble, monosaccharide or disaccharide CARBOHYDRATES, e.g. GLUCOSE, FRUCTOSE, Sucrose (CANE SUGAR), etc. It is commonly applied to the latter, common table sugar, which is obtained from sugar cane and SUGAR BEET.

Sugar Beet. A rootcrop of the genus *Beta* which also includes FODDER BEET and MANGEL, grown for its high content of sucrose (◗ CANE SUGAR). The sugar content varies, subject to the variety grown, soil type, seasonal characteristics, etc., but averages about 16%. In Britain the crop is grown on contract to the BRITISH SUGAR CORPORATION under E.E.C. quotas. It is mainly grown in East Anglia and the East and West Midlands within easy transporting distances of processing factories and on deep, well-drained, non-acid soils, usually as a root break. In dry years it may be irrigated. The residual tops (crowns and leaves) following mechanical harvesting are used, after wilting, as a feedingstuff for cattle and sheep (usually by FOLDING for the latter) or for SILAGE. Spent sugar beet pulp, a factory by-product, is also valued as a feedingstuff as a replacement for roots.

A sugar beet harvester.

429

Sugar Beet Harvester. A machine (either tractor-drawn and p.t.o.-driven, or self-propelled) used to harvest SUGAR BEET, which cuts off the beet tops (crowns and leaves), lifts the roots (◗ LIFTING SHARES) from the ground, cleans off adhering soil, and deposits them onto an elevator mechanism which carries them into an adjacent trailer or storage tank. The majority of beet is now harvested by a complete harvester which carries out the above operations in one unit. Some two-stage harvesters are still used which have one topping unit and a separate lifting and cleaning unit (*see p. 429*).

Suint ◗ YOLK.

Sulphate of Ammonia. ($(NH_4)_2SO_4$.) Ammonium sulphate, a nitrogenous FERTILIZER occurring as a soluble salt in the form of white needle-like crystals once produced as a by-product of coal-gas manufacture but now synthesised. In common with other non-metallic ammonium salts, when applied to fields the sulphate radical combines with calcium in the soil and is leached out, depleting the soil of LIME and increasing acidity. It contains about 21% NITROGEN.

Sulphate of Potash. (K_2SO_4.) Potassium sulphate, a FERTILIZER containing about 50% POTASH, occurring as a white crystalline powder, produced by the action of sulphuric acid on MURIATE OF POTASH. Mainly used by fruit growers and in market gardens, and also on potatoes which produce tubers with a higher DRY MATTER content than when muriate of potash is applied.

Sulphur. (S.) A non-metallic chemical element, occurring as pale-yellow crystals in several forms, required by living organisms as a constituent of PROTEIN (particularly KERATIN in animals) and some oils.

Sulphuric Acid. An extremely corrosive, colourless, oily liquid, which chars ORGANIC MATTER and is sometimes used as a CONTACT HERBICIDE, e.g. to burn off diseased potato HAULM. Also called oil of vitriol.

Summer Fallow ◗ BASTARD FALLOW.

Summer Feeding. 1. The supplying of water to marshland in the summer, usually by impounding river water to prevent natural drainage and the lowering of the water table.
2. The feeding regime of cattle in the summer months involving outdoor grazing with or without SUPPLEMENTARY RATIONS. (◗ WINTER FEEDING)

Sunflower. A composite plant the seeds of which are rich in edible oil, used for margarine, cooking oil and in medicine. The residue, after oil extraction, is used to produce CAKE or MEAL with a high FIBRE content, sometimes used in CONCENTRATES for cattle and sheep but not fed to pigs. The seed is sometimes fed to poultry, particularly during moulting.

Supers. 1. ◗ BEEHIVE.
2. A term often used for SUPERPHOSPHATE.

Superphosphate. A type of FERTILIZER once widely used but now largely superseded by the more concentrated ammonium phosphate and TRIPLE SUPERPHOSPHATE. It is a greyish, granular or powdery substance, produced by treating finely ground ROCK PHOSPHATE with a restricted amount of sulphuric acid, yielding mono-calcium phosphate, calcium sulphate and some unchanged rock phosphate. Superphosphate contains 18–21% water soluble phosphate.

Supplementary Rations. The CONCENTRATES fed to livestock in addition to bulk foods such as hay, straw and roots.

Support Buying ◗ INTERVENTION BUYING.

Surface Caterpillar ◗ CUTWORM.

Surface Cultivations. Those TILLAGE operations which are of a shallow nature and only affect the topsoil, e.g. to kill weeds, as distinct from those which involve deeper ploughing.

Surface Drainage. The practice of removing surplus water from land by surface channels, e.g. ridges and furrows, as distinct from via subsoil DRAINS.

Surface Water. Water unable to penetrate and drain through the soil and which finds its way to drainage channels, streams and rivers via the surface of the soil.

Sussex. 1. A breed of beef cattle, closely related to the smaller DEVON breed and native to Sussex where it was developed from old red cattle. Until this century the breed was used for draught purposes. Characteristically dark cherry red coloured (of a deeper shade than the Devon) with a short, broad head and medium-sized horns which grow outwards with forward curling tips. Some polled cattle have now been accepted in the HERD BOOK. A hardy breed which can exist on poor or rough pasture, and renowned for producing high quality, lean carcases. It has been exported to South America and Southern Africa.
2. A heavy breed of poultry, variously coloured including brown, grey, red, white and barred, with white legs, a single comb and producing tinted eggs. (◗ LIGHT SUSSEX)

Sussex Ground Oats. A FEEDINGSTUFF given to poultry consisting of coarsely ground oats and including the HUSKS. Also called Sussex ground oatmeal. (◗ OATMEAL)

Swaledale. A very hardy moorland breed of sheep originating in Swaledale in the northern Pennines, now replacing the SCOTTISH BLACKFACE in southern Scotland and also found in the South of England. Characterised by a fleece of long outer coarse wool covering an inner layer of dense fine wool, long curled horns, double-spiralled in rams, a black face, mealy grey nose, and mottled or grey legs. Ewes are crossed with the BLUEFACED LEICESTER and WENSLEYDALE rams to produce the crossbreds known respectively as the MULE and MASHAM.

Swan Goose ◗ CHINESE GOOSE.

Sward. The carpet of grasses, clovers and other GRASSLAND SPECIES covering the ground in a pasture.

Swath. A band of grass, corn or other crop cut by a MOWER or SCYTHE, lying on the ground.

Swath Turner. A tractor-drawn implement used in HAYMAKING which gently inverts a SWATH exposing the underside to enable it to dry. Two types of swath turner are available, (a) those incorporating rake bars (four horizontal bars attached at each end to a disc) and (b) those using spring-tined finger or spider wheels, usually 4 or 6, which rotate on independent mountings. These machines can easily be converted into SIDE

DELIVERY RAKES. Swath turners are often used in conjunction with a TEDDER and some machines combine the action of both.

Swayback. 1. A brain disease of lambs due to an inadequacy of copper in the ewe's diet, which causes paralysis and is often fatal. Lambs are seen to stagger or are unable to walk. 2. A horse with a sagging back.

Sweat-box House ♦ PIGGERY.

Swede. A ROOTCROP (*Brassica rutabaga*) similar to the TURNIP, characterised by smooth, ashy-grey leaves growing from an extended stem or 'neck', marked by scars. By contrast, turnips have hairy, grass-green leaves arising direct from the bulb itself. Swedes have a higher nutritional value than turnips, with a DRY MATTER content of 10–13%. They also have a longer growing period, are more hardy and are easily stored for winter feeding to livestock. They are sometimes pulped and mechanically fed to mangers. They are also sometimes grazed off in the field by folding (♦ FOLD). White and yellow-fleshed types exist, and of the latter, purple, bronze and green-skinned varieties are available. They are well adapted to cooler and damper climate conditions of the north and west of the U.K., and are often grown as a root break following cereals.

Sweep ♦ HAY SWEEP.

Sweet Corn. Varieties of MAIZE, the cobs of which are sweet and harvested for human consumption. Demand in the U.K. has increased in recent years with better varieties and farm gate sales, whereby the cobs can be blanched and frozen to preserve their sweet flavour. A large area of the crop is now grown in the South-East where favourable climate conditions are found. Areas grown vary from about 1 ha (2–3 acres) for the pick-your-own or farm shop trade to a few growers having 40–50 ha (100–125 acres). Commonly called 'corn on the cob'.

Sweet Cure. A cure for bacon increasingly used as a replacement for the WILTSHIRE CURE, involving the injection of brine as distinct from soaking in brine.

Sweet Soil. A fertile ALKALINE SOIL, as distinct from a sour or infertile ACID SOIL.

Swill. WASTE FOOD from kitchens used for feeding to pigs. By law swill must be sterilised before feeding by thorough boiling, for at least 1 hour, to prevent the transmission of infectious diseases, particularly SWINE VESICULAR DISEASE, SWINE FEVER and FOOT AND MOUTH DISEASE. The boiling also softens fibrous materials and separates meat from bones. All swill production plants must be licensed by the Ministry of Agriculture, Fisheries and Food under the Diseases of Animals (Waste Food) Order, 1973.

Swim Bath ◗ DIPPING BATH.

Swine. An old-fashioned term for pigs.

Swine Erysipelas ◗ ERYSIPELAS.

Swine Fever. An infectious NOTIFIABLE DISEASE of pigs caused by a virus. It is characterised by fever, refusal to eat, foul-smelling diarrhoea, discharges from the eyes, distressed breathing and general weakness. The disease may be acute, particularly in young pigs, and death may occur in a few days or it may take a chronic form, mainly in older pigs, which remain ill for a long period and lose condition, but may not die. There have been no cases in Britain since 1971. Also called hog cholera and pig typhoid.

Swine Vesicular Disease. A NOTIFIABLE DISEASE of pigs, caused by a virus, which first appeared in the U.K. in 1972. The symptoms are identical to those of FOOT AND MOUTH DISEASE and it is only distinguishable by virological testing in the laboratory. Its spread is associated with the feeding of SWILL. It is easily transmitted by untreated animal waste, and swill must, by law, now be boiled before feeding. S.V.D. is controlled by compulsory slaughter of infected pigs, the declaration of CONTROLLED AREAS, and the requirement for a MOVEMENT LICENCE for all movements of pigs (other than those direct to a slaughterhouse or slaughter market from premises where waste food is not fed to the pigs) under the Movement and Sale of Pigs Order, 1975. All swill-fed pigs may only be moved to a slaughterhouse. All pigs sent for slaughter are required to be marked with a red cross.

Swing, Swingle-tree ◗ WHIPPLE-TREE.

Swing Plough. An old type of plough which lacked wheels and had a shortish MOULDBOARD, so that it produced a greater digging action than ordinary ploughs. It was generally used in the North of England and in Scotland on the boulder clay soils.

Swingle ▶ FLAIL.

Switch. 1. A long, flexible twig or shoot.
2. To prune or trim a hedge, tree, etc.

Symbiosis. The phenomenon of two different organisms living together for mutual benefit, e.g. the association of nitrogen-fixing bacteria with leguminous plants in the root NODULES. The bacteria obtain CARBOHYDRATES and other foods from the plants and fix atmospheric nitrogen (▶ NITROGEN FIXATION) which is made available to the plants in various compounds. Lichens consist of fungi and algae living together symbiotically.

Systematic Ploughing. The ploughing of fields in LANDS.

Systemic Compound. A chemical which, when applied to foliage or the soil, is absorbed by a plant and moved in the sap to all parts of the plant. Insects sucking the sap of plants treated with systemic insecticides are poisoned. TRANSLOCATED-HERBI-CIDES are systemic compounds.

T

2,4,5–T. An abbreviation for 2,4,5–trichlor phenoxyacetic acid, the active ingredient in HERBICIDE products used in forestry and other applications to control woody weeds such as brambles. All products containing 2,4,5–T used in the U.K. have been cleared under the PESTICIDE SAFETY PRECAUTIONS SCHEME but must be used strictly in accordance with recommended precautions. In 1979 the United States Environmental Protection Agency imposed an emergency ban on 2,4,5–T on grounds of a report concluding a correlation between its usage and miscarriage in women in Oregon. The U.K. Advisory Committee on Pesticides did not accept the correlation as proven and the Government accepted its advice not to impose any restrictions on the use of 2,4,5–T. It is closely related chemically to 2,4–D.

Tabanid. A FLY of the genus *Tabanus*. (♦ GADFLY)

Table Birds. Poultry reared for their meat as distinct from for egg production.

Tack. 1. Pasture rented for grazing livestock. (♦ AGISTMENT)
2. Harness equipment.

Tag. 1. ♦ DAG.
2. A term for SHEEP SCAB.

Tag-lock ♦ DAG.

Tail Corn. Small, undersized or very light grains of corn in a sample. Also called tailings. (♦ HEAD CORN)

Tailings. 1. A type of wool oddment purchased by the BRITISH WOOL MARKETING BOARD comprising very small bits of fleece detached during shearing, generally derived from HILL BREEDS of sheep in Wales and North England. (♦ BROKES)
2. ♦ TAIL CORN.

Tailpiece. An extension sometimes attached to the rear end of a MOULDBOARD. It can be set to push the turning furrow further over.

436

Take. 1. The taking possession of buildings and land by a farmer when he assumes a tenancy. Also called entry.
2. A crop which germinates or a successful GRAFT are said to take.

Take-all. A disease of cereals due to a soil-borne fungus (*Gaumannomyces graminis*) which infects the roots and stem base, causing discoloration, stunted growth, premature ripening and the production of bleached EARS containing little or no grain. One strain of the disease affects only wheat and barley, whilst another also affects oats but is mainly confined to the wetter areas in the North and West. Control is by ROTATION. Also called whiteheads.

Tall Fescue. A tufted perennial grass (*Festuca arundinacea*) which is persistent and resistant to drought, but is more coarse and generally less palatable than MEADOW FESCUE. Although slow to establish it grows early in the spring and is often used to produce an EARLY BITE and also for winter grazing due to its hardy nature (*see p. 211*).

Tamworth. A breed of pig, characteristically golden-red coloured with a long narrow body, a long straight snout, and large, rigid, forward projecting ears. A hardy breed, slow to mature and fatten, producing a quality lean carcase. The breed has been widely exported and is popular in tropical climates due to its resistance to sun scald. In America it is valued as a bacon pig whilst in Britain it is mostly found in the Midlands where it is kept as a pork pig. The breed does well under outdoor systems of management and is sometimes used to reclaim rough pasture and scrubland.

Tandem Parlour ◗ MILKING PARLOUR.

Tank Mixing. A relatively new practice of mixing crop-protection chemicals (and also liquid fertilizers, adjuvant oils, trace elements and foliar feeds) before spraying on field crops, particularly cereals. The benefits include savings in time, labour, water and fuel, reduced wear and tear on machinery and less crop damage through wheelings. Manufacturers recommended mixes are cleared and approved under the PESTICIDES SAFETY PRECAUTIONS SCHEME and the AGRICULTURAL CHEMICALS APPROVALS SCHEME.

Tankage ♦ MEAT MEAL.

Tankard Mangel. A type of MANGEL which is tankard-shaped and classified with the intermediate types which have less of their swollen root above ground than the globe types.

Tapeworms. A class of parasitic flatworms which inhabit the small intestines of animals and are named from their dirty white tape-like appearance. They consist of a ribbon-like chain of segments anchored at one end and trailing through the gut, reaching several feet in length in some cases. They seldom require treatment in livestock.

Tapioca ♦ MANIOC.

Tares ♦ VETCH.

Target Price. The wholesale price which it is hoped will be achieved in the E.E.C. for certain agricultural products. The COMMON AGRICULTURAL POLICY is designed to keep import prices as close as possible to the target price. THRESHOLD PRICES and INTERVENTION PRICES are linked to it. For cereals, the target price is varied monthly to allow for storage costs throughout the year.

Tassel. 1. The male flower of the MAIZE plant, responsible for pollen production.
2. A black hairy tuft on a turkey's neck.
3. ♦ DODDLE.

Tear. The soil not cut on the COULTER FACE during ploughing.

Teart Land. The rich pastures of Somerset on heavy soils containing an excess of MOLYBDENUM, renowned for causing SCOUR in sheep and cattle grazing on them.

Teat. A nipple, particularly one of the four protuberances on a cow's udder from which milk may be sucked by a calf or removed by MILKING MACHINE.

Teat Cup. One of the four tube-like devices forming the cluster of a MILKING MACHINE. Each teat cup consists of an outer metallic shell lined with synthetic rubber, which fits over the teat and into which milk is drawn by the application of a partial vacuum.

Teat Dipping. A method of applying disinfectant to a cow's teats as a precaution against the spread of MASTITIS, involving the dipping of each teat into a cup containing the disinfectant. The three conventional designs of cup are (*a*) a straight cup, (*b*) a cup on top of a 'squeeze' bottle, and (*c*) a non-squeeze, two chamber model. The most effective cups are at least 7.5 cm (3 in) deep, which ensure complete immersion. Disinfectant may also be applied by sprayers but this is regarded as less effective. The three main disinfectants used are IODOPHORS, hypochlorite and chlorhexidenes, and they are applied after washing the udder, but prior to milking.

Ted. To toss swaths of new-mown grass in HAYMAKING in order to expose more grass to the sun and air and quicken the drying process.

Tedder. A tractor-drawn HAYMAKING machine, usually used soon after a grass crop has been cut to loosen and aerate the SWATH. The crop is picked up by tines rotating in either a vertical or horizontal plane, and is thrown backwards onto the ground. Tedders may be either wheel- or power-driven and may be designed to ted a single row or to combine two swaths into one WINDROW. Modern machines have spring tines, the angle of attack of which can be adjusted to vary the throw of the crop.

Teeswater. 1. A breed of sheep revived after the Second World War in Upper Teesdale from survivors of the old Teeswater 'Mug' (a large long-woolled breed). The modern Teeswater is a smaller, polled breed with a white or greyish face. It is used to provide rams for cross-breeding, e.g. with SWALEDALE ewes to produce the MASHAM.
2. ◗ DURHAM.

Teg ◗ HOGG.

Temporary Grassland. Arable land sown to a LEY for a limited period of years, as distinct from permanent GRASSLAND.

Tenant. A person who, subject to certain conditions, has temporary possession of a farm, buildings and land owned by a landlord, for which a rent is paid. Under the Agriculture

(Miscellaneous Provisions) Act, 1976, a farmer and his near relatives have security in the tenancy of a farm for 3 generations, but subject to stringent eligibility tests.

According to the 1980 ANNUAL REVIEW White Paper about 35% of agricultural holdings in Britain were tenanted in 1979.

Tendon. The strong, fibrous cord or band of connective tissue by which a muscle is attached to a bone or other part of the body. It consists mostly of the protein collagen. Also called sinew.

Tendril. A modified leaf or stem in the form of a coiling thread by which certain climbing plants adhere to a support. The upper leaflets of the sweet pea are modified as climbing tendrils.

Terminal Bud. A BUD growing at the tip of a stem or branch. (♦ LATERALS)

Terrier. A register or inventory of the farms of an estate including such details as field boundaries, acreages, rents, etc.

Teschen Disease. A NOTIFIABLE DISEASE of pigs due to a virus, resulting in an excited condition with fever and paralysis, and often death. It was first recognised in Czechoslovakia but has not yet appeared in the U.K.

Testa. The usually dry, hard protective coat enclosing the embryos of seed plants. Also called seed coat.

Testicle. The reproductive gland of male animals occurring in pairs, producing SPERM. Also called testis (plural, testes).

Tetanus. A bacterial disease (*Clostridium tetani*) of livestock and man which enters the body via wounds, particularly from the soil. Muscles become stiff, spasms may occur, and the muscle which closes the jaws may become continuously contracted. Death usually follows. Few animals recover even if treated. Also called lockjaw.

Tether. A rope, chain or halter used to confine or restrain an animal within certain limits, e.g. by tying it to a post in the ground.

Tetraploid. Having four times the HAPLOID number of CHROMOSOMES.

Texel. A breed of sheep originating in Holland and widely distributed in Europe. It is renowned for its milk production, and also noted for its rate of growth and potential for meat production (particularly in France where a distinct meat type has been developed).

Texture ◗ SOIL TEXTURE.

Theave ◗ SHEEP NAMES.

Thin. To remove some of the plants in a crop so that it is less crowded and those remaining have room to develop.

Thin Crop. A poorly developed crop with plants growing relatively sparsely.

Thin Land. Shallow soil.

Thirds. A term once given to a category of wheat offal resulting from milling. All wheat offals, apart from BRAN, are now sold as WEATINGS for use in FEEDINGSTUFFS.

Thole ◗ NIB.

Thorax. That part of an animal's or insect's body between the head and ABDOMEN. In insects, the legs and wings are borne on the three segments which comprise the thorax.

Thorter-ill ◗ STURDY.

Thousand-headed Kale. A very branched variety of KALE which generally grows 2–3 ft or more tall, bearing many plain, small, uncurled leaves, and is grown as a FODDER CROP. It is more frost tolerant than MARROW-STEM KALE and is normally kept for feeding in the late winter. Canson kale is a dwarf variety of Thousand-headed Kale sometimes fed to sheep early in the year.

Thrash ◗ THRESH.

Thrave, Threave. An old measure of straw and fodder roughly equivalent to 2 STOOKS, usually of 12 SHEAVES each.

Thread Worm. Any member of the Nematoda or ROUND WORMS, but particularly of the genus *Oxyuris*.

Three-point Linkage. The three hydraulically operated arms fitted to the rear of a tractor on which implements are mounted.

Thresh. To separate the grain or seeds from crops (e.g. cereals, beans, peas, clovers, etc.) by beating.

Thresher. A THRESHING MACHINE.

Threshing Drum ♦ DRUM.

Threshing Machine. A machine once commonly used to THRESH cereals and other seed crops. Most cereal crops are now harvested and threshed by COMBINE HARVESTER.

Threshold Price. The minimum import price into the E.E.C. for cereals, milk products and sugar from non-E.E.C. countries. It is fixed in relation to the TARGET PRICE so that after the addition of transport costs from the port of import the commodity is marketed at or above the target price. Commodities imported into the E.E.C. below the threshold price are subject to a variable level to raise the price to the threshold price.

Thrifty. A term applied to animals which are developing well, gaining weight, producing milk or eggs well, etc., and generally thriving.

Thrips. An order of small, sap-sucking flies (Thysanoptera), characterised by narrow wings with fringes of long hairs. Very many species exist, some of which are pests of crops, e.g. Onion Thrips (*Thrips tabaci*) and Pea Thrips (*Kakothrips robustus*). Also called Thunder Flies. (Thrips is both singular and plural.)

Through Stone. A stone in a DRY STONE WALL, where it is more than one stone thick, laid across the wall to bond the stones together.

Throw. 1. To give birth to.
2. To sire.

Throw-back ♦ ATAVISM.

Thrunter. A three-year-old ewe.

Thrush. An inflammation of the FROG of a horse's hoof, accompanied by the production of a foul-smelling discharge.

Thunder Flies ♦ THRIPS.

Tick. A sub-order of large blood-sucking insects (Acarina) of the class ARACHNIDA, allied to the MITES. They are serious parasites of livestock, and when present in large numbers they cause 'worrying' and lack of THRIFT. The best known is the sheep tick (*Ixodes ricinus*), responsible for transmitting LOUPING-ILL, REDWATER and TICK-BORNE FEVER. Ticks are controlled by DIPPING, washing or spraying with insecticides. Also called faggs.

Tick Bean ▶ BEANS.

Tick-borne Fever. A disease of cattle and sheep caused by the micro-organism *Rickettsia* (intermediate between bacteria and viruses) which infects the white blood cells and is transmitted by the sheep tick (▶ TICK). In cattle it causes fever, reduced milk yield for a period and increased susceptibility to other diseases (e.g. REDWATER). In sheep the fever is accompanied by listlessness and loss of appetite, and pregnant ewes may abort.

Tied Cottage ▶ SERVICE HOUSE.

Tile Drain. A baked, extruded clay pipe used for underground DRAINS in LAND DRAINAGE systems. Such pipes replaced the original bent or horse-shoe clay tiles (from which the term 'tile-draining' was derived) in the mid-nineteenth century.

Till. To work land with implements in preparation for a crop, e.g. ploughing, cultivating, harrowing, manuring, sowing, etc.

Tillage. 1. The act or practice of tilling the land. (▶ TILL)
2. Land lying fallow or under a crop, as distinct from under a LEY.

Tiller. A side-shoot from the base of a plant near to the ground, e.g. from the bottom of the stalk or stem of cereals and grasses.

Tilth. The physical condition of the topsoil after TILLAGE. A fine tilth consists of small clods and loose, crumbling soil particles or mould. In a coarse tilth, comparatively large clods constitute most of the broken material.

Timothy. A loose to densely tufted perennial grass (*Phleum pratense*), with hairless, broad, rough, bluish-green leaf blades and flowerheads in the form of spike-like, cylindrical PANICLES. Timothy is a very palatable species which flowers late in the spring, being popular for summer grazing. It is a hardy grass which persists in winter and thrives on heavy soils, particularly in cool and wet areas. It makes excellent HAY. The name is derived from Timothy Hanson who introduced the grass to the U.S.A. in about 1720 (*see p. 211*). Also called Cat's Tail.

Tine. A spike or prong on a CULTIVATOR, HARROW, RAKE or fork. Also called point.

Tine Harrow ◗ DRILL HARROW.

Tipping Plough ◗ BALANCE PLOUGH.

Tithe. A term once applied to a tenth of the produce of land and stock on a farm given to the church, and later to a cash sum of approximately equal value. A tithe barn was one in which corn paid as tithe was stored.

Tod. An old measure once used in wool buying, equal to 28 lbs.

Toggenburg. A small breed of GOAT imported from Switzerland, brown in colour with two white stripes on the face, white legs and white marks on the tail. They have small, pricked ears. Females weigh about 45–50 kg (100–110 lbs) and produce about 1 450 l (320 gal.) of milk per lactation containing in excess of 4% BUTTERFAT. Males weigh about 65 kg (140 lbs).

Tolpuddle Martyrs. Six Dorset farmworkers who were prosecuted and transported to Australia for trade union activities in 1834.

Tom ◗ TURKEY COCK.

Tongue Graft ◗ WHIP GRAFT.

Tool Bar. A metal frame fixed usually to the rear of a tractor on which the cultivating tools (e.g. hoes) for inter-row work are mounted.

Top. 1. The tallest growing grasses in a SWARD.
2. To lightly mow grassland to encourage more leafy growth.
3. To cut off the crowns and leaves of sugar beet. (◗ SUGAR BEET HARVESTER)

Top Dressing. FERTILIZER applied to a crop after it has emerged from the ground.

Top Fruit. FRUIT with relatively firm flesh, as distinct from the soft flesh of SOFT FRUIT. Divided into hard or pome fruits (e.g. APPLES, PEARS, quinces, etc.) and stone fruits or DRUPES (e.g. CHERRIES, plums, etc.). Cherries and plums are grown in Kent, and plums also in the Vale of Evesham and parts of East Anglia.

Top Wire. One of the parallel wires at the top of the WIREWORK in a HOP GARDEN to which string is attached, up which the BINES grow.

Tops. The unwanted portions of SUGAR BEET plants cut off during harvesting, comprising the crown of the root and the leaves, used for feeding to cattle and sheep.

Topsoil. The surface soil disturbed during cultivation and rich in ORGANIC MATTER. (♦ SOIL HORIZON)

Total Digestible Energy Value. A percentage value which is a summation of the digestible nutrients (i.e. digestible crude protein, digestible crude fibre, digestible oil × 2.3, and digestible nitrogen-free extract) in a FEEDINGSTUFF, determined by a DIGESTIBILITY TRIAL. T.D.N. values are used to assess the energy value of pig and poultry diets. They are slightly higher than D-VALUES since the high energy value of fats (oils) is taken into account.

Total Exchange Capacity ♦ CATION EXCHANGE CAPACITY.

Total Herbicide. A non-selective HERBICIDE used to kill all types of vegetation, applied prior to crop planting, and also used on footpaths, roads, waste land, etc.

Toulouse. A large grey breed of goose with an orange beak and feet, originating in France. Egg production is relatively poor. Adults weight about 9–13.6 kg (20–30 lbs).

Toxic. A term usually applied to a chemical capable of killing, injuring or impairing an organism.

Toxin. A TOXIC substance, especially one produced by bacteria infecting part of an animal's body.

Trace Element. A chemical ELEMENT required by a plant or animal in very small quantities for its METABOLISM. Such elements are often essential for the functioning of ENZYMES, HORMONES and VITAMINS. They include iron, zinc, boron, manganese, cobalt, molybdenum and copper and a lack of these elements may cause a DEFICIENCY DISEASE (e.g. HEART ROT in sugar beet and PINING in sheep, due to insufficient boron and cobalt respectively). Also called micronutrient.

Traces. The ropes, chains or straps attached to the collar of a draught animal (usually a trace horse) by which it pulls an implement or cart.

Tracklayer. A type of TRACTOR propelled by many small wheels running on two parallel, closed, articulated tracks. Also called a caterpillar or crawler tractor.

Tractor. A vehicle which has now almost entirely replaced the horse for pulling implements and trailers. The modern tractor is also able to operate various implements attached to its three-point-linkage and front-loader arms, and to provide power to trailed and mounted machinery via the p.t.o. and hydraulic drive system. Some stationary machinery (e.g. sawbenches, CRUSHING MILLS, etc.) are also powered by tractors by belt pulley.

A wide range of tractors are available and are usually classified according to engine power as indicated below. Tractors have increased in size and power over the years so that models over 100 h.p. are much more common now than in the late 1960s. The 75 h.p. machine is now regarded as a 'common workhorse' on arable farms. Two- and four-wheel drive types exist, and the latter may have equal or unequal sized drive tyres.

(a) *Less than 10 h.p. (7.6 kw):* Usually single-axle and hand operated. Used in market gardens for hoeing and cultivating.

(b) *10–40 h.p. (7.6–30 kw):* Usually two-wheel drive, manoeuvrable, with good ground clearance, sometimes with narrow wheels. Used for row-crop work.

(c) *40–75 h.p. (30–57 kw):* Mostly two-wheel drive but some four-wheel drive and TRACKLAYERS. The general-purpose tractor used for a wide range of jobs.

(d) *75–150 h.p. (57–114 kw):* Large models, both two- and four-wheel drive, and tracklayers. Used on large farms and by contractors for heavy work.

(e) *More than 150 h.p. (114 kw):* Large four-wheel drive types and tracklayers.

Since 1970 new tractors have been required to be fitted with a SAFETY CAB or safety frame (unless exempted).

A land tractor is defined precisely under Regulation 3 of the Motor Vehicles (Construction and Use) Regulations 1978 as: 'a tractor, having an unladen weight not exceeding 7 370 kg (7¼ tons), designed and used primarily for work on the land in connection with agriculture, grass cutting, forestry, land levelling, dredging or similar operations, which is:

(a) the property of a person engaged in agriculture or forestry . . . and

(b) not constructed or adapted for the conveyance of a load other than –

(i) water, fuel, accumulators and other equipment used for the purpose of propulsion, loose tools and loose equipment,

(ii) . . . goods . . . or appliances . . . for which a goods vehicle licence does not apply, and

(iii) an implement fitted to the tractor and used for work on the land on farms or forestry estates. . . . . . .'

Tractor Cab ◆ SAFETY CAB.

Tractor Hoe. A type of HOE consisting of rigid or spring-loaded blades attached to the toolbar of a row-crop tractor. It incorporates a set of A-shaped blades which loosen the soil between the rows, behind which are fitted a set of L-shaped blades which cut the roots of weeds. Shallow-rooting weeds are disturbed and die on drying. Discs are sometimes mounted between the two sets of blades, set at an angle, to prevent damage to crop seedlings by soil covering.

Tractor Vaporising Oil. A type of fuel at one time widely used in tractors, consisting of paraffin with added aromatic hydrocarbons to prevent pre-ignition, enabling its use in the internal combustion engine. Almost all tractors now use diesel fuel although some old models are still run on t.v.o. Some farmers prepare t.v.o. on-farm by adding petrol to paraffin approximately in the ratio 1 : 9.

Traditional Farm-fresh Poultry. Poultry plucked but with the VISCERA, head and feet intact. Formerly referred to as New York dressed poultry.

Trailer. An agricultural trailer is defined under Regulation 3, Motor Vehicles (Construction and Use) Regulations, 1978 as 'a trailer the property of a person engaged in agriculture which is not used on a road for the conveyance of any goods or burden other than agricultural produce or articles required for the purpose of agriculture.'

Trailer-spreaders ▶ MANURE SPREADERS.

Tramlines. Accurately spaced pathways left in a growing crop or field area to provide wheel guide marks for a tractor driver or machine operator to follow during subsequent operations. Tramlines enable the accurate application of pesticides, fertilizers or other products, facilitate precision during cultivation and lighten the load or task of the operator. They can be produced by omitting to sow seed along particular 'paths' during drilling (▶ DRILL), by crop removal after emergence by chemical or other means, and by continuous wheeling.

Translocated Herbicide. A HERBICIDE which when absorbed by a plant via the leaves (following foliar spraying) or via the roots (after application to the soil) moves in the sap to all parts of the plant, eventually killing it, e.g. 2,4-D. Translocated herbicides are SYSTEMIC COMPOUNDS.

Transpiration. The evaporation of water vapour from the surface of plant leaves, mainly through stomata (▶ STOMA). (▶ EVAPO-TRANSPIRATION)

Transplanter. A tractor-drawn machine used mainly for transplanting BRASSICA crops. Various designs exist. In the simple hand-fed machine the operator is seated close to the ground and places the plants into a furrow opened by the planter. The soil is firmed around the roots by a pair of following, inclined press wheels. An assistant is sometimes required to prepare and hand plants to the operator from a supply tray fitted to the machine. In other types the plants may be conveyed to the furrow in fingers on a continuous conveyor belt, or in other 'feeding' devices.

Transported Soil. A soil comprised of material carried from elsewhere (e.g. by wind, water, ice, gravity, etc.) and deposited, such as ALLUVIUM, COLLUVIUM, etc.

Trap Nest. A type of nesting box which a hen or duck can enter via a trap door but cannot leave until released. Such devices are used to record the number and size of eggs laid, to assist in selecting hens for breeding.

Trave. A STOOK, or row of stooks.

Tray Drier ▶ GRAIN DRIER.

Treaty of Rome. A treaty signed by the original 6 members of the E.E.C. in Rome on 1 January 1958, by which the E.E.C. was established, and which, amongst other things, set out the main objectives of the COMMON AGRICULTURAL POLICY.

Tree Fruit. FRUIT which grows on a tree (e.g. apples, pears, cherries, etc.) as distinct from those growing on bushes, canes or low-growing plants.

Trefoil. An annual leguminous plant (*Medicago lupulina*), with trifoliate leaves, small yellow flowers and twisted, black pods. Abundant on lime-rich soils. It is usefully used for short-term LEYS and for CATCH CROPS and produces a useful EARLY BITE for sheep on chalk land.

Trenching. A method of deep digging in which adjacent trenches are dug two spits deep, with the top spit being turned into the bottom of the last (adjacent) trench dug, and the bottom spit being turned on top. Soil layers are reversed, as opposed to being replaced in their natural order as in BASTARD TRENCHING.

Tribunal ▶ AGRICULTURAL LAND TRIBUNAL.

Trichinosis. A disease of pigs and other animals, including man, in which muscles become infected with the LARVAE of the small ROUNDWORM *Trichinella spiralis*. Pigs are usually infected after eating infected raw SWILL.

Trichomonas. A genus of protozoa, *T. foetus*, which infects the genital passages of cows, causing abortion, pus formation, and sometimes sterility.

Trifolium. The genus of CLOVER. Particularly applied to Crimson Clover, *Trifolium incarnatum*, a tall, erect annual plant with crimson flowers. It is frost-sensitive and cultivated only in the South and South-East of England, mainly as a CATCH CROP. It is sown in lightly harrowed corn stubble and sheep are grazed on it in late April and May in the following year. Early, medium and late strains are available, which can be grazed in sequence.

Tripod Hay ◖ HAYMAKING.

Trocar ◖ CANNULA.

Trotter. The foot of a pig or sheep.

Trough. A long, narrow vessel in which water or feed is provided to animals.

Trough Room. The amount of space required by each animal in a group feeding from a TROUGH.

True Protein. An assessment of the actual PROTEIN content of an animal FEEDINGSTUFF derived by subtracting the non-protein nitrogen from the total nitrogen content. Roots and SILAGE contain significant amounts of non-protein nitrogen. (◖ CRUDE PROTEIN, DIGESTIBLE CRUDE PROTEIN and PROTEIN EQUIVALENT)

Trug. A fruit basket made of plaited wooden strips.

Truss. 1. A bundle of HAY or STRAW.
2. A cluster of fruit or flowers, e.g. tomatoes.
3. To prepare a bird for cooking after killing and plucking, by removing the head and viscera, and tying or skewering.

Trusser. A mechanism sometimes fitted to the back of a COMBINE HARVESTER which trusses the straw leaving the combine into bundles. It avoids the need to subsequently use a BALER on the straw deposited on the field. Also called a buncher or straw press.

Trypsin. A proteolytic ENZYME which breaks down PROTEINS and PEPTONES to PEPTIDES. Found in PANCREATIC JUICE.

Tuber. A swollen underground stem (stem tuber) bearing buds in the axils of rudimentary leaves or scales, e.g. potato. Also a swollen root (root tuber), e.g. dahlia. Tubers are organs of

VEGETATIVE REPRODUCTION and contain stored food, e.g. in the form of STARCH in the potato.

Tubercle. 1. A small TUBER.
2. A small, rough bump on the surface of a bone.
3. A nodule-like swelling on a plant.
4. A small, barely visible swelling in a lung or other organ of the body, caused by bacteria. The first stage of TUBERCULOSIS.

Tuberculin. A liquid preparation derived from a culture of tubercle bacilli, containing tuberculo-protein, injected into the skin of animals to determine if animals are infected with TUBERCULOSIS (the Intradermal Comparative Test). Infected animals react by producing a swelling at the site of injection.

Tuberculosis. A contagious disease of many animals including man and all domesticated animals, particularly cattle and pigs, caused by the bacterium *Mycobacterium tuberculosis*. Characteristic tubercles or nodular swellings develop in one or more of the organs of the body and degenerate, becoming cheesy, so that affected organs are destroyed. T.B. of the lungs is the most common and is usually associated with coughing (chronic in cattle) and loss of condition. (♦ BOVINE TUBERCULOSIS)

Tull, Jethro (1674–1740). A Berkshire landowner and farmer, inventor of the SEED DRILL. He also developed the horse hoe and advocated thorough cultivation of the soil, drilling of seed as opposed to broadcasting (♦ BROADCASTER), and the practice of wide drilling and inter-row cultivations, with a consequent great reduction in the amount of seed required. These methods allowed the abolition of the need for a periodic one-year fallow which had been a feature of the medieval open-field system and of subsequent crop ROTATIONS.

Tumbril. A tip-cart used for carrying dung.

Tump. A hillock or mound. Also a clump.

Tunnel Drying. A method of drying BALES of HAY built into a stack with a central tunnel blocked at one end, so that air blown into it with a fan is forced up through the stack. The stack can be constructed in the field or more usually under a DUTCH BARN when it is also called barn hay drying.

Tup. 1. An uncastrated male sheep. Also called ram. (◆ SHEEP NAMES)

2. A term applied to a ram when it mates with a ewe.

Tup Lamb. An uncastrated male lamb between birth and weaning. Also called ram lamb, heeder or pur lamb. (◆ SHEEP NAMES)

Tupping Time. The mating season for sheep.

Turbary. A place where PEAT is dug. Also a legal right to cut peat on another person's land.

Turbidity Test. A statutory test which milk must pass before it can be sold for public consumption as STERILISED MILK. The principle of the test is that when milk is boiled all the ALBUMEN in it is precipitated. It involves the addition of ammonium sulphate to precipitate other substances, e.g. casein, which are then filtered out. The filtrate is heated and any albumen in the milk is revealed by turbidity. If the milk had been adequately sterilised (by heating to in excess of 100°C) then the albumen would all have been previously precipitated, and the test would reveal no turbidity.

Turf. The carpet of grasses covering the surface of the land in a pasture or meadow, the roots of which are matted with the soil. Also a cake or sod of such turf sliced off and removed.

Turkey. A large poultry bird (*Meleagris gallopavo*) of the pheasant family native to North America, several varieties of which are kept for their meat. Nowadays the main producers are specialists keeping large numbers of birds, although some small-scale producers still keep turkeys, themselves undertaking the killing, plucking and dressing of oven-ready birds (mainly at Christmas).

Turkey Grower. A turkey between 8 and 26 weeks of age. (◆ POULT, TURKEYCOCK)

Turkey Hen. A female turkey older than 26 weeks. (◆ TURKEY GROWER)

Turkeycock. An adult male turkey older than 26 weeks. Also called turkey stag, or tom. (◆ TURKEY GROWER)

Turn. 1. A term used of a female animal which comes on HEAT again after being mated. Also called return.
2. ♦ STURDY.

Turn Up. To lift a sheep so that it sits on its rump to facilitate shearing, veterinary treatment, etc.

Turnabout Plough. An old type of horse-drawn one-way plough with two bodies (♦ BODY), one right-handed, the other left-handed, mounted on opposite sides of a single beam. At the end of the furrow the other body was brought into action by 'unlocking' the beam from a plate fixed to the wheels, and rotating it through 180°. Also called turnover plough.

Turnip. A ROOTCROP (*Brassica rapa*) similar to the SWEDE, characterised by hairy, grass-green leaves which arise direct from the swollen root or bulb. By contrast, swedes have smooth, ashy-grey leaves growing from an extended stem or 'neck'. Turnips have a lower nutritional value than swedes, with a DRY MATTER content of $7\frac{1}{2}$–10%. They also have a shorter growing period, are less hardy, and less easily stored. Turnips, like swedes, are extremely valuable for feeding to sheep and cattle, and some varieties are used as table vegetables. They may be grown as a CATCH CROP and grazed *in situ* or grown as a break crop in cereal ROTATIONS (they were the root break in the NORFOLK ROTATION). The roots may be stored for winter feeding (often with the tops left on) or removed from the ground as required. They are sometimes pulped and mechanically fed to mangers. Turnips are well adapted to the cooler and damper climatic conditions of the north and west of the U.K. White and yellow-fleshed varieties exist, the former being faster-growing initially, and maturing earlier, and often grazed by folding. (♦ FOLD)

Turnover plough ♦ TURNABOUT PLOUGH.

Turnsick ♦ STURDY.

Turnwrest Plough. An old type of horse-drawn, one-way plough with a single BODY on which the MOULDBOARD and SHARE had right- and left-hand faces. Either set of faces could be brought into action by partially rotating the body beneath the beam from one side to the other.

453

Tussock. A tuft or bunched clump of grass, rushes, etc.

Twine ◗ BALER-TWINE.

Twinter. A two-year-old sheep or other animal. Derived from the Old English term *twi-wintre* meaning two-winter.

Twitch. 1. A device sometimes used to hold animals still, particularly horses, consisting of a rod with a hole in one end, through which a piece of soft cord passes, which can be tightened round the upper lip or ear by twisting the rod.
2. A term for COUCH GRASS.

Two Sward Grazing. A system of grassland management in which areas intended for grazing and for cutting for conservation are treated separately. It is more efficient than systems which integrate the two functions, in that varieties specifically suited to either grazing or cutting can be selected, particular fertilizers can be used, and the harvest can be planned by the use of varieties which reach the cutting stage of growth at different times. Grazing areas, particularly for large herds, can be arranged close to the farm buildings.

Two-crop Ewe. A EWE that has borne two crops of lambs.

Two-tooth Sheep. Sheep of about 12–15 months of age. (◗ FOUR-TOOTH, SIX-TOOTH, FULL MOUTH and SHEEP NAMES)

U

Uberous. Yielding milk in abundance.

Udder. The organ containing the MAMMARY GLANDS of animals. It is usually applied to the pendulous bag hanging beneath a cow just in front of the rear legs, bearing four TEATS. Also called bag.

Udder-clap ▶ MASTITIS.

Ulcerative Stomatitis ▶ ORF.

Ulster Bacon Agency Ltd. A company, on the Board of which the PIGS MARKETING BOARD (N. IRELAND) and independent curers have equal representation, responsible for the promotion and pricing of Ulster bacon in Great Britain. It also carries out trade and cookery demonstrations, and advertising and promotional work in Northern Ireland (other than brand advertising).

Ulster Curers Association. An organisation composed of all companies and individuals engaged in the processing of pigmeat in Ulster, including factories owned by the PIGS MARKETING BOARD (N. IRELAND). The Association has a technical division responsible for advisory work. Close relationships are maintained between the Association and the P.M.B., with representatives of the two bodies meeting to agree arrangements for sales of pigs to curers, and to discuss other matters affecting the industry. The two bodies also jointly sponsor promotional work in the province, bacon carcase competitions at agricultural shows, demonstrations for the trade and cookery demonstrations for various organisations, advertising and other promotional work in Ulster.

Ultra Heat Treated Milk. A designation of MILK sold for liquid consumption (▶ MILK PRODUCTS) after subjection to ultra high temperature to sterilise it by killing BACTERIA and other organisms, including spores, present in it. The milk is flow heated and held at a temperature of at least 270°F (132°C) for one second before cooling. This treatment destroys some vitamins and probably some proteins and affects the palatability of the milk. Before sale U.H.T. milk must pass the statutory Colony Count Test (to determine the possible presence of bacteria). It is sold in sterile packages, often cartons, to prevent recontamination, and can be stored indefinitely if kept sterile. (♦ PAS-TEURISED MILK, STERILISED MILK and UNTREATED MILK)

Umbel. A type of flat-topped inflorescence in which the main axis of the stem stops growing and a number of flower stalks develop at one level, like the spokes of an umbrella, e.g. carrot, parsley.

Unbroken Furrow. A continuous, smooth type of furrow slice such as that produced by a Lea Plough.

Undecorticated Seeds. Seeds which have not been subjected to DECORTICATION.

Underbeam Clearance. The measurement from the underside of the SHARE to the underside of the BEAM of a plough.

Under Drain. A type of drain constructed beneath the ground, e.g. MOLE DRAIN, TILE DRAIN, etc., in a LAND DRAINAGE system, as distinct from an open ditch or surface drain. The water draining into such pipes is carried away by gravity flow and is usually discharged into a ditch. Before the advent of tile drains the ducts were often formed from locally available materials, e.g. thorn or other brushwood faggots, stones, dried peat or turf tiles, or straw bundles.

Undersow. To sow a grass SEEDS MIXTURE with another crop (a COVER CROP), usually a cereal crop.

Undulant Fever. A recurrent feverish disease of man which can last for a long period, sometimes many months. It is caused by the bacteria *Brucella abortus* which is responsible for BRUCELLOSIS and is usually contracted by contact with infected cattle or by drinking unpasteurised (♦ PASTEURISED MILK) infected milk. Also called Human Brucellosis.

Unexhausted Manurial Value. The amount of FERTILIZER, MANURE or LIME remaining in the soil, and usefully available to a crop, after one or more crops have drawn upon it after its application.

Under the Agriculture (Calculation of Value for Compensation) Regulations 1978, provision is made for the compensation of an outgoing tenant for fertilizers, manures, magnesium and copper, and lime applied to the soil, and for purchased feeding-stuffs, including corn produced on the holding, consumed by livestock (producing SLURRY or FARMYARD MANURE for storage) or fed directly on the land. The compensation is calculated according to reducing unexhausted manurial values of the materials over specified numbers of growing seasons since their application to the soil or consumption by livestock.

Ungulates. Herbivorous hoofed animals, e.g. cattle, deer, horses, pigs and sheep.

Unisexual Flower. A flower bearing either only STAMENS (male flower) or only a PISTIL (female flower).

Unit. 1. An implement directly mounted on a tractor as distinct from one trailed behind.
2. ▶ LIVESTOCK UNIT.
3. ▶ UNIT OF ACCOUNT.
4. ▶ UNIT OF FERTILIZER.

Unit of Account. The common 'artificial' currency of the E.E.C. used for accounting purposes, and in terms of which INSTITUTIONAL PRICES are expressed. For trading it is translated into the national currency of each member state at a fixed exchange rate, the Representative Rate (▶ GREEN POUND). The unit of account now in use is the European Currency Unit (E.C.U.) – the central unit of the European Monetary System – which is defined in terms of a weighted average, or 'basket', of the national currencies of all nine E.E.C. member states. In January 1980 1 ecu = 61.87p.

Unit of Fertilizer. 1% of 1 cwt (i.e. 1.12 lb.) Thus a nitrogen FERTILIZER with 15% nitrogen in each cwt is said to contain 15 units of nitrogen. A COMPOUND FERTILIZER containing say 10, 15 and 20 units of nitrogen, phosphate and potash, respectively, contains 10%, 15% and 20% respectively of each per cwt (shown as 10:15:20). With metrication fertilizer application is increasingly expressed in kilograms per hectare.

United Kingdom Agricultural Supply Trade Association. The national trade association of the U.K. agricultural supply industry, representing both private and co-operative firms which market and distribute animal feedingstuffs, market and process home-grown cereals and pulses in the U.K. and abroad, process and distribute seeds and supply farmers with agrochemicals, fertilizers and other requisites. U.K.A.S.T.A. was formed in 1977 following the merger of the British Association of Grain, Seed, Feed and Agricultural Merchants (B.A.S.A.M.) and the Compound Animal Feedingstuffs Manufacturers' National Association (C.A.F.M.N.A.). The Association monitors U.K. and E.E.C. legislation, representing its members views at all levels, reports and advises on the implications of proposed changes and works closely with producers' organisations and other bodies, e.g. N.F.U., A.D.A.S., etc.

United Kingdom Seeds Executive. A statutory Committee of the INTERVENTION BOARD FOR AGRICULTURAL PRODUCE, established in 1973 to co-ordinate plant variety and seeds administration throughout the U.K. with particular reference to the adoption and operation of the E.E.C. plant variety and seeds regime. (◗ COMMON PRICES)

Unkindly. A term applied (*a*) to soils which lack natural fertility or are cold, late, wet, etc., and (*b*) to crops or animals which lack THRIFT or are in poor condition.

Untreated Milk. Milk sold for liquid consumption direct from farms as opposed to being sent to dairies, and without any heat treatment (◗ HEAT TREATED MILK). In the past producer-retailers and milk dealers were able, under licence, to sell such milk freely and it was identified by the green foil top used on the bottles (and thus also called Green Top Milk).

On 1 May 1980, new regulations came into operation which introduced 30 April 1985, as a common expiry date for all producer–retailer licences. From that date sales to shops, schools, hotels and other catering establishments and institutions will no longer be permitted.

Urban Fringe. The zone where the built-up area of a city, town or other urban area merges with the surrounding rural areas. Sometimes called the rural-urban interface or urban fringe. (◗ GREEN BELT)

'**Urea.** ($CO(NH_2)_2$.) A white, crystalline, organic compound which occurs in urine and was the first organic compound to be artificially synthesised. It is the most concentrated solid nitrogenous FERTILIZER, containing 46% nitrogen, and is sold in a granulated form. It decomposes rapidly giving off AMMONIA which is converted to NITRATE in the soil. Urea is very soluble and is often used in liquid fertilizers and in concentrated COMPOUND FERTILIZERS. It is also fed to stock as a source of non-protein nitrogen, mixed with a digestible CARBOHYDRATE source, e.g. MOLASSES, sugar-beet pulp, cereals, maize, SILAGE, etc.

V

Vaccinate. To inoculate an animal with a preparation containing dead or living, but weakened, antigens (e.g. bacteria or viruses) so that the animal produces antibodies (◗ ANTIBODY) in sufficient numbers to protect itself against the specific disease caused by the antigen concerned. The preparation is called a vaccine.

Vacuum Silage. SILAGE prepared in a large, sealed, polythene 'bag'. The 'bag' is evacuated of air by a pump through a valve in the sheeting. This removal of air and the build up of carbon dioxide reduces RESPIRATION losses and aids fermentation, producing high quality silage. (◗ SILO)

Vale. A broad, level valley.

Variegated. Of more than one colour, often in patches, e.g. the leaves of certain plants.

Variety. A group of animals or plants within a species or sub-species which share similar characteristics, but which differ in respect of those characteristics from other groups or varieties within the species. Used loosely to mean a variation of any kind within the species. Also called BREED, race or strain. (◗ PLANT BREEDING, IN-BREEDING)

Vascular Bundle. A longitudinal strand of tissue in a plant, consisting of PHLOEM and XYLEM, through which water and food materials are conducted.

Veal. The meat from calves slaughtered at less than 15 weeks old. (◗ SLINK CALF)

Vegetable. A general term applied to plants as distinct from animals. Normally applied to plants or parts of plants cultivated for human consumption or for stock-feeding, e.g. potatoes, carrots, cabbages, etc. Some fruit (e.g. tomatoes, cucumbers) and some seeds (e.g. peas, beans) are also considered as vegetables. Most vegetables contain useful amounts of Vitamin C and MINERALS. Root vegetables contain stored CARBOHYDRATES and seed vegetables are rich in PROTEIN.

Vegetables and salad crops have traditionally been grown as horticultural crops in the Thames Valley, North Kent, Vale of Evesham, South Lancashire and Humberside. Certain other areas have become prominent for particular crops due to favourable climatic conditions or soil type, e.g. the light soils of Norfolk and Cambridgeshire for carrots and onions, Lincolnshire for summer and autumn cauliflowers, etc.

Field-scale production of vegetables for the processing industries has become significant in recent years, particularly in the case of PEAS for canning, drying and quick-freezing. Peas for processing now represent about a quarter of the area of vegetables grown, other than potatoes. Most peas and dwarf BEANS are grown close to processing factories in the East of England and in Scotland.

Vegetable Classes ◗ FRUIT AND VEGETABLE CLASSES.

Vegetative Reproduction. Reproduction in plants not involving the flower or any sexual process, by the detachment of part of the plant (e.g. RHIZOME, TUBER, BULB, etc.), which then develops into a new plant. (◗ PROPAGATION)

Vein. 1. A tubular vessel which carries blood back to the heart in an animal.
2. A VASCULAR BUNDLE or rib forming part of the 'skeleton' of a leaf.

Venison. Meat derived from deer.

Verandah. A type of housing for poultry with a slatted or wire-mesh floor raised above the ground, allowing the faeces to fall through. Most designs incorporate a sleeping area containing nest boxes and perches and a covered run in which troughs for food and water are located. The walls of the run are of wire mesh although the side walls may sometimes be partly boarded for protection.

Verandah House Piggery ◗ PIGGERY.

Vermicide. A worm-killing agent.

Vernalisation. The subjection of plants to a period of cold conditions to cause flowering. Varieties of certain plants (e.g. cereals), if sown in spring, will fail to flower in the same year and continue vegetative growth. Such 'winter varieties' are sown in the autumn so that the cold stimulus received during the winter enables them to flower in the following year. By contrast 'spring varieties' sown in spring will flower in the same year, requiring no cold period. By vernalisation winter cereals are treated so that they acquire the properties of spring cereals and will flower in the summer following spring sowing. The process involves the slight moistening of the seeds to stimulate GERMINATION and the exposure of the emerging RADICLES to a temperature just above freezing for a few weeks.

Vet. An abbreviation for VETERINARY SURGEON.

Vetch. A leguminous plant, *Vicia sativa*, with a slender, square, climbing stem, tendrilled pinnate leaves and blue or purple flowers. It is sown with rye to provide spring KEEP for sheep and with oats, or beans and oats, for SILAGE. It may also be made into HAY. Sometimes it is grown as a pure crop following cereals to produce seed. Winter and spring varieties are available. Also called tares.

Veterinary Service. A division of the AGRICULTURAL DEVELOPMENT AND ADVISORY SERVICE with many functions including the control and eradication of NOTIFIABLE DISEASES, national disease control programmes, import and export of animals and animal products, investigation of new diseases, meat hygiene, animal welfare, licensing of medicines and liaison with private veterinary surgeons in the investigation of disease and production problems in livestock. There are Regional Veterinary Officers throughout the U.K. assisted by staff in the field and in Veterinary Investigation Centres.

Veterinary Surgeon. A person qualified in veterinary science (the study of animal diseases), and able to diagnose diseases, prescribe medicines and carry out surgery. Only veterinary surgeons and practitioners registered with the Royal College of Veterinary Surgeons are legally able to practice veterinary surgery.

Vine. A climbing or trailing stem, e.g. that of the grape.

Vineyard. A plantation of grapes

Vining Peas. PEAS grown for picking green, before they are fully ripe.

Vinquish ▶ PINING.

Virgin Land. Land which has never been cultivated.

Virus. One of a group of sub-microscopic, self-reproducing, proteinous agents, which infect plants and animals causing disease (e.g. MOSAIC DISEASES, FOOT AND MOUTH DISEASE). They are transmitted between plants mainly by insects, particularly APHIDS, and by EELWORMS, and between animals by insects, contact and the inhalation of mucus droplets expelled by coughing and sneezing.

Virus Yellows. A mosaic virus disease of SUGAR BEET or MANGEL plants due to a virus complex, *Beet Milk Yellow Virus* (B.M.Y.V.) and *Beet Yellow Virus* (B.Y.V.) spread by certain APHIDS. The disease is characterised by the outer and middle leaves turning yellow from the tips and upper margins, and becoming thicker and brittle. Significant reductions in the sugar content and yield of crops are caused.

Viscera. The abdominal organs of animals, e.g. liver, kidneys, spleen, etc.

Vitamins. A class of organic substances required by animals in small amounts, essential to METABOLISM and thought to act together with ENZYMES. Animals are able to synthesise several vitamins in their bodies but most must be supplied in their food. Vitamins A, D, E and K are fat-soluble and are stored in the FAT compounds of the body. Vitamins B and C are water-soluble and cannot be stored and are regularly required in the diet. The known vitamins are listed in the table *The Physiological Functions and Chief Sources of the Vitamins.* All farm animals are able to synthesise sufficient vitamin C in their bodies. RUMINANTS and horses contain gut bacteria capable of producing sufficient supplies of the B group vitamins. Livestock kept outdoors are able during summer to produce adequate supplies of vitamin D through the irradiation of their coats and skin.

Deficiency in particular vitamins can cause ill-health and

livestock diets usually include synthetic vitamin supplements. The latter are normally added to CONCENTRATE rations in balanced amounts according to the animals' needs.

The Physiological Functions and Chief Sources of the Vitamins.

Name	Bodily functions for which required	Dietary sources
FAT-SOLUBLE		
Vitamin A	Vision, health of skin and walls of respiratory and urogenital tracts, growth and reproduction	Some fish liver oils. Carotene (pro-vitamin A), fresh green leaves and carrots
Vitamin D	Absorption and deposition of bone, growth and reproduction	Some fish liver oils, irradiated yeast
Vitamin E	Normal reproduction. Muscle health	Wheatgerm, whole cereal grains, some vegetable oils
Vitamin K	Clotting of blood	Liver, egg, fresh green leaves
WATER-SOLUBLE		
Vitamin B group		
B_1 (thiamine)	Carbohydrate metabolism, functioning of central nervous system	Dried yeast, soya bean meal, liver, heart
B_2 (riboflavin)	Oxidation processes in the body	Dried yeast, dried milk, liver, heart
Nicotinic acid	Oxidation processes in the body	Dried yeast, dried milk, liver, heart, wheat and barley
B_6 (pyridoxine)	Amino acid metabolism. Of greatest importance in meat eating species	Cereals, dried yeast, dried milk, liver, egg yolk
Pantothenic acid	Oxidation of fats	Dried liver, dried yeast, dried milk, egg yolk, liver
Biotin	Amino acid metabolism	Wide distribution
Choline	Fat formation, amino acid metabolism	Most foods contain reasonable amounts
B_{12} (cobalamin)	Amino acid metabolism. Normal growth, reproduction and blood formation	Cheese, dried milk, liver, kidney
Vitamin C	Normal digestion and carbohydrate metabolism	Liver, green leaves, some fruits

Source: *An Introduction to Animal Husbandry* by J.O.L. King. Blackwell Scientific Publications 1978. Reproduced with permission.

Vitriol ◗ SULPHURIC ACID.

Viviparous. A descriptive term for animals which give birth to live offspring. (◗ OVIPAROUS)

Volunteer. A plant which grows spontaneously in a crop of different plants, from seed present in the soil.

W

Wages Committee, Wages Order ♦ AGRICULTURAL WAGES
BOARD.

Walnut Comb. An additional fleshy COMB on the heads of some
breeds of poultry, resembling a half walnut, also called a
strawberry comb.

War Agricultural Executive Committees. County-based committees established during the First and Second World Wars with
legal powers to enforce full and efficient use of all land suitable
for agriculture. The Committees had powers to direct farmers
and landowners to plough-up and crop their land, and had
facilities to supply services of labour (♦ LAND ARMY), machinery, etc., which were necessary to comply with their directions,
but which farmers could not always provide for themselves.
They had power to requisition and supervise land under the
Defence Regulations. The Committees were also concerned
with promoting technical efficiency, directed towards the rapid
increase of food production. After the Second World War, due
to continued food difficulties, they continued to operate until
the establishment of COUNTY AGRICULTURAL EXECUTIVE
COMMITTEES in 1947.

Warble Fly. One of two species of fly, *Hypoderma bovis* and *H.
lineata,* which cause great annoyance to cattle in the spring and
summer when they lay their eggs on the underparts and legs.
Cattle often rush about (known as gadding) to avoid attack,
sometimes injuring themselves, and there may be a loss of
condition and reduced milk yield. On hatching, the eggs burrow into the skin causing sores at the points of entry. The
maggots migrate through the body and reach the back in
January–June of the following year, where they cause swellings
(warbles) about the size of a small walnut, sometimes pus-filled
and each with a perforation or breathing hole. On maturity the
maggots escape through the hide and fall to the ground to
pupate (♦ PUPA). During migration in the body, the maggots

466

cause damage to the tissues of the gullet and back, causing severe irritation and inflammation and loss of health and condition. The value of the hides is also reduced due to the perforations. The condition is most serious in young animals and death sometimes occurs. Treatment is by the use of 'pour-on' insecticides applied to the back which are absorbed and kill the migrating maggots.

Ware Potatoes. The largest potatoes in a crop sold for human consumption, as distinct from CHATS and SEED POTATOES. They are dressed by registered producers and licensed Ware Merchants to a standard quality set by the POTATO MARKET-ING BOARD (P.M.B.). Ware potatoes must (*a*) be virtually clean, sound and not shrivelled or wizened, (*b*) not contain tubers which pass (without pressure) through a 40 mm horizontal square mesh or stand on an 80 mm mesh, or exceed 165 mm in length, (*c*) be free from a range of specified diseases, growth faults and damage, and (*d*) be practically free from adhering growing shoots.

Registered producers and Licensed Ware Merchants may also sell potatoes for human consumption without a P.M.B. permit which are classified as BAKERS or MIDS. The P.M.B. has also formulated a higher or 'Premium Grade' which it recommends producers and merchants to adopt.

Warp. 1. To lay eggs or to give birth, especially prematurely. Also to miscarry, particularly in cattle and sheep.
2. Alluvial sediment. (◗ WARPING)

Warping. 1. The natural build-up of a saltmarsh by the deposition of silt.
2. The controlled flooding of agricultural land adjoining a river with silt-charged river water so as to build up fertile soil.

Warren. The burrows of a colony of RABBITS.

Wart Disease. A NOTIFIABLE DISEASE of the POTATO, due to the fungus *Synchytrium endobioticum*, which causes the development of rough-surfaced, dark brown or black, tumorous growths on the tubers which, under favourable conditions, release motile spores which are able to infect young tubers, but which can persist indefinitely in the soil. Most varieties of potato grown nowadays are immune.

467

Waste Food. Defined by the Diseases of Animals (Waste Food) Order, 1973, as 'any meat, bones, blood, offal, or other part of the carcase of any livestock, or of any poultry, or product derived therefrom, or hatchery waste or eggs or egg shells; or any broken or waste foodstuffs (including table or kitchen refuse, scraps or waste), which contain or have been in contact with any meat, bones, blood, offal, or with any other part of the carcase of any livestock or of any poultry.' (◗ SWILL)

Water. Water is vital to all forms of farming. In 1980, in its report, 'Water For Agriculture; Future Needs', the ADVISORY COUN-CIL FOR AGRICULTURE AND HORTICULTURE estimated the annual intake of water by agriculture to be 65 000 million gallons (300 Mm^3), 60% derived from public mains and the remainder from private abstractions. Livestock husbandry was found to be the largest consumer accounting for 35 000 million gallons (160 Mm^3), followed by IRRIGATION using 20 000 million gallons (90 Mm^3) in a dry year. Average daily consumption by livestock is as follows:

	Daily Requirements
In pig sows	at least 4.55 litres
Suckling sows	up to 22.75 litres
Fattening pigs	0.9–1.35 litres water per kg
Sheep	approx. 4.55 litres
Fattening cattle	27–33 litres in winter
	33–36 litres in summer
Milking cows	4.55 litres per 50 kg body weight
	+ 4.55 litres per litre milk produced

Water Meadow ◗ MEADOW.

Water Table. The surface at which GROUND WATER has settled in the ground and below which fissures and pores in the rock strata or soil are saturated with water. This surface is uneven and its position varies according to the amount of rain which has fallen. Where a water table rises to intersect the ground surface a spring arises.

Water-holding Capacity. A general term for the ability of a soil to retain water. (◗ FIELD CAPACITY, PERMANENT WILTING POINT, READILY AVAILABLE MOISTURE and SOIL MOIS-TURE DEFICIT)

Waterlogged Soil. A soil which is saturated with water so that the PORE SPACE is completely filled with water, and conditions are unsuitable for plant growth since the roots are unable to obtain oxygen for RESPIRATION.

Watershed. The ridge of high ground separating the CATCHMENT areas of two distinct river systems.

Wattle. 1. The coloured fleshy skin beneath the throat of some birds. Also the DODDLE of a goat.
2. A twig or flexible stick, woven with others to make a framework for fences, roofs, hurdles, etc.

Waymark. To mark the course of a RIGHT OF WAY, path or trail at points along it to enable users to use it accurately.

Weak Wheat ▶ STRONG WHEAT.

Wean. To separate a young animal from its mother so that it no longer has access to its mother's milk and must rely on other food supplied.

Weaner. A piglet which has been weaned (nowadays usually at 3–5 weeks) but has not reached the age of about 10 weeks when it becomes a feeding pig or 'fattener' for a variety of purposes. (▶ PIG)

Weathering. The process by which rocks disintegrate and decompose, eventually producing soil particles, by exposure to the physical and chemical effects of atmospheric agents, e.g. rain water, frost, wind, temperature changes, plants and animals. Soils when formed, continue to degrade under the influence of such agents.

Weatings. A term now given to all wheat offals, other than BRAN, containing not more than 6% FIBRE, and used as a FEEDINGSTUFF. At one time such offals were grouped in three main categories; pollards, coarse middlings or sharps, and fine middlings or thirds.

Web Punching. The punching of holes in the webs between the toes of FOWLS and other POULTRY for identification purposes.

Wedder ▶ WHETHER.

Wedge Shape. The CONFORMATION regarded as ideal for dairy cattle, the hindquarters being broader and deeper than the forequarters.

Weeds. A term applied loosely to any plants growing where they are not wanted by man, and specifically to unwanted plants growing in cultivated land where they compete with crops for light, water and nutrients, reducing yields, hindering harvesting, and contaminating or tainting produce. They may also harbour pests and diseases which can spread to the crop. Weeds characteristically produce plentiful seed and rapidly colonise open ground. Most are ANNUALS or PERENNIALS, the former growing quickly from seeds and troublesome mostly on arable land, the latter reproducing by seeds and vegetative organs such as RHIZOMES (e.g. COUCH GRASS) and troublesome in arable and grassland. Weeds may be controlled by good ROTATIONS and tillage treatments (◗ BARE FALLOW, BASTARD FALLOW, SPRING CLEANING, STUBBLE CLEANING) or by the use of HERBICIDES. Some weeds were introduced from abroad in contaminated grain and other seeds, but this is now less likely since the introduction of SEED CERTIFICATION. (◗ INJURIOUS WEEDS)

Weedkiller. A term for a HERBICIDE.

Weeds Act, 1959 ◗ INJURIOUS WEEDS.

Weevils. A large family of small beetles (Rhynchophora) characterised by a beak-like prolongation of the head (the rostrum) and elbowed antennae, which cause damage, either as larvae or adults, to fruit, nuts, grain (e.g. the Grain Weevil, *Sitophilus granarius*) or trees. Pea and Bean Weevils (*Sitona sp.*) are unrelated (family, Bruchidae) and, as adults, cut notches in the leaves of the plants.

Well-sprung. A descriptive term for an animal with well-arched ribs so that it is not flat-sided.

Welsh. A white, lop-eared breed of pig, in appearance very similar to the LANDRACE. It is sometimes used instead of the latter in cross-breeding programmes (*see p. 327*).

Welsh Black. A dual-purpose breed of cattle. The Welsh Black Cattle Society was formed when the old North Wales or Anglesey and the Castlemartin types were amalgamated in 1904. The breed is black, sometimes with white hairs on the udder. Some types with white markings and occasionally red types are seen. The horns sweep outward, turning forward, and are black-tipped. It is a hardy and thrifty breed often used for SINGLE SUCKLING on exposed upland pastures. MILK YIELDS are low and carcases have a high proportion of lean meat (*see p. 87*).

Welsh Half-bred. A cross-bred type of sheep derived from a BORDER LEICESTER ram mated with a WELSH MOUNTAIN ewe.

Welsh Mountain. A hardy mountain breed of sheep including a number of distinct varieties (e.g. Aberystwyth, Hardy, South Black and Speckleface), the most common being the Aberystwyth Welsh Mountain. It is the smallest British breed characterised by a grey-white face and legs, sometimes with tan markings, and a fleece of fine, soft wool, often containing red hairs (kemp). Rams only are horned. The breed is extensively found on the Welsh mountains, tolerating high rainfall conditions and sparse, poor quality grazing. DRAFT ewes crossed with BORDER LEICESTER rams produce the Welsh Half-bred used in the lowlands (*see p. 393*).

Welsummer. A light laying breed of fowl, brown in colour, with yellow legs, a single comb, and producing dark brown eggs.

Wensleydale. A large, hardy, long-woolled breed of sheep which originated in North Yorkshire, with a distinctive bluish-grey face, ears and legs. Some lambs are born black. The fleece is silky, finely purled, bright and lustrous. The breed has a restricted distribution and is mainly found in its native area. It has been popularly used for crossing with hill breeds to produce cross-bred ewes (e.g. the MASHAM), although it is now increasing in popularity as a sire of heavyweight lambs.

Went. A single journey across a field in cultivating, the return journey being known as a bout.

Wessex ♦ BRITISH SADDLEBACK.

471

Western ♦ WILTSHIRE HORN.

Westerwolds Ryegrass. An annual type of ITALIAN RYEGRASS (*Lolium multiflorum* var *Westerwoldicum*) which, when sown in spring or summer, rapidly sets seed and flowers. It is usefully used to provide HAY in the year of seeding. DIPLOID and TETRAPLOID varieties are available. Sometimes spelt Westerwolths.

Wet Feeding. The feeding to livestock of MEAL mixed with water in the form of a porridge. (♦ DRY FEEDING)

Wet Fencing. Water-filled ditches and dykes in MARSH areas which livestock will not or cannot cross, and which therefore act like fences in retaining animals.

Wet Mash. MEAL fed to livestock after water has been added to reduce it to a crumbly state.

Wet Spell. A period (in the U.K.) of at least 15 consecutive days on each of which at least 1 mm (0.04 in) of rain has fallen. Not an internationally accepted definition. (♦ DRY SPELL, ABSOLUTE DROUGHT and PARTIAL DROUGHT)

Wether. A castrated adult male sheep. Also called wedder. (♦ BELL WETHER and SHEEP NAMES)

Wetting Agent. A substance which lowers the surface tension of water so that it spreads out rather than remains in droplet form when sprayed. Wetting agents are added to detergents to increase their efficiency and to sprays of insecticides, fungicides, herbicides or mineral sprays to improve the cover achieved when applied to leaf surfaces.

Whale Meat Meal. A FEEDINGSTUFF with a high PROTEIN content, consisting mainly of whale flesh, processed in a similar manner to MEAT MEAL.

Wheat. A cereal (*Triticum vulgare*) grown for its grain, used mainly for flour making (♦ MILLING WHEAT). Many varieties are available, the grains of which differ in size, shape and colour (♦ STRONG WHEAT), and which exhibit different characteristics in respect of time of ripening, disease resistance, hardiness and adaptation to soil type and fertility. They fall into two groups, autumn- and winter-sown varieties (◆ VERNALISATION). The flowering head of wheat is a four-sided SPIKE bearing tightly-set spikelets, the PALEA of each ending in a point, and in some varieties (bearded wheats) extending into long bristly AWN so that they resemble BARLEY. Wheat is mainly grown in the drier eastern half of Britain, doing best on well-drained, fertile soils. It is usually grown following a LEY or after a root or legume break in a ROTATION (*see p. 95*).

Wheat Bulb Fly. A two-winged fly (*Hylemyia coarctata*), the larvae of which bore into the central shoots of wheat in the spring (and also winter barley and rye) causing them to turn yellow, and die if TILLERS have not developed. Control is by SEED DRESSING using chlorfenvinphos and carbophenothion (ORGANOPHO-SPHOROUS COMPOUNDS) and sometimes by spray treatment in the spring.

Whey. The residue from milk after the CASEIN and most of the fat has been removed as CURD in CHEESE making. Dried whey is added to processed cheese. Most whey is fed to pigs in liquid form. Its composition is as follows:

	%
Water	93.04
Total solids	6.96
Fat	0.36
Proteins	0.84
Lactose, salts, etc.	5.67

Whip. 1. A long twig or slender branch.
2. A young unpruned fruit tree.

Whip Graft. A GRAFT in which a tongue cut on the SCION is fitted into a slit cut slopingly in the STOCK. Also called a tongue graft.

Whipple-tree, Whippance. The crossbar of a horse-drawn implement or cart, pivoted in the middle, to which the TRACES are fixed. Also called bullet, bodkin, swing or swingle-tree.

Whitbred Shorthorn. A white, fast-growing variety of the BEEF SHORTHORN, mainly found in the English-Scottish border country, and crossed with GALLOWAY cows, producing BLUE-GREY suckler cows.

White Clover. A perennial type of CLOVER (*Trifolium repens*), characterised by a creeping growth habit which knits the sward together, keeps out weeds and unproductive grasses, and minimises the destructive effect of poaching by livestock. Several varieties are available, all of which are persistent in the sward but are less bulky than RED CLOVER. White clovers tolerate grazing well and are used in all long LEYS. (♦ DUTCH WHITE CLOVER and WILD WHITE CLOVER)

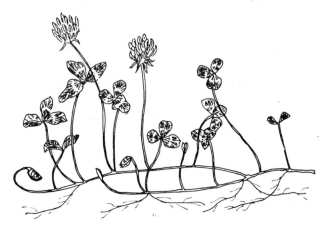

White Clover.

White Grubs ♦ COCKCHAFER.

White Mustard. A tall, branched, yellow-flowered, cruciferous plant (*Sinapsis alba*), producing long curved pods. It is grown as a GREEN MANURE crop and sometimes for folding (♦ FOLD) sheep on.

White Straw Crops. A term used for wheat, barley, oats and rye.

Whitefaced Woodland. A horned, mountain breed of sheep mainly found in the Pennines where Derbyshire borders with West and South Yorkshire. It has a white face and pinkish nostrils. Rams are considerably larger than ewes and possess a strong, muscular tail. They are crossed with ewes of other blackfaced, horned mountain breeds to improve the size of the latter, producing cross-bred ewes characterised by speckled faces and possessing considerable HYBRID VIGOUR. Also called Penistone.

Whiteheads ♦ TAKE-ALL.

Whole. A term used to mean complete, e.g. (*a*) whole animals, which are entire and have not been castrated, (*b*) whole corn, which is unmilled, or (*c*) whole milk, which has not been separated. (♦ SEPARATED MILK)

Whole Work. Ploughing which leaves the furrow unbroken. (♦ BROKEN WORK)

Whorl. A ring of plant organs (e.g. leaves, petals, etc.) around a stem, all attached to it at the same level.

Wick. 1. A dialectal term sometimes used for a farm.
2. A term for a MAGGOT.

Wild Oats. A tall, annual weed (*Avena fatua*) related to cultivated oats, but distinguished by having long, yellow or brown, hairy grains, each bearing a strong, twisted AWN. Wild oat infestations in cereal crops are a serious problem, increasing competition and reducing grain yield, and making harvesting, cleaning and drying more difficult and costly. They also lower the value of the crop. The National Wild Oat Advisory Programme, before it was officially wound up in 1978, estimated that 95% of English farms had wild oats. An Industrial Wild Oat Group, consisting of representatives of certain agrochemical companies, is continuing the work of the Programme to promote awareness of the problem and the need for control measures.

Wild White Clover. A small-leaved, vigorous, truly perennial form of WHITE CLOVER (*Trifolium repens*), the STOLONS of which spread close to the ground keeping out weeds. It is drought resistant, persistent, and remains continuously palatable to grazing animals. It is an important constituent of long LEYS, and most seed is produced in Kent.

Wilting. A condition of plants due to loss of cell turgour as a result of water loss, characterised by the leaves and young stems drooping and becoming limp. Wilting can result from lack of water supply when TRANSPIRATION exceeds root intake, and can also be induced by certain fungal diseases (wilt diseases). In HAYMAKING and SILAGE making, plants are often crimped or lacerated to increase wilting in order to reduce moisture content.

Wilting Point. The point at which the water content of a soil reaches such a level on drying out that it is all firmly held by the soil and is unavailable to plant roots, so that the plants wilt permanently and die.

Wiltshire Bacon Pig ♦ PIG.

Wiltshire Cure. The curing of traditional British bacon by injecting prepared carcases (sides) with brine and stacking them in brine-filled tanks for several days, and then smoking them over smouldering wood chippings.

Wiltshire Horn. A distinctive, white-faced breed of sheep, characterised by a very short fleece of thickly matted hair, and curled horns in both sexes. It is a robust breed producing active, rapid-growing lambs. It is mainly found in Northamptonshire and in North Wales and Anglesey where rams are used as meat sires for crossing with draft ewes of the WELSH MOUNTAIN breed. Also called Western.

Wiltshire Side. A SIDE of bacon prepared by cutting a pig carcase along the backbone and removing the head, feet, tail, shoulder and vertebrae.

Windbreak. A screen, particularly a line or clump of trees or a piece of woodland, providing protection from the wind to an area of land growing crops or livestock. (♦ SHELTER BELT)

Windgall. A puffy swelling in the region of a horse's FETLOCK.

Windrow. A row of HAY (or other cut crops), consisting of two or more SWATHS combined into one in preparation for picking up. (♦ WINDROW PICK-UP)

Windrow Pick-up. A mechanism attached to the front of various harvesting machines (e.g. COMBINE HARVESTER, pea harvester) to pick up a crop left in a WINDROW after previous cutting by a MOWER.

Wine Lake. A loose descriptive term for the E.E.C. structural surplus of wine, purchased as an INTERVENTION STOCK under the COMMON AGRICULTURAL POLICY. (♦ MOUNTAIN)

Winnow. To separate the CHAFF from the GRAIN of threshed corn by blowing it away in a draught from a fan.

Winter Feeding. The feeding regime of livestock, mainly cattle, in the winter months, involving the feeding of HAY, SILAGE, other bulk foods and CONCENTRATES. (♦ SUMMER FEEDING)

Winter Green. A term applied to plants which remain green and fresh through the winter.

Winter Proud ♦ PROUD.

Winter Varieties ♦ VERNALISATION.

Winter Wash. A spray applied to fruit trees and bushes during the winter months which soaks into crevices in the bark and kills the overwintering stages of pests and diseases. Usually a tar oil or similar organic compound.

Wirework. A term applied to the system of wires supported on a 'grid' of poles, erected in hop gardens (♦ HOP), to which 'stringing wires' are attached, up which the hop bines grow. Various designs of wirework are in use.

Wireworms. The smooth, thin, yellow larvae of the CLICK BEETLE which inhabit grassland and attack various crops, particularly cereals which are usually sown following a LEY. Wireworm attack is greatest in spring and autumn when they eat into plants just below the soil surface, causing foliage to turn yellow before the plants die. Wireworms take 4–5 years to mature after which they pupate (♦ PUPA) in the soil. Control is by the use of SEED DRESSINGS containing B.H.C.

Withers. The ridge between the shoulder blades of a horse, its highest part.

Withy ♦ OSIER.

Wold(s). An open, rolling, upland tract of country, e.g. the CHALK hills of Yorkshire and Lincolnshire, and the LIMESTONE hills of the East Midlands and Cotswolds.

Wooden Tongue ♦ ACTINOBACILLOSIS.

Wool. A soft, modified form of hair in which the fibres are shorter and curled, and which has an imbricated surface of minute, overlapping, interlocking scales which hold the wool fibres together. Wool is found on various mammals but the term is usually restricted to the fleeces of sheep. The annual wool yield per sheep varies from as low as 1 kg for Welsh short-woolled breeds to 7 or 8 kg for certain long-woolled breeds. All wool in Britain is sold by registered producers through the BRITISH WOOL MARKETING BOARD. About 20 million sheep are shorn annually in the U.K. (the remaining 10 million mostly being sold for meat before growing a full fleece), producing a national 'clip' of 30–35 million kg of wool. The average clip per farm is about 400 kg.

Wool Ball. A mass of wool, tangled into ball, found in the first or fourth stomach (◗ RUMINANT) of lambs. Wool is swallowed from the mothers' fleece when suckling and accumulates, sometimes blocking the stomach outlet and causing death.

Wool Grades. In Britain each producer's clip is graded by merchants into one or other of the 300-odd grades of British wool for which the BRITISH WOOL MARKETING BOARD has laid down detailed specifications. The latter are mainly concerned with the following characteristics:

(a) *Degree of Fineness.* The diameter of the wool fibre influences to a large extent the use to which the wool can be put and also the length of yarn that can be spun from a given weight of wool (◗ BRADFORD WORSTED COUNT). Users and manufacturers normally use a series of 'quality numbers' to indicate fibre diameter. These numbers in ascending order of fineness, are 28s (the thickest, coarsest fibres), 32s, 36s, 40s, 44s, 46s, 48s, 50s, 56s, 58s, 60s, 64s, 66s, 70s, 90s, 100s (the very thinnest, finest fibres). Most British wools are within the range 28s to 58s. Quality numbers are not based on any particular unit of measurement; they are standards handed down from generation to generation of woolmen and can be learned only by practical experience and handling of wools.

(b) *Length of Staple.* This is the measurement of the unstretched staple from tip to base.

(c) *Handle of Wool.* The softness or harshness of the wool when handled.

(d) *Degree of Lustre.* The amount of gloss or sheen visible on the fibres. Degree of lustre varies greatly between different types. A bright lustre is an asset for certain manufacturing processes.

(e) *Degree of Springiness*. The extent to which a handful of wool will expand again after being compressed and released. Springiness is a valuable quality of most British wools.

(f) *Colour*. The nearness of the wool to white (or black in the case of black fleeces).

(g) *Strength*. The ability of the staple to resist breakage during manufacture.. The word 'sound' is used to describe wool of satisfactory strength rather than 'strong' which, in wool terminology, refers to thick or coarse fibres. The opposite to a sound wool is a 'tender' wool. Tender wools may have a break in all the fibres at one point in the staple as a result of illness or drought.

The main groups of wool grades in the U.K. are:

Down wools	Cheviot, Radnor and Welsh
Fine wools	Swaledale, Blackface and
Medium wools	Herdwick
Masham, Cross and Leicester	Lamb wools
Lustre wools	Northern Ireland grades

Wool Marketing ◗ BRITISH WOOL MARKETING BOARD.

Wool Oddments. Various types of wool pieces detached or separately clipped from fleeces and purchased by the BRITISH WOOL MARKETING BOARD from registered producers, including BRANDS, BROKES, CLAGS, DAGGINGS, FALLEN WOOL, LOCKS and TAILINGS.

Wool Rot. A condition of sheep due to a fungus (*Dermatophilus dermatonomus*) which causes irritation to the skin in wet weather and the development of hard, yellowish scabs. The sores heal and the growing wool carries the scabs away in the fleece. Also called Lumpy Wool.

Wool Sheet. An oblong sack, generally measuring about 2 m (7 ft) long and about 1.4 m (4 ft 6 in.) high, in which fleeces are packed, after which it is sown up for delivery to a wool merchant. At one time they were mainly made of jute but polypropylene is now becoming more common. In some areas, such as South-East England, fleeces are predominantly delivered in bulk, loosely piled in lorries.

Workers ◗ BEE.

Worms. A common name for a range of elongate, cylindrical, legless invertebrates, many of which are parasitic. (♦ EARTH-WORM, EELWORMS, FLATWORMS, ROUNDWORMS and TAPEWORMS)

Worrying. The tormenting or harassing of livestock, usually by uncontrolled dogs. This often involves the biting and tearing of flesh. Sheep, which become particularly anxious when worried, are frequently killed by dogs, and pregnant ewes often abort. Farmers are entitled to shoot dogs found worrying livestock.

Wrees. A term used in South-West Scotland for sheep gathering pens.

Wrest. An old term for the MOULDBOARD of a plough. (♦ TURNWREST PLOUGH)

Wry Tail. A term applied to poultry when the tail is carried turned to one side.

Wurzel ♦ MANGEL.

Wyandotte. A heavy, dual-purpose breed of FOWL of American origin, variously coloured including white, silver or gold laced, blue, buff, partridge or barred, etc., with yellow legs, a rose comb and producing light to dark brown eggs.

X

Xanthophyll. A carotenoid plant pigment (◗ CAROTENE), yellowish in colour, present in CHLOROPLASTS and other plant parts where CHLOROPHYLL is absent. It assists in PHOTO-SYNTHESIS.

Xylem. Vascular tissue in plants which carries water and dissolved mineral salts from the roots, where they are taken in, to the various parts of the plant. It also provides mechanical support to plants, the cells being impregnated with LIGNIN, and is located to the inside of the CAMBIUM. Secondary xylem comprises the bulk of the woody tissue in woody plants and is laid down by the process of SECONDARY THICKENING.

Y

Yean ♦ EAN.

Yearling. An animal between 1 and 2 years of age.

Yeast. A single-celled ASCOMYCETE fungus, economically important in being capable of FERMENTATION and an important source of VITAMINS.

Yeld. Dry or barren, not producing MILK.

Yellow Rust. A fungal disease (*Puccinia striiformis*) of barley and wheat, forming deep yellow pustules in closely parallel lines on the infected parts of leaves and stems. The disease commences on VOLUNTEER plants and spreads to both spring and winter crops, particularly when conditions are cool and moist. Spores are able to survive the winter. (♦ RUSTS)

Yellows ♦ VIRUS YELLOWS.

Yelm. A bundle of straw used for thatching.

Yelt ♦ GILT.

Yeoman Farmer. One of a class of small freeholders who emerged as a result of the enclosure movement (♦ AGRICULTURAL REVOLUTION) which gave them the ability to cultivate and improve their land throughout the year. This ability enabled the improvements of the 'revolution' to be introduced effectively into agriculture. The term is also applied to any small farmer or countryman. During the wars of the French Revolution, county-based volunteer cavalry forces were raised, under the lords-lieutenant, from yeomen or smaller farmers, the men providing their own horses and uniforms, and known as Yeomanry.

Yilt ♦ GILT.

Yoke. 1. A frame joining a pair of draught oxen. Also any pair of animals working together.
2. A collar placed round an animal's neck to keep it in a stall.
3. ♦ YOLK.

482

Yolk. 1. The food store containing PROTEIN and FAT granules in the eggs of most animals. The yellow central part of the eggs of birds.
2. The grease in wool. Also called euk, suint or yoke.

Yorkshire. 1. ⧫ DURHAM.
2. ⧫ LARGE WHITE.

Yorkshire Boarding. A type of boarding used to clad the walls of livestock buildings, particularly those housing cattle, allowing ventilation. It consists of vertical wooden slats set apart by gaps of varying width, according to the exposure of the site (*see p. 131*).

Yorkshire Fog. A perennial, soft, downy grass (*Holcus lanatus*), common in poor pastures and an indicator of LIME deficiency in the soil. Also called Meadow Soft Grass.

Young Farmers' Clubs. Clubs which have the aim of encouraging an interest in agriculture and an appreciation of country life. They also provide training in the arts of citizenship and develop abilities to serve the community. In England and Wales clubs are affiliated to a National Federation of Young Farmers' Clubs. There is also a Scottish Association and an Ulster Federation. The clubs have a total U.K. membership of approximately 60 000, aged between 10 and 26 years.

Z

Zebu ♦ CATTLE.

Zero Grazing. A GRAZING SYSTEM in which grass (or other FODDER) is cut daily and then taken to cattle kept in a yard or in a small exercise paddock near the buildings. The system obviates problems of poaching (particularly on heavy land), fouling, over-eating and wastage due to selective grazing. More grass can be provided from a given area by the system than by grazing. Also called mechanical grazing or soilage.

Zig-Zag Harrow. A type of HARROW in which the frames are zig-zag in shape, bearing short pointed tines, spaced so that each covers different ground. Used usually in the final stages of SEED BED preparation, and also to remove tractor and DRILL wheel marks after sowing and to cover the seeds fully.

Zinc. (Zn.) A metallic TRACE ELEMENT required by plants and animals for the correct functioning of ENZYMES.

Zygote. A fertilised OVUM before it starts to divide and form an EMBRYO.

Zymogenous Flora. Organisms found in the soil in large numbers following the addition of readily decomposable ORGANIC MATTER.

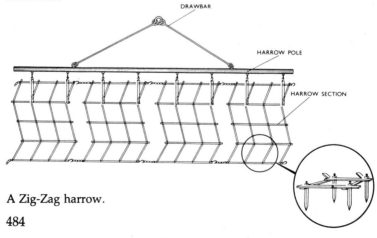

A Zig-Zag harrow.

Abbreviations, Acronyms and Initials

a.	Acre
A.A.B.	Association of Applied Biologists
A.A.C.	Agricultural Advisory Council for England and Wales
A.A.E.S.	Association of Agricultural Education Staffs
A.A.P.P.	Average All Pigs Price
A.B.A.O.	Association of British Abattoir Owners
A.B.B.B.C.P.	Association of Butter Blenders and Butter and Cheese Packers
A.B.I.M.	Association of British Insecticide Manufacturers
A.B.M.A.C.	Association of British Manufacturers of Agricultural Chemicals
A.B.M.M.P.	Association of British Manufacturers of Milk Powder
A.B.P.I.	Association of the British Pharmaceutical Industry
A.B.R.O.	Animal Breeding Research Association
A.B.V.	Apparent Biological Value
ac.	Acre
A.C.A.	1. Accession Compensatory Amount
	2. Advisory Council for Agriculture
	3. Agricultural Cooperative Association
A.C.A.S.	1. Agricultural Chemicals Approval Scheme
	2. Arbitration, Conciliation and Advisory Service
A.C.C.	1. Agricultural Credit Corporation
	2. Agricultural Correspondence College
	3. Association of County Councils
A.C.M.S.	Agricultural Cooperation and Marketing Services
A.C.P.	Association of Cheese Processors
A.C.T.	Agricultural Central Trading Ltd
A.C.W.W.	Associated Country Women of the World
A.D.A.	Association of Drainage Authorities
A.D.A.S.	Agricultural Development and Advisory Service
A.D.C.	Association of District Councils
A.D.E.	Apparent Digestible Energy
A.D.H.A.C.	Agricultural Dwelling House Advisory Committee
A.D.R.I.	Animal Disease Research Institute
A.D.S.	Animal Defence Society
A.E.A.	1. Agricultural Education Association
	2. Agricultural Engineers' Association
A.E.C.	1. Agricultural Education Council
	2. Association Européene pour la Coopération

A.E.R.I.	Agricultural Economics Research Institute
A.E.S.	Agricultural Economics Society
A.F.M.M.	Association of Fish Meat Manufacturers
A.G.C.D.	Association of Green Crop Driers
A.G.P.D.	Association Générale des Producteurs de Blé et autres Céréales (French, Association of Wheat and other Cereal Producers).
agr., agric.	Agriculture, Agricultural
A.H.D.	Animal Health Division
A.H.Q.S.	Asymmetric Hindquarters Syndrome (condition of pigs)
A.H.T.	Animal Health Trust
A.I.	Artificial Insemination.
a.i.	1. Artificial Insemination
	2. Active Ingredient
A.I.D.	Association Internationale de Développement
A.I.V.	A silage making process (see dictionary text)
Al	Aluminium
A.L.A.M.	Association of Lecturers in Agricultural Machinery
A.L.P.C.	Agricultural Lime Producers' Council
A.L.T.	Agricultural Land Tribunal
A.M.A.	Association of Metropolitan Authorities
A.M.C.	Agricultural Mortgage Corporation Ltd
A.M.D.	Amalgamated Master Dairymen
A.M.I.	Association of Meat Inspectors
A.M.T.D.S.	Agricultural Machinery Training Development Society
A.O.N.B.	Area of Outstanding Natural Beauty
A.P.C.A.	Assemblée Permanente des Chambres d'Agriculture
	(French, Permanent Assembly of the Chambers of Agriculture)
A.P.D.C.	Apple and Pear Development Council
A.P.F.	1. Animal Protein Factor
	2. Association of Professional Foresters
arbor.	Arboriculture
A.R.C.	Agricultural Research Council
A.S.A.	Apprenticeship Scheme for Agriculture
A.S.A.O.	Association of Show and Agricultural Organisations
A.S.C.C.	Agricultural Statistics Consultative Committee
A.S.E.A.	Agricultural Show Exhibitors' Association
A.T.B.	Agricultural Training Board
A.V.R.O.	Animal Virus Research Organisation
A.W.B.	Agricultural Wages Board
A.W.C.	1. Available Water Content
	2. Agricultural Wages Committee
A.W.T.	Animal Welfare Trust
A.Y.R.	All Year Round

486

B.A.A.	British Agrochemicals Association
B.A.A.O.	British Association of Abattoir Owners
B.A.B.	Le Bureau de l'Agriculture Britannique (the N.F.U. in Brussels)
B.A.C.	British Agricultural Council
B.A.E.C.	British Agricultural Export Council
B.A.G.C.D.	The British Association of Green Crop Driers Ltd
B.A.G.M.A.	British Agricultural and Garden Machinery Association
B.Agr.	Bachelor of Agriculture
B.A.H.P.A.	British Agricultural and Horticultural Plastics Association
B.A.H.S.	British Agricultural History Society
B.A.L.I.	The British Association of Landscape Industries
B.A.M.R.G.	British Agricultural Marketing Research Group
B.A.P.B.	British Association of Plant Breeders
B.A.S.I.S.	British Agrochemical Supply Industry Scheme Ltd
B.B.	Boerenbond Belge (Belgian Farmers' Union)
B.B.A.	British Bee-keepers' Association
B.C.B.C.	British Cattle Breeders' Club
B.C.P.C.	British Crop Protection Council.
B.D.A.M.A.	British Distributors of Animal Medicines Association
B.D.S.	British Deer Society
B.E.A.	British Egg Association
B.E.I.S.	British Egg Information Service
B.E.P.A.	British Edible Pulses Association
B.E.U.C.	Bureau Européen des Unions de Consommateurs
b.f.	Butter fat
B.F.M.I.R.A.	British Food Manufacturing Industries Research Association
B.F.P.C.	British Farm Produce Council
B.F.S.S.	British Field Sports Society
B.G.S.	1. British Goat Society
	2. British Grassland Association
B.H.A.	Butylated hydroxyanisole
B.H.C.	1. British Herdsmen's Club
	2. Benzene hexachloride
B.H.S.	British Horse Society
B.H.S.M.A.	British Hay and Straw Merchants' Association
B.I.A.C.	British Institute of Agricultural Consultants
B.I.C.	Butter Information Council
B.M.B.	Bottled Milk Buyers
B.M.M.A.	Bacon and Meat Manufacturers' Association
B.M.Y.	Beet Milk Yellow Virus
Bo	Boron
B.O.D.	Biochemical Oxygen Demand
B.O.G.A.	British Onion Growers' Association Ltd

487

B.O.V.	Brown Oil of Vitriol
B.P.A.	British Ploughing Association
B.P.B.H.A.	British Poultry Breeders and Hatcheries Association Ltd
B.P.M.A.	British Poultry Meat Association Ltd
B.P.P.F.	British Pig Producers' Federation
B.R.	Benefit Ratio
B.R.C.	British Rabbit Council
B.S.A.P.	British Society of Animal Production
B.S.B.P.A.	British Sugar Beet Producers' Association
B.S.C.	1. British Seeds Council
	2. British Sugar Corporation
B.S.E.	British Semen Exports
B.S.P.	Base Saturation Percentage
B.S.R.A.E.	British Society for Research in Agricultural Engineering
B.T.F.	British Turkey Federation Ltd
Bu., Bus., bush.	Bushel
B.V.A.	British Veterinary Association
B.V.D.	Bovine Virus Diarrhoea
B.V.Sc.	Bachelor of Veterinary Science
B.W.D.	Bacillary White Diarrhoea
B.W.M.B.	British Wool Marketing Board
B.Y.V.	Beet Yellow Virus
C.	1. Centigrade/Celsius
	2. Carbon
Ca	Calcium
C.A.A.V.	Central Association of Agricultural Valuers
C.A.B.	Commonwealth Agricultural Bureau
C.A.E.C.	County Agricultural Executive Committee
C.A.P.	Common Agricultural Policy
C.B.H.P.C.	Commonwealth Bureau of Horticulture and Plantation Crops
C.B.I.	Confederation of British Industry
C.C.	1. Countryside Commission
	2. Contemporary Comparison
C.C.A.	Chrome Copper Arsenate
C.C.A.H.C.	Central Council for Agricultural and Horticultural Cooperation
C.D.A.	Controlled Drop Application
C.E.	Council of Europe
C.E.A.	Confédération Européenne de l'Agriculture
C.E.C.	Cation Exchange Capacity
C.E.N.E.C.A.	Centre National des Exportateurs et Concours Agricoles (French)
C.E.P.F.A.R.	Centre Européen pour la Promotion et la Formation dans le Milieu Agricole et Rural
C.F.	Crude Fibre
cf.	Calf

C.F.A.	1. Confédération Française de l'Aviculture
	2. Commonwealth Forestry Association
C.F.M.	Centre for Farm Management
C.F.T.	Complement Fixation Test
C.G.A.	1. Confédération Générale de l'Agriculture
	2. Country Gentlemen's Association Ltd
C.I.A.A.	Commission des Industries Agricoles et Alimentaires
C.I.B.	Channel Islands Breed
C.I.B.E.	Confédération Internationale des Betteraviers Européens (International Confederation of Sugar Beet Growers)
C.I.C.R.A.	Centre International pour la Co-ordination des Recherches Agricoles
C.J.A.H.S.A.	Council of Justice to Animals and Humane Slaughter Association
Cl	Chlorine
C.L.A.	Country Landowners' Association
C.L.W.B.	Committee of London Wool Brokers
Cm	Centimetre
C.M.P.P.	2,methyl-4-chlorophenoxyproprionic acid
C.M.R.	Controlled Milk Recording
C.M.T.A.	Cooperative Milk Trade Association
C : N	Carbon : Nitrogen ratio
Co	Cobalt
CO_2	Carbon dioxide
C.O.B.	Cost of Building
C.O.G.E.C.A.	Comité Général des Co-opératives Agricoles des Pays de la C.E.E. (Committee of E.E.C. Farmers' Co-operatives)
C.O.P.A.	Comité des Organisations Professionnelles Agricoles de la C.E.E. (Committee of E.E.C. Farmers' Unions)
C.O.S.F.P.A.	Commons, Open Spaces and Footpaths Preservation Society
Co.S.I.R.A.	Council for Small Industries in Rural Areas
C.P.	Crude Protein
C.P.A.	Creamery Promotion Association
C.P.E.	Commercial Product Evaluation
C.P.R.E.	Council for the Protection of Rural England
C.P.R.W.	Council for the Protection of Rural Wales
C.Q.P.P.A.	Council of Quality Pig Producers' Association
C.R.A.	Commercial Rabbit Association
C.R.D.	Chronic Respiratory Disease
C.S.T.S.	Craft Skills Training Scheme
Cu	Copper
cwt.	A hundredweight

2,4-D.	2,4,dichloro-phenoxyacetic acid
D.A.F.S.	Department of Agriculture and Fisheries for Scotland
D.A.N.I.	Department of Agriculture for Northern Ireland
d.b.W.	Drawbar power in Watts
D.C.F.	Digestible Crude Fibre
D.C.P.	Digestible Crude Protein
D.D.T.	Dichlor-diphenyl-trichlor-ethane
D.E.	Digestible Energy Value
D.E.A.	Dairy Engineers' Association
D.I.T.E.C.	Dairy Industry Training and Education Committee
D.M., d.m.	Dry Matter
D.M.P.	Di-methyl-phthalate
DNA	Deoxyribonucleic acid
D.N.C., D.N.O.C.	Di-nitro-ortho-cresol
D.N.B.P.	Di-nitro-ortho-secondary butyl-phenol
D.O.M.	Digestible Organic Matter
D.P.A.	Duck Producers' Association
D.S.W.A.	Dry Stone Walling Association
D.T.F.	Dairy Trade Federation
D.V.O.	Divisional Veterinary Officer (of A.D.A.S.)
d.w.	Deadweight
E.	Energy
E.A.C.	European Association for Cooperation
E.A.G.G.F.	European Agricultural Guarantee and Guidance Fund
E.B.L.	Enzootic Bovine Leucosis
E.C.C.C.	English Country Cheese Council
E.C.F.	Eastern Counties Farmers Ltd
E.C.U.	European Currency Unit
E.D.F.	European Development Fund
E.D.T.A.	Ethylene-diamine-tetra-acetic acid
E.E.C.	European Economic Community
E.F.D.R.	European Federation of Diary Retailers
E.F.T.A.	European Free Trade Association
E.H.F.	Experimental Husbandry Farm
E.L.F.	European Landworkers' Association
E.M.L.A.	East Malling-Long Ashton (see dictionary text)
E.M.P.	Ethylmercury phosphate
E.M.R.S.	East Malling Research Station
E.M.S.	European Monetary System
Est.d.c.w.	Estimated Dressed Carcase Weight
E.U.A.	European Unit of Account
E.V.A.	English Vineyards Association
E.W.R.C.	European Weed Research Council
F1., F2	First, and Second, Filial Generation

F.A.C.	Federation of Agricultural Co-operatives (U.K.) Ltd
F.A.F.S.	Farm and Food Society
F.A.O.	Food and Agricultural Organisation (of the United Nations)
F.B.A.	Farm Buildings Association
F.B.G.	Farm Buildings Group (of ADAS Land Service)
F.B.R.I.C.	Farm Based Recreation Information Centre
F.C.G.S.	Farm Capital Grant Scheme
F.C.R.	Food Conversion Ratio
f.c.s.	filled, carted and spread (of manure)
F.D.I.C.	Food and Drink Industries Council
Fe	Iron
F.E.I.	Fédération Equestre Internationale
F.E.O.G.A.	Fonds Européen d'Orientation et de Garanties Agricoles
F.F.M.W.	Federation of Fresh Meat Wholesalers
F.F.V.I.B.	Fresh Fruit and Vegetable Information Bureau
F.H.D.S.	Food and Horticultural Development Scheme
Fl	Fluorine
F.L.K.S.	Fatty Liver/Kidney Syndrome (of chickens)
F.M.A.	1. Fertilizer Manufacturers' Association Ltd
	2. Farm Management Association
F.M.B.R.A.	Flour Milling and Baking Research Association
F.M.C.	Fatstock Marketing Corporation Ltd
F.M.D.	Foot and Mouth Disease
F.M.F.	Food Manufacturers' Federation
F.M.I.G.	Food Manufacturers' Industrial Group
F.P.C.	Flowers and Plants Council (formerly the Flowers Publicity Council)
F.R.W.	Federation of Rural Workers
F.S.	Foundation Seed
F.U.W.	Farmers' Union of Wales
F.W.A.G.	Farming and Wildlife Advisory Group
fym	Farmyard Manure
g	Gram
G.A.F.T.A.	Grain and Feed Trade Association Ltd
G.A.J.	Guild of Agricultural Journalists
gal., gall.	Gallon
G.C.R.I.	Glasshouse Crops Research Institute
G.E.	Gross Energy
G.E.F.A.P.	Groupement Européen des Associations Nationales des Fabricants de Pesticides
G.F.W.	General Farm Worker
G.R.I.	Grassland Research Institute
G.W.D., g.w.d.	Gang-Work-Day
H	Hydrogen
ha	Hectare

H.C.G.S.	Horticultural Capital Grant Scheme
H.D.	Heavy Duty
H.D.R.I.	Hannah Dairy Research Institute
H.E.	Hay Equivalent
H.E.A.	Horticultural Education Association
H.E.T.P.	Hexa-ethyl-tetraphosphate
H.F.R.O.	Hill Farming Research Organisation
H.G.C.A.	Home Grown Cereals Authority
H.I.	Heat Increment
H.M.B.	Hops Marketing Board
H.M.I.	Horticultural Marketing Inspectorate
hort., hortic.	Horticulture, Horticultural
H_2O	Water
h.p.	Horse-power
H.S.E.	Health and Safety Executive
H.T.A.	Horticultural Trades Association
H.T.S.T.	High Temperature, Short Time (pasteurisation method)
H.V.S.	Higher Voluntary Standard (for seed certification)

I	Iodine
I.A.A.C.	International Agricultural Aviation Centre
I.A.E.	Institute of Agricultural Engineers
I.A.F.M.M.	International Association of Fish Meal Manufacturers
I.A.H.P.	International Association of Horticultural Producers
I.A.S.	Institute of Agricultural Secretaries
I.B.	Infectious Bronchitis
I.B.A.P.	Intervention Board for Agricultural Produce
I.B.R.	Infectious Bovine Rhinotracheitis
I.B.R.A.	International Bee Research Association
I.C.A.C.	International Confederation for Agricultural Credit
I.C.C.	1. International Chamber of Commerce
	2. Improved Contemporary Comparison
I.D.B.	Internal Drainage Board
I.D.D.	Internal Drainage District
I.E.C.	International Egg Commission
I.F.A.	Irish Farmers' Association
I.F.A.D.	International Fund for Agricultural Development
I.F.A.J.	International Federation of Agricularal Journalists
I.F.A.P.	International Federation of Agricultural Producers
I.F.G.B.	Institute of Foresters of Great Britain
I.L.A.M.	Institute of Lecturers in Agricultural Machinery

I.L.M.B.	Irish Livestock and Meat Board
I.M.T.A.	Imported Meat Trade Association
I.O.M.	Institute of Meat
I.O.M.F.	Isle of Man Farmers Ltd
I.P.B.C.	Irish Pigs and Bacon Commission
I.R.A.D.	Institute for Research in Animal Diseases
I.S.D.S.	International Sheep Dog Association
I.S.O.	International Sugar Organisation
I.S.V.A.	Incorporated Society of Valuers and Auctioneers
I.W.C.	International Wheat Council
I.W.S.	International Wood Secretariat
J.	Joule
J.C.C.	Joint Consultative Council of the Fresh Fruit, Vegetable and Flower Industry
J.F.U.	Jersey Farmers' Union
K	Potassium
K₂O	Potash
kg, kilo.	Kilogram
km	Kilometre
K.O.	Killing Out
K-Slag	Potassic Basic Slag
kw	Kilowatt
kw/h	Kilowatt Hour
l	Litre
L.A.	Local Authority
L.A.M.C.E.W.	Livestock Auctioneers' Market Committee for England and Wales
L.C.P.	Low Cost Production
L.I.	Landscape Institute
L.I.C.	Lands Improvement Company
L.M.C.N.I.	Livestock Marketing Commission for Northern Ireland
L.S.A.	Land Settlement Association, Ltd
L.T.A.	Livestock Traders' Association
L.V.G.A.	Lea Valley Growers' Association
L.V.I.	Licensed Veterinary Inspector
lw	Liveweight
L.W.G., l.w.g.	Liveweight gain
L.W.I.	Liveweight Increase
M.	Maintenance
m	Metre
M.A.F.F.	Ministry of Agriculture, Fisheries and Food
m.c.	Moisture Content
M.C.A.	Monetary Compensatory Amount
M.C.P.A.	2-methyl-4-chloro-phenoxyacetic acid

M.C.P.B.	2-methyl-4-chloro-phenoxybutyric acid
M.D.A.	Maize Development Association
M.E., m.e.	1. Metabolisable energy
	2. Mechanical efficiency
M.E.A.T.	Meat Eating Advisory Team (of M.L.C.)
m.eq.	Milligram equivalent
M.F.A.	Maize and Forage Association Ltd
Mg	Magnesium
M.G.A.	Mushroom Growers' Association
M.J.	Mega Joule
M.L.C.	Meat and Livestock Commission
ml	Millilitre
m/m	Milking machine
M.M.A.	Meat Manufacturers' Association
M.M.B.	Milk Marketing Board
M.M.M.A.	Milking Machine Manufacturers' Association
Mn	Manganese
Mo	Molybdenum
M.P.E.	Meat Promotion Executive
M.R.I	1. Meuse-Rhine-Ijssel
	2. Meat Research Institute
M.R.S.	Monthly Recording by Statement (of milk)
N	Nitrogen
Na	Sodium
N.A.A.C.	National Association of Agricultural Contractors
N.A.B.I.M.	National Association of British and Irish Millers (Inc.)
N.A.B.M.A.	National Association of British Market Authorities
N.A.C.	1. National Agricultural Council
	2. National Agricultural Centre
N.A.C.P.	National Association of Creamery Proprietors and Wholesale Dairymen (Inc.)
N.A.P.A.E.O.	National Association of Principal Agricultural Education Officers
N.A.S.P.M.	National Association of Seed Potato Merchants
N.A.S.S.	National Association of Semen Suppliers
N.C.A.F.M.W.	National Council for the Associations of Fresh Meat Wholesalers
N.C.B.A.	National Cattle Breeders' Association
N.C.C.	1. Nature Conservancy Council
	2. National Consumer Council
N.D.	Newcastle Disease
N.D.A.	1. National Diploma in Agriculture
	2. National Dairymen's Association (Inc.)
N.D.B.	National Diploma in Beekeeping
N.D.C.	National Dairy Council
N.D.G.P.A.	National Dairy Goat Produce Association
N.D.P.A.	National Dairy Produce Association

N.E.	Net Energy
N. Em., N. Ef.	Net Energy (i) for Maintenance, and (ii) for Fattening
N.A.A.S.	National Agricultural Advisory Service (now A.D.A.S.)
N.E.P.A.	National Egg Packers' Association Ltd
N.E.P.R.A.	National Egg Producers' and Retailers' Association
N.F.A.	National Farmers' Association (Eire)
N.F.A.P.C.S.	National Federation of Agricultural Pest Control Societies Ltd
n.f.e.	Nitrogen–free extract
N.F.F.P.T.	National Federation of Fruit and Potato Trades Ltd
N.F.M.T.	National Federation of Meat Traders
N.F.S.I.	National Farm Seeds Industries Ltd
N.F.T.	Nutrient Film Technique
N.F.U.	National Farmers' Union
N.F.Y.F.C.	National Federation of Young Farmer' Clubs
NH_3	Ammonia
N.I.	Non Immune (to Wart Disease). Used of potato varieties
N.I.A.B.	National Institute of Agricultural Botany
N.I.A.E.	National Institute of Agricultural Engineering
N.I.D.A.	Northern Ireland Department of Agriculture
N.I.P.H.	National Institute of Poultry Husbandry
N.I.R.D.	National Institute for Research in Dairying
N : K.	Nitrogen : Potassium ratio
N.M.F.B.A.E.A.	National Master Farriers, Blacksmiths and Agricultural Engineers' Association
N.M.P.C.	National Milk Publicity Council (Inc.)
N.M.R.	National Milk Records
N.M.T.F.	National Market Traders' Federation
NO_3, NO_2.	Nitrate, Nitrite
N.P.B.A.	National Pig Breeders' Association
N.P.K.	Nitrogen, Phosphorus and Potassium
N.P.N.	Non Protein Nitrogen
N.P.T.C.	National Proficiency Tests Council
n.p.w.	Net present value
N.R.	Nutritive ratio
N.S.A.	1. National Sheep Association
	2. Nuclear Stock Association
N.S.D.O.	National Seed Development Organisation Ltd
N.S.L.G.	National Society for Leisure Gardeners
N.U.A.A.W.	National Union of Agricultural and Allied Workers
n.v.	Neutralising value (of a liming material)
N.V.R.S.	National Vegetable Research Station
N.Y.D.	New York Dressed (poultry)

O_2	Oxygen
O.E.C.D.	Organisation for Economic Co-operation and Development
O.E.E.C.	Organisation for European Economic Cooperation
O.M.	Organic Matter
O.P.	Organo Phosphorus compound
P	Phosphorus
P.B.I.	Plant Breeding Institute
P.C.P.	Pentachlorophenol
P_2O_5	Phosphoric Acid
P.E.	Protein Equivalent
p.g.e.	Parasitic gastro-enteritis
P.G.R.O.	Processors and Growers Research Organisation
pH	A measure of acidity (see dictionary text)
P.H.C.A.	Pig Health Control Association
P.H.S.	Pig Health Scheme
P.H.S.I.	Plant Health and Seeds Inspector (of M.A.F.F.)
P.I.D.B.	Pig Industry Development Board (now defunct)
P.M.B.	1. Potato Marketing Board
	2. Pigs Marketing Board (Northern Ireland)
P.M.F.I.C.	Potash and Magnesium Fertilizer Information Centre
P.M.P.	Potassic Mineral Phosphate
P.M.T.V.	Potato Mop-Top Virus
PO_4	Phosphate
P.P.D.C.	Pig Production Development Committee
P.P.G.A.	Pot Plant Growers' Association
p.p.m.	Parts per million
P.S.P.S.	Pesticides Safety Precautions Scheme
p.t.o.	Power take off
P.V.G.A.	Processed Vegetable Growers' Association
P.Y.O.	Pick Your Own
Q.M.P.	Quality Milk Producers Ltd
Q.P.P.A.	Quality Pig Producers Association
R.A.B.D.F.	Royal Association of British Dairy Farmers
R.A.B.I.	Royal Agricultural Benevolent Institution
r.a.m.	Readily available moisture
R.A.N.A.	Registered Animal Nursing Auxiliary
R.A.S.E.	Royal Agricultural Society of England
R.B.S.	Rare Breeds Society
R.B.S.T.	Rare Breeds Survival Trust
R.B.V.	Relative Breeding Value
R.C.V.S.	Royal College of Veterinary Surgeons
R.E.M.I.	Regional Egg Marketing Inspector (of M.A.F.F.)
r.h.	Relative humidity

R.H.A.	Road Haulage Association Ltd
R.H.A.S.S.	Royal Highland and Agricultural Society of Scotland
R.H.M.I.	Regional Horticultural Marketing Inspector (of M.A.F.F.)
R.H.S.	Royal Horticultural Society
R.I.C.S.	Royal Institute of Chartered Surveyors
R.I.R.	Rhode Island Red (fowl)
RNA	Ribonucleic acid
R.O.P.S.	Roll Over Protection Structure
R.P.G.	Resource Planning Group
Ro.S.P.A.	Royal Society for the Prevention of Accidents
R.S.A.B.I.	Royal Scottish Agricultural Benevolent Institution
R.S.C.	Royal Smithfield Club
R.S.P.C.A.	Royal Society for the Prevention of Cruelty to Animals
R.U.A.S.	Royal Ulster Agricultural Society
R.V.C.	Royal Veterinary College
R.W.A.S.	Royal Welsh Agricultural Society
S	Sulphur
S.A.A.A.	Scottish Agricultural Arbiters' Association
S.A.G.A.	Sand and Gravel Association
S.A.O.S.	Scottish Agricultural Organisation Society
S.A.T.	Serum Agglutination Test
S.A.W.M.A.	Soil and Water Management Association
S.B.A.	Scottish Beekeepers' Association
S.C.A.T.S.	Southern Counties Agricultural Trading Society
S.D.	Semi-Digger (mouldboard)
S.D.T.	Society of Dairy Technology
S.E.	Starch Equivalent
S.E.T.A.	Scottish Egg Trade Association
S.F.A.	Small Farmers' Association
S.F.B.I.U.	Scottish Farm Buildings Investigation Unit
S.F.W.	Standard Feed Wheat
S.G.A.	Sand and Gravel Association
S.I.	Système International d'Unités (international system of units for measurement)
Si	Silicon
S.L.F.	Scottish Landowners' Federation
S.M.M.B.	Scottish Milk Marketing Board
S.M.D.	1. Soil Moisture Deficit
	2. Standard Man Day
S.M.P.	Skim Milk Powder
S.M.R.A.	Scottish Milk Records Association
s n f	Solids-not-fat
s.p.f.	Specific pathogen free
Sp. gr.	Specific gravity
S.P.M.B.N.I.	Seed Potato Marketing Board for Northern Ireland

S.P.V.S.	Society of Practising Veterinary Surgeons
S.S.	Stock seed
S.S.N.T.A.	Scottish Seed and Nursery Trade Association
S.S.S.I.	Site of Special Scientific Interest
S.V.D.	Swine Vesicular Disease
S.W.M.A.	Soil and Water Management Association Ltd
2,4,5,T	2,4,5,-trichloro-phenoxyacetic acid
t	Tonne
TB	Tuberculosis
T.C.P.A.	Town and Country Planning Association
T.D.N.	Total Digestible Nutrients
T.E.P.P.	Tetra-ethyl-pyrophosphate
T.F.F.	Traditional Farm Fresh (poultry)
T.G.E.	Transmissible gastro-enteritis
T.G.O.	Timber Growers' Organisation
T.G.W.U.	Transport and General Workers' Union
T.P.	True Protein
T.P.I.	Town Planning Institute
T.R.S.	Tough Rubber Sheathed
T.R.V.	Tobacco Rattle Virus
T.T.C.	Triphenyltetrazolium chloride
T.T.F.	Timber Trade Federation
T.V.I.	Temporary Veterinary Inspector (of M.A.F.F.)
t.v.o.	Tractor vaporising oil
U.A.	Unit of Account
U.B.A.	Ulster Bacon Agency Ltd
U.C.A.	Ulster Curers' Association
U.C.F.	United Co-operative Farmers Inc.
U.F.A.W.	Universities Federation of Animal Welfare
U.F.U.	Ulster Farmers' Union
U.K.A.S.T.A.	United Kingdom Agricultural Supply Trade Association
U.K.S.A.	United Kingdom Sponsoring Authority for the Exchange of Young Agriculturists
U.K.S.C.S.	United Kingdom Seed Certification Schemes
U.K.W.G.F.	United Kingdom Wool Growers' Federation
U.L.V.	Ultra low volume
U.M.V.	Unexhausted manurial value
U.S.D.A.	United States Department of Agriculture
Var.	Variety
v.f.a.s.	Volatile fatty acids
V.I.C.	Veterinary Investigation Centre
V.O.	Vaporising Oil
V.P.C.	Veterinary Products Committee
V.P.P.	Virus Pneumonia of Pigs
V.T.S.C.	Virus Tested Stem Cuttings (a grade of seed potato)

498

W.A.E.R.S.A.	World Agricultural Economic and Rural Sociological Abstracts
W.A.G.B.I.	Wildfowlers' Association of Great Britain and Ireland
W.A.O.S.	Welsh Agricultural Organisation Society Ltd
W.C.F.	Worshipful Company of Farmers
W.C.F.T.S.	West Cumberland Farmers' Trading Society Ltd
W.F.G.A.	Women's Farm and Garden Association
W.J.F.I.T.B.	Wool, Jute and Flax Industries Training Board
W.P.	Wilting Point
W.P.B.S.	Welsh Plant Breeding Station
W.P.O.	World Ploughing Organisation
W.P.S.A.	World Poultry Science Association
W.R.O.	Weed Research Organisation
W.S.G.F.	Welsh Seed Growers' Federation Ltd
Zn	Zinc